Handbook of
Microcircuit Design
and Application

OTHER McGRAW-HILL HANDBOOKS OF INTEREST

Handbook of Microcircuit Design and Application

DAVID F. STOUT
Sr. Engineering Specialist
Ford Aerospace and Communications Corporation
Palo Alto, California

Edited by **MILTON KAUFMAN**
President, Electronic Writers and Editors

W. E. MOERNER

McGRAW-HILL BOOK COMPANY

New York   St. Louis   San Francisco   Auckland
Bogotá   Hamburg   Johannesburg   London
Madrid   Mexico   Montreal   New Delhi
Panama   São Paulo   Singapore
Sydney   Tokyo   Paris
Toronto

Library of Congress Cataloging in Publication Data

Stout, David F

  Handbook of microcircuit design and application.

  Includes bibliographical references and index.

  1. Integrated circuits—Handbooks, manuals,
etc. 2. Microelectronics—Handbooks, manuals,
etc. 3. Electronic circuit design—Handbooks,
manuals, etc. I. Kaufman, Milton, joint author.
II. Title.

TK7874.S83   621.381′73   79–19999

ISBN 0–07–061796–1

1234567890   KPKP   89876543210

The editors for this book were Harold B. Crawford and Ruth L. Weine,
the designer was Naomi Auerbach, and the production supervisor was
Teresa F. Leaden. It was set in Baskerville by The Kingsport Press.

Printed and bound by The Kingsport Press.

Contents

Preface

Analog and digital microcircuits constitute a very significant portion of the rapidly growing field of electronics. This Handbook provides a comprehensive presentation of all major classes of microcircuits. The material is organized to facilitate rapid solutions to many specific problems confronting the design engineer. And yet the level of presentation is useful to technicians as well.

All chapters of this Handbook are organized for quick reference to the important material without excessive background theory or proofs. An easily followed format is used for all digital chapters, and a slightly different format is strictly adhered to for the analog chapters.

Each digital integrated circuit (IC) is presented by using most of the following: alternate names for the device, principles of operation for the device, circuit diagram of a typical IC in that class, boolean equations if applicable, timing diagrams if applicable, and several application examples per chapter. More complex digital ICs such as the microprocessor are presented via the building block approach. First, the elemental portion of the device is presented, then peripheral circuitry is gradually added and explained, until finally, the device is completely described.

Since the microprocessor is such a widely useful and general-purpose device, five chapters are devoted to this IC. The first (Chapter 13) describes the device in detail and Chapters 14 through 17 present four diverse applications.

Each analog integrated circuit is presented using most of the following design aids: alternate names for the device, principles of operation, a list of design parameters, a list of important design equations, recommended design steps, and an example using the design steps. The parameter list is extremely useful since it carefully explains each parameter or circuit element and states what effect it has on the circuit.

Much of the handbook emphasizes microprocessors and microcomputers and problems related to interfacing with these LSI components.

Other LSI bus-oriented devices discussed are read-write random-access memories, read-only memories, serial and parallel communication devices, A/D and D/A converters, direct-memory-access chips, number crunchers, and coder-decoders.

The chapter on operational amplifiers contains many of the unique features appearing in the author's *Handbook of Operational Amplifier Circuit Design*. These include the two fundamental rules needed to design op-amp circuits, a comprehensive list of op-amp error sources and methods by which they can be minimized, and an example of inverting amplifier design that considers all error sources.

Eight types of active filters are presented in Chapter 20. These include low-pass, high-pass, bandpass, and bandstop. In most cases design procedures and a numerical example are also provided. Sensitivity parameters and other sources of filter errors are quantitatively discussed.

The author wishes to acknowledge the helpful support of his wife, Mildred, who meticulously typed the 1000-page manuscript. The author also appreciates the support of his fellow workers and supervisors at Ford Aerospace and Communications Corp. Those who offered suggestions and provided moral support include Bill Slivkoff, Ed Petroka, Layard Kirby, and Dave Westcott. We also wish to thank Fairchild Semiconductor, Harris Semiconductor, Intersil, Motorola Semiconductor, National Semiconductor, RCA Solid State, Signetics, Teledyne Semiconductor, and Texas Instruments for allowing us to use some of their material.

David F. Stout

Chapter 1

Introduction to Microcircuits

1.1 INTRODUCTION

The microcircuit appeared first in the early 1960s. Since that time the electronics industry has experienced tremendous changes, both in manufacturing methods and in the theory of electronics design. The use of microcircuits has made possible both great reductions in the size and cost of complex electronics circuitry and the development of new products, such as compact calculators and microprocessors.

1.2 TYPES OF MICROCIRCUITS

Microcircuits may be classified as either monolithic or hybrid types. *Mono-* means single and *-lithic* refers to rock or crystal. Thus, a monolithic microcircuit is a single crystal (chip) of silicon which contains all the elements of an electronic circuit.

MONOLITHIC INTEGRATED CIRCUITS A monolithic integrated circuit, usually referred to simply as an *integrated circuit* (IC), is another name for a monolithic microcircuit.

A single "slice" of silicon may contain hundreds of ICs, as shown in Fig. 1.1. The figure shows a typical silicon slice, 1.5 to 4 in in diameter and 6 mils in thickness. Each chip (die) is typically $\frac{1}{16}$ by $\frac{1}{16}$ in. In processing, the individual chips are cut from the slice and individually mounted in a special package, as shown in Fig. 1.2. Connecting leads are bonded to the chip, and the package lid is sealed on. The completed microcircuit is then ready for test.

HYBRID MICROCIRCUITS A hybrid microcircuit is a packaged device containing one or more ICs plus resistors, capacitors, and conductors. The last three items are deposited on a substrate (such as ceramic)

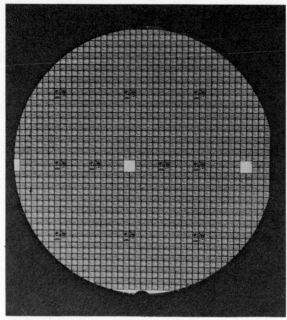

Fig. 1.1 A wafer of silicon containing hundreds of separate microcircuit dice.   *(Motorola, Inc.)*

and interconnected to the IC or ICs and each other.   In general, monolithic microcircuits are preferred where mass production of ICs is required.   For lesser requirements, hybrid production may currently be
more profitable.   Figure 1.3 shows how a hybrid microcircuit is mounted
in its package.

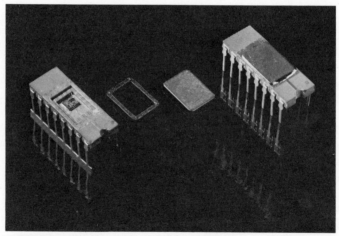

Fig. 1.2 A microcircuit die placed in a header.   Wires are attached and a lid is sealed
into place to complete the package.   *(Burr-Brown, Inc.)*

Fig. 1.3 Exploded view of the major parts of a hybrid microcircuit. *(Burr-Brown, Inc.)*

SINGLE-CHIP DEVICES It is felt that single-chip devices (monolithic ICs) are the ultimate destiny of all useful microcircuits. The hybrid microcircuit is merely one step on the road to this configuration.

The popularity of monolithic ICs is based on a very practical reason. Electronic equipment which required many cubic feet of space several decades ago has been reduced by the use of monolithic ICs to only a small fraction of its former dimensions.

Nearly every analog or digital circuit conceived can be reduced to a microcircuit. The only important limitations are power-handling capability and the required quantity of external connections, but even these constraints are being conquered as multiwatt ICs and 60-pin (or more) ICs are introduced.

1.3 THE PURPOSE OF THIS HANDBOOK

This handbook has been written to assist the working engineer, as well as instructors and students, in the design and practical application of ICs. It is frequently difficult to obtain all the desired information for ICs. Although for certain ICs good data sheets and useful application notes are available, in other cases, particularly with new ICs, there may be very little useful literature.

Typically it may be found that many required pieces of information are missing, difficult to understand, or even contradictory. In magazine articles and application notes, sometimes three or four contradictory equations have appeared for the same parameter. It is rare that an article will provide all the information required for the complete design of a particular circuit. Although this handbook cannot provide 100 percent of the required equations and design information for a specific IC, it attempts to approach this goal as closely as possible.

THE SCHEME OF THIS HANDBOOK Throughout this handbook, a systematized presentation of each microcircuit is followed. For example, one section describes how the particular device operates in typical applications, and another section thoroughly discusses each parameter of the microcircuit. In addition, application sections provide complete lists of design equations, design steps, and numerical illustrations of the use of the design steps.

Throughout the handbook, the major emphasis is on microprocessors. It is believed that these devices will become the central controllers for most electronic systems in the foreseeable future. All other electronic circuits will eventually become peripheral devices to microprocessors. Consequently, anyone interested in the study and application of electronics circuitry must become intimately familiar with the standard analog and digital ICs and how they communicate with or support microprocessors.

1.4 THE BASIC CONTENT OF THIS HANDBOOK

Although we live in an analog world, most large electronic systems which assist people in complex tasks are controlled by digital computers. Typical of these are data storage and retrieval systems.

Digital systems are ideal for most of these complex tasks since they are nearly absolute in handling numbers and making decisions. Many of the interfaces of these systems must be analog because most processes and material phenomena are analog in nature. The ultimate system will have an analog input and output and a central processing system which is purely digital.

THE ULTIMATE SYSTEM In this handbook we are attempting to anticipate this ultimate system. In future designs the analog input and output and the digital processing will all be performed on a single silicon chip. While most of the electronic circuits on this ultimate single silicon chip are now separate ICs, they are gradually being combined into complex multifunction devices. Many chips are currently available which contain both analog and digital circuitry.

SECTIONS OF THE HANDBOOK This handbook is divided into three basic sections, as follows:

Chaps. 2 to 18: digital ICs
Chaps. 19 to 23: analog ICs
Chaps. 24 to 27: analog-digital ICs

The emphasis is on digital microcircuits, especially microprocessors, since the majority of the ICs in large electronic systems are digital and the ratio of digital to analog ICs becomes greater each year.

1.5 THE FORMAT

Basically, this handbook is composed as follows:

1. Each chapter covers a different type or class of microcircuit.

2. Each chapter has a discussion section which describes the IC and presents any general notation applicable to the entire chapter.

3. A list of parameters with complete descriptions is provided either for the chapter or for each specific device.

4. Since many devices have more than one name, a list of alternate names is provided for each device.

5. For each digital microcircuit there is a functional description, timing diagrams, logic symbols, and the boolean equation.

6. Analog ICs are described with internal schematics, typical external connections, design equations, waveforms, and a functional description.

7. Each chapter contains two or three application sections, where common uses of the devices are described. These sections contain a list of alternate names, a section on principles of operation, a list of parameters, and a list of design equations. All these data will be put to use in a design-procedure section, which contains a recommended set of design steps. A numerical example follows to show how easily the design steps are implemented.

8. Several theory chapters are included to assist the reader in using the design chapters. Chapter 2 discusses all the common digital families and makes comparisons between them in many key areas. Microprocessors are covered in Chaps. 13 to 17, and a theoretical introduction to these devices is provided in Chap. 13.

Digital Families

INTRODUCTION

The selection of an optimum digital device family for an electronic circuit or system is one of the most important trade-offs the designer must make. In this chapter we provide all the necessary tables, curves, and comparative data necessary to complete this trade-off successfully. Many systems are designed around several logic families. Often the system has fast sections, requiring fast logic devices, and slow sections, where the designer has some latitude in the selection of device families.

2.1 TYPICAL GATE CIRCUIT FOR EACH FAMILY

Figure 2.1 shows a typical two-input gate for each of the digital families evaluated in this chapter. Each manufacturer uses a slightly different approach for each digital family. We will show some of the more popular circuits used by several large manufacturers. Important options, such as tristate or open-collector outputs, should also be considered when making this part of the evaluation.

2.2 VOLTAGE TRANSFER FUNCTIONS

The voltage transfer function of a digital gate contains several items of information useful to the circuit designer. From this transfer-function plot one can evaluate such things as:
1. Output ON and OFF voltages relative to input ON and OFF voltages
2. Behavior of transfer function in linear region
3. Changes in transfer function for various supply voltages

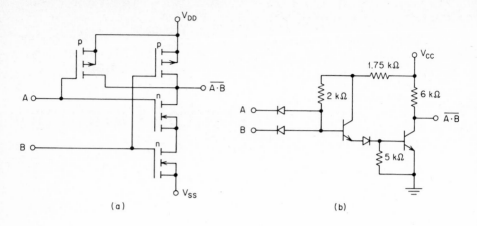

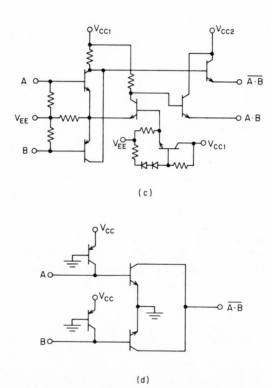

Fig. 2.1 Typical two-input gates for several digital IC families: *(a)* CMOS (RCA CD4011A), *(b)* DTL (National DM946), *(c)* ECL (National DM10105), *(d)* I²L (*Electronics,* February 21, 1974, p. 93), *(e)* MOS (National MM480), *(f)* RTL (Motorola MC914), *(g)* TTL (Texas Instruments SN7400), *(h)* high-speed Schottky TTL, and *(i)* low-power Schottky TTL.

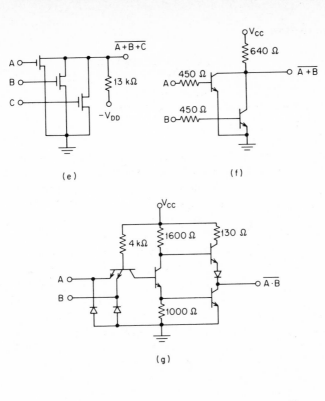

(e)

(f)

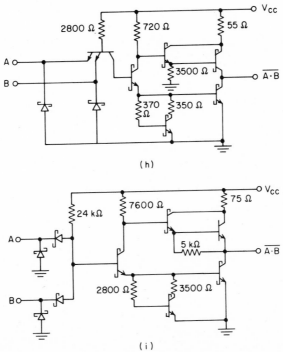

(g)

(h)

(i)

4. All the above as a function of temperature

5. Power dissipation at various points in the transfer function if the supply current is plotted on same figure

6. The optimum (and sometimes the maximum) input and output voltage ranges

7. DC noise margin

The voltage transfer functions shown in Fig. 2.2 are typical at 25°C. To characterize this parameter adequately a family of curves should be shown for each device. Parameter variations over temperature and guaranteed maximum or minimum limits would assist the designer. Many of these data are available in tabular form but rarely given in voltage plots, as in Fig. 2.2. These tabular data, showing worst-case input-output voltage ranges, are extremely important when interfacing one logic family with another. Problems seldom arise when interfacing devices within the same logic family, but some combinations of different logic families will not interface under any conditions. Some will interface only if the device parameters are close to their typical specification-sheet values but will not interface when worst-case variations are considered. The best combinations are between most of the TTL and CMOS families which will usually interface under worst-case situations.

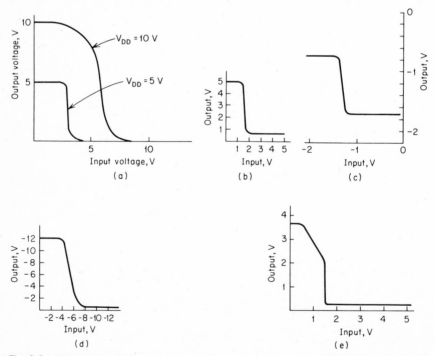

Fig. 2.2   Voltage transfer functions of several digital families: (a) CMOS (RCA CD4011A), (b) DTL, (c) ECL, (d) MOS, and (e) TTL.

Four parameters must be considered when an output terminal of one device is to be connected to the input terminal of another device. These parameters are pictorially represented in Fig. 2.3 and described as follows:

V_{IH} = minimum HIGH input voltage
V_{IL} = maximum LOW input voltage
V_{OH} = minimum HIGH output voltage
V_{OL} = maximum LOW output voltage

The devices will interface properly if the inequalities

$$V_{IH} \leq V_{OH} \quad \text{and} \quad V_{OL} \leq V_{IL}$$

are satisfied. The inequalities must remain satisfied for all expected temperatures and for all worst-case combinations of input and output

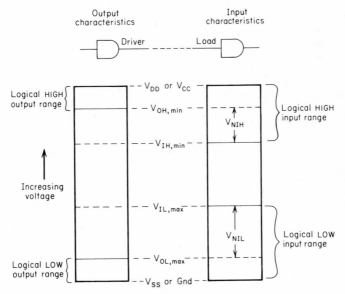

Fig. 2.3 Graphical description of the HIGH dc noise immunity V_{NIH} and the LOW dc noise immunity V_{NIL}.

currents. For example, if an output device has a minimum HIGH output voltage V_{OH} of 2.7 V while supplying 400 μA, the input device must have a minimum HIGH input voltage V_{IH} of 2.7 V or less while drawing no more than 400 μA from the output device. If the output device has a maximum LOW output voltage V_{OL} of 0.5 V while sinking 20 mA, the input device must have a maximum LOW input voltage V_{IL} of 0.5 V or more while supplying no more than 20 mA back to the output device. In practice, when devices in the same family are interfaced, the output device is capable of handling at least 20 times the current required by the input device.

NOISE IMMUNITY   Noise immunity (or noise margin) is a parameter closely related to input-output voltage characteristics.   Using this parameter, one can determine the allowable noise voltage on the input terminal of a device.   Noise immunity is specified using two parameters. As shown in Fig. 2.3, the LOW noise immunity V_{NIL} is the difference in magnitude between the maximum LOW output voltage of the driving device and the maximum input LOW voltage recognized by the receiving device.   Thus,

$$V_{NIL} = |V_{IL,\max} - V_{OL,\max}|$$

The HIGH dc noise immunity V_{NIH} is the difference in magnitude between the minimum HIGH output voltage of the the driving device and the minimum input HIGH voltage recognized by the receiving device.   Thus,

$$V_{NIH} = |V_{OH,\min} - V_{IH,\min}|$$

Some digital microcircuit data sheets provide only one number for noise margin, i.e., the smaller of V_{NIL} or V_{NIH}.   Most data sheets do not specify noise margin, but the input-output parameters needed to calculate V_{NIL} and V_{NIH} are always given.   Table 2.1 compares the dc noise immunity of various microcircuits.   The noise immunity is listed as a percentage of supply voltage for easy comparison.

TABLE 2.1   DC Noise Immunity for Various Digital ICs

Logic family	DC noise immunity, % (minimum if the two are different)
Standard TTL	8
L-TTL	8
S-TTL	6
LS-TTL	6
CMOS	30
ECL	3
DTL	8

The parameters V_{NIL} and V_{NIH} are dc noise margins.   Anyone testing a digital system is aware that ac noise and transients are always present on signal lines between microcircuits.   Careful board layout and good bypassing will minimize this noise.   The ac noise margin is never provided on data sheets because it depends on the speed of the device and the total capacitance (to ground) for the input considered.   Figure 2.4 indicates how TTL and CMOS noise immunity depends on the pulse width of the input. These curves should not be taken literally because the stray capacitance and input capacitance of every setup will be different.

In low-impedance logic circuits, like those using TTL and ECL, most noise originates through inductive coupling. In high-impedance circuits, like those using CMOS devices, mutual capacitance dominates the noise sources. Switching rates in logic circuits are of the order of 10^7 A/s and/or 10^8 V/s. In a low-impedance circuit a high di/dt will induce a noise voltage on adjacent lines. In high-impedance circuits a high dv/dt will induce a noise voltage on adjacent lines. It is therefore important to not allow long runs of closely spaced parallel lines having different signals.

Noise can also be effectively transferred to an input line if the ground or power lines contain noise. It is wise to design the system so that the power and ground have a low dc resistance back to the power supply

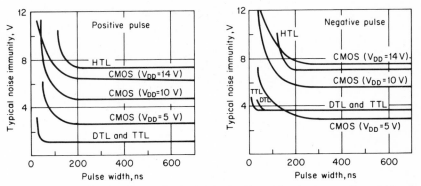

Fig. 2.4 Curves showing how noise immunity is affected by the input pulse width of a digital microcircuit [6].

for all points in the circuit. Likewise, the ac impedance between the two lines should be low at all points in the circuit for as wide a frequency range as possible. This is usually approximated by installing one or two tantalum capacitors of 4.7 μF or larger per board and also a 0.1-μF ceramic capacitor for every five or ten ICs between the supply voltage and ground.

2.3 POWER DISSIPATION

Power supplies represent an expensive item in most digital systems. Any design trade-offs which reduce system power consumption will also reduce the size and cost of power-supply equipment. For this reason many new systems which do not require high-speed operation have been designed around CMOS or I²L devices. The designer must be aware, however, that at frequencies above 2 MHz the low-power Schottky devices consume less power than CMOS and at 20 MHz all low-power devices begin to approach 20 mW per gate or more.

Figure 2.5 demonstrates that all logic devices except ECL show an increase in power consumption as frequency increases. This is due primarily to the totem-pole output stages, where two transistors are in series directly between the power terminal and ground. Since transistor turn-off times are normally longer than turn-on times, a spike of current

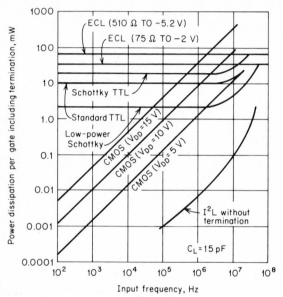

Fig. 2.5 Curves of power dissipation per logic gate as a function of frequency for several logic families.

is pulled from the power supply during the short period when both output devices are ON. At low frequencies these spikes represent a small portion of the standby current. However, as the switching frequency increases, the current-spike contribution becomes significant and causes total device power consumption to increase.

The turn-off time of a transistor increases as its load current increases or as temperature increases. Thus, one can expect power consumption of a digital device to increase drastically if it is heavily loaded and operating at an elevated temperature. If these conditions are likely to occur in a system, the ability of the power supply to handle current spikes should be thoroughly analyzed. Extensive use of bypass capacitors throughout the system will also be mandatory.

2.4 SWITCHING AND PROPAGATION TIMES

Several parameters are used to indicate the speed of a logic device. The most common parameter, and in some cases the only one given, is the *propagation delay*. This parameter, as shown in Fig. 2.6, is defined as the time required for an output to change state after an input has changed state. The measurement is usually taken at 50 percent amplitude. Some devices respond at different speeds when changing states in one direction compared with the other direction. Two parameters are therefore used:

t_{PLH} = propagation delay for output changing from LOW to HIGH

t_{PHL} = propagation delay for output changing from HIGH to LOW

Propagation delay can be defined for any output which depends on an input. For example, multiplexers such as the 74LS151 have three different kinds of inputs for which propagation delay is defined. In all cases the parameter is called t_{PHL} or t_{PLH}, depending on which direction the output is changing. Some data sheets call t_{PHL} the *turn-on delay* and t_{PLH} the *turn-off delay*. This naming arises from the idea that the output-current-sinking transistor (the one whose emitter is tied to ground) is the device which has the greatest effect on switching speed. Thus, when it turns ON, the output falls and the delay is called t_{PHL}.

Propagation delay must be accounted for in high-speed circuits. For example, in a microcomputer system, the data bus has valid data present for only a very short period of time. During this short interval the address bus must be decoded and the results from it used to control the destination or source of the data bus. Some MOS microprocessors allow only a few tens of nanoseconds to perform this address decoding. Needless to say, one cannot use CMOS devices to decode the address bus. Instead, TTL is often required since it has a delay per gate of

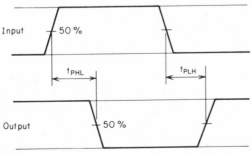

Fig. 2.6 Definition of propagation delay (t_{PLH} or t_{PHL}) as applied to an inverting buffer microcircuit.

less than 10 ns. Figure 2.7 shows the approximate propagation delay for a number of widely used digital microcircuits. Since CMOS devices are high-impedance devices, capacitive loading has a marked effect on propagation delay. For example, the 4049 has a t_{PHL} of 20 ns (typical) if $V_{DD} = 10$ V and the load is 50 pF. If $V_{DD} = 10$ V and a variable capacitive load is attached to the output, the propagation delay is specified at

$$t_{PHL} = (0.1 \text{ ns/pF})(C_L) + 15 \text{ ns}$$

where C_L is in picofarads.

Thus, if $C_L = 50$ pF,
$$t_{PHL} = (0.1 \text{ ns/pF})(50) + 15$$
$$= 5 + 15 = 20 \text{ ns}$$

A second common switching parameter for digital microcircuits is the *maximum toggle frequency* (or the *maximum clock frequency*). This parameter, called f_{max}, applies only to devices with storage elements, e.g., flip-flops, registers, counters, and latches. All static digital storage elements require positive feedback of one or more output signals to one or more of the inputs. This causes regeneration to occur, thereby making the device "snap" from one state to the other faster than it would without the positive feedback. As a consequence of this feedback, the clock input of a storage device cannot receive clock pulses with spacing closer than the time required for the entire feedback loop to be traversed. This time, in some cases, is the delay of only two gates. If each gate

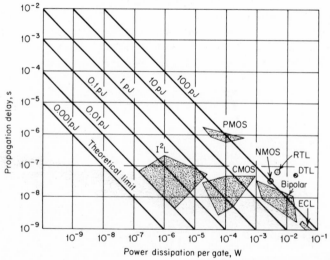

Fig. 2.7 Typical propagation delay per gate as a function of gate power dissipation [4].

has a propagation delay of 10 ns, a total loop would require approximately 20 ns (neglecting rise times). This corresponds to a maximum toggle frequency of 1/(20 ns) = 50 MHz.

Another switching parameter related to maximum clock frequency is the *minimum clock-pulse width* (t_w or t_{pw}). The clock pulse must stay ON long enough for the previously mentioned positive-feedback loop to be traversed. Assume the clock input is a square wave. We then observe that the maximum clock frequency is approximately

$$f_{max} \approx 1/(2\,t_w)$$

A few devices are also characterized for minimum HIGH clock-pulse width $t_{w,H}$ and for minimum LOW clock-pulse width $t_{w,L}$. From these we can also find the maximum clock frequency to be approximately

$$f_{max} \approx \frac{1}{t_{w,H} + t_{w,L}}$$

Many of the newer MSI and LSI microcircuits containing storage elements specify parameters called *setup time t_S* and *hold time t_H*. In most cases these parameters describe a relationship between a data input and a control input. In general, t_S and t_H are used for describing the minimum time by which one signal must precede another signal. The time required at the front edge of the signals is the *setup time*, and the time required at the back edge is the *hold time*. For example, the 4076 CMOS device (a quad D register) has t_S and t_H specified for clocking in data with respect to the rising edge of the clock. As shown in Fig. 2.8, if the 4076 is operated at $V_{DD} = 5$ V dc, the data must precede the rising edge of the clock by no less than 30 ns. It can be any value more than 30 ns, but it should not be less if data are to be latched with 100 percent certainty. We also note that the data must be held for at least 130 ns after the rising edge of the clock.

As noted in Fig. 2.8, the maximum t_S is 30 ns and the maximum t_H is 130 ns. These are device minimum time requirements. The system providing inputs to the 4076 must transmit new data a minimum of 30 ns before the clock rises. It must also hold data stable for a minimum time of 130 ns after the clock rises. Thus, the system minimum t_S must be larger than the device maximum t_S. Likewise, the system minimum t_H must be larger than the maximum t_H of the 4076. Therefore, we note that there exist two definitions for each of these parameters. In this type of design problem, many designers use small subscripts such as t_s and t_h for the device parameters and large subscripts t_S and t_H for the required system parameters. The same procedure is useful in many other timing parameters in microprocessor and memory-system design. One should always be able to separate system minimum and

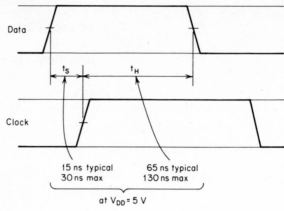

Fig. 2.8 Definition of the setup time t_S and the hold time t_H for the data input of the 4076 quad CMOS register.

maximum parameters from device minimum and maximum parameters. A system minimum parameter should always have a margin of safety (worst case) above a device maximum parameter.

Rise time t_r and *fall time t_f* are specified for some digital devices. In most cases these two parameters indicate the time required for the output to switch between 10 and 90 percent of full amplitude. Some ECL devices also provide data for transitions between 20 and 80 percent.

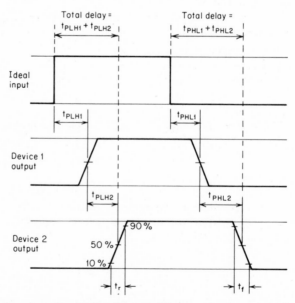

Fig. 2.9 Timing chart showing how system delays can be determined from individual propagation delays independent of rise and fall times.

These parameters are also called the *transition time,* LOW to HIGH or HIGH to LOW, that is, t_{TLH} or t_{THL}.

Rise and fall time are not generally provided for TTL digital ICs. These data are not required for most system analysis since the propagation delay is defined from the 50 percent point of the input amplitude to the 50 percent point of the output amplitude. This method of defining propagation delay effectively includes the delay contribution of rise and fall times. Figure 2.9 clarifies the meaning of these statements.

Memory devices have a complex set of timing relationships which must be carefully evaluated in order to obtain reliable writing and subsequent reading of data. This subject is covered in Chap. 5.

REFERENCES

1. RCA COS/MOS Integrated Circuits Manual, *RCA Corp. Tech. Ser.* CMS-271, 1972, p. 77.
2. Garrett, L. S.: Integrated Circuit Digital Logic Families, *IEEE Spectrum,* November 1970, p. 63.
3. Franson, P.: Logic Family Update: SSI/MSI Still Thrive in the World of LSI, *EDN,* Feb. 20, 1977, p. 79.
4. Altman, L.: Logic's Leap Ahead Creates New Design Tools for Old and New Applications, *Electronics,* Feb. 21, 1974, p. 94.
5. Ormond, T.: IC Logic Families: They Keep Growing in Numbers and Size, *EDN,* Oct. 5, 1974, p. 30.
6. Boaen, V.: Designing Logic Circuits for High Noise Immunity, *IEEE Spectrum,* January 1973, p. 53.

Chapter 3

Combinational Circuits

INTRODUCTION

A circuit which contains no storage elements and performs logical arithmetic operations is called a *combinational circuit.* Other common names given these devices are arithmetic circuits or arithmetic elements, names which take on more significance when we realize that the basic arithmetic circuits discussed in this chapter are the building blocks of the largest of all arithmetic circuits, the microprocessor (see Chap. 13).

When memory (storage elements) is added to a combinational circuit, the result is a *sequential circuit.* Combinational circuits and sequential circuits are thus defined as follows:

Combinational circuit. The outputs of the device are completely defined by the present state of its inputs.

Sequential circuit. The outputs of the device are defined by the present state of its inputs and the state of its memory devices.

The relationship between the set of inputs and the set of outputs in a combinational circuit can be expressed in several ways. In this chapter we will use the three most popular methods:

Truth table. A table showing the logic level of all outputs for every combination of the logic levels of all inputs. Truth tables are also called function tables.

Boolean equation. An equation for each output relating the logic level of that output with the logic level of all inputs.

Timing chart. A chart showing output voltage waveforms relative to input voltage waveforms.

The truth table and boolean equation of a device are true for all time except for a small period after the inputs change state. This short time, called the *propagation delay,* is usually measured in nanoseconds. The timing charts show this delay.

The read-only memory (ROM) is also a combinational circuit. Since ROMs are such an important class of device, we devote all of Chap. 6 to them.

Table 3.1 lists many of the common parameters used for combinational circuits. This table is not exhaustive since each particular circuit also has its own specialized parameters.

TABLE 3.1   Parameters of Combinational Circuits

Parameter	Description
BV_{CEO}	Breakdown voltage of output (open-collector) transistor
C_i	Input capacitance of any single input
$I_{CC}, I_{DD}, I_{EE}, I_{SS}$	Device power-supply current (from $V_{CC}, V_{DD}, V_{EE},$ or V_{SS})
I_{IH}	Input current of any single input in HIGH state
I_{IL}	Input current of any single input in LOW state
I_{OH}	Output source current of any single output in HIGH state
I_{OL}	Output sink current of any single output in LOW state
I_{OS}	Output short-circuit current (to ground)
P_D	Power dissipation in device
t_{PHL}	Propagation-delay time of output compared to input (output changing from HIGH to LOW)
t_{PLH}	Propagation-delay time of output relative to input (output changing from LOW to HIGH)
t_{THL}	Transition time (90 to 10%) of output (changing from HIGH to LOW)
t_{TLH}	Transition time (10 to 90%) of output (changing from LOW to HIGH)
V_{CC}	Collector supply voltage
V_{DD}	Drain supply voltage
V_{EE}	Emitter supply voltage
V_{IH}	Minimum allowed HIGH input to hold output at required state
V_{IL}	Maximum allowed LOW input to hold output at required state
V_{OH}	Minimum HIGH-level output voltage
V_{OL}	Maximum LOW-level output voltage
V_{SS}	Source supply voltage
V_T	Threshold voltage, i.e., input voltage at which output voltage changes state
Y	Output in boolean algebra expression

3.1   BUFFER GATE

ALTERNATE NAMES   Driver, power booster, fan-out booster, hex buffer (six to a package), noninverting buffer.

PRINCIPLES OF OPERATION   These circuits are usually furnished in quantities of six per package. The logic symbol and timing diagram for this device are shown in Figs. 3.1 and 3.2, respectively. Buffer gates are useful in logic design for several important reasons:

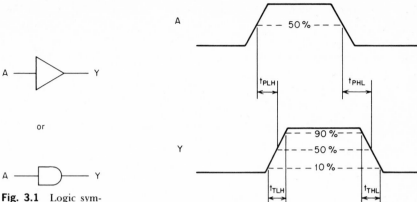

A ─────▷───── Y

or

A ─────⊐───── Y

Fig. 3.1 Logic symbol for the buffer gate.

Fig. 3.2 The timing chart for a buffer gate.

1. They increase the fan-out of a circuit without inverting the logic level.

2. They provide substantially more output current (source or sink) than other devices in their logic family.

3. They are useful as interface circuits where the load may not always be carefully controlled.

4. They also come with high-voltage open-collector outputs. The breakdown voltage rating of the output transistor is often 3 to 10 times greater than the nominal supply voltage for that particular logic family. This is useful when interfacing with other logic families, displays, electromechanical devices, or analog circuits.

The truth table and boolean equation for a buffer gate are shown in Table 3.2.

TABLE 3.2   Truth Table and Boolean Equation for a Buffer Gate

Truth table		Boolean equation
Input A	Output Y	
0	0	Positive logic: $Y = A$
1	1	Negative logic: $Y = A$

3.2 NOT GATE

ALTERNATE NAMES   Inverter, hex inverter or hex buffer (if six to a package), driver, power booster, fan-out booster, inverting buffer, inverting gate, complementing gate.

PRINCIPLES OF OPERATION   The NOT gate is commonly called the hex inverter since it usually comes six to a package. This circuit has the same four uses tabulated for the buffer gate (Sec. 3.1) except that the device inverts the logic level. This inverting function is the most popular use of the NOT gate.

It is very useful to know that the NAND gate and the NOR gate can also be used as a NOT gate. One need merely tie all the inputs of a NAND gate or NOR gate together to make this conversion, shown in Fig.

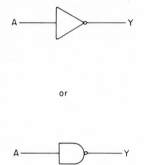

or

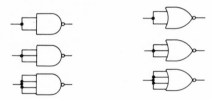

Other ways to make a NOT gate:

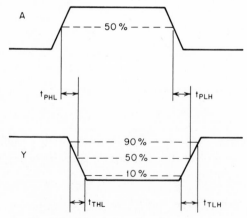

Fig. 3.3 Logic symbol for the NOT gate.

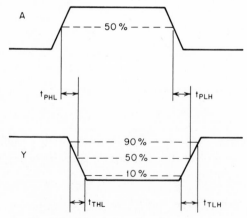

Fig. 3.4 The timing chart for a NOT gate.

3.3. The timing diagram for the NOT gate is shown in Fig. 3.4. The truth table and boolean equation for this device are shown in Table 3.3.

TABLE 3.3   Truth Table and Boolean Equation for a NOT Gate

| Truth table | | Boolean equation |
Input A	Output Y	
0	1	Positive logic: $Y = \overline{A}$
1	0	Negative logic: $Y = \overline{A}$

NOT-GATE APPLICATION; THE ASTABLE MULTIVIBRATOR One of the common applications [1][1] for the NOT gate is the astable multivibrator. As shown in Fig. 3.5, two NOT gates are required. We will present design equations and an example for an astable multivibrator using CMOS NOT gates. CMOS gates are especially suitable for this application because their high input resistance has almost zero effect on the circuit

[1] References appear at the ends of chapters.

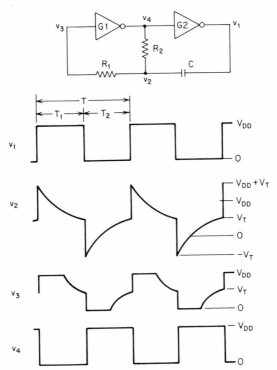

Fig. 3.5 An astable multivibrator using CMOS NOT gates and the waveforms at indicated locations in the circuit.

time constant R_2C. This particular circuit is also quite insensitive to changes in V_T, the threshold voltage, i.e., the input voltage at which the output changes state. A 30 percent change in V_T typically produces only a 9 percent change in the multivibrator period. Resistor R_1 makes the circuit reasonably insensitive to power-supply variations. If $R_1 \geq 2R_2$, a 100 percent change in V_{DD} produces only a 1 to 2 percent change in the multivibrator period.

DESIGN PARAMETERS

Parameter	Description
C	Determines multivibrator period along with R_2
f_o	Frequency of multivibrator
G1	Gate which senses charging and discharging of C and trips whenever v_2 passes through V_T
G2	Gate which changes states when G1 trips and then overdrives G1 through C and R1; this assures a clean transition into each half cycle of operation
R_1	Resistor which isolates the charge-discharge waveform of R_2 and C from the input clamp diodes of G1; otherwise the v_2 waveform would look like v_3 and the period would depend on the magnitude of V_{DD}
R_2	Determines multivibrator period along with C
R_{in}	Input resistance of G1 or G2
T_o	Period of multivibrator
T_1	Half period when C is discharging
T_2	Half period when C is charging

DESIGN EQUATIONS

Description	Equation	
Period of one cycle	$T_o = R_2C \left(\ln\dfrac{V_{DD}+V_T}{V_T} + \ln\dfrac{2V_{DD}-V_T}{V_{DD}-V_T} \right)$	(1)
Period of one cycle assuming $V_T = 0.5V_{DD}$	$T_o = 2.2R_2C = \dfrac{1}{f_o}$	(2)
Time period when v_1 is HIGH	$T_1 = R_2C \ln\dfrac{V_{DD}+V_T}{V_T}$	(3)
Time period when v_1 is HIGH assuming $V_T = 0.5V_{DD}$	$T_1 = 1.1R_2C$	(4)
Time period when v_1 is LOW	$T_2 = R_2C \ln\dfrac{2V_{DD}-V_T}{V_{DD}-V_T}$	(5)
Time period when v_1 is LOW assuming $V_T = 0.5\,V_{DD}$	$T_2 = 1.1R_2C$	(6)
Optimum R_1	$2R_2 \leq R_1 \ll R_{in}$	(7)
Optimum R_2	$R_2 \ll R_{in}$	(8)

ASTABLE MULTIVIBRATOR DESIGN EXAMPLE Suppose we require a 3000-Hz multivibrator using two gates of a 74CO4 hex inverter. We choose 10 V for V_{DD}. Assuming $V_T = 0.5 V_{DD}$, we get $T = 2.2 R_2 C$. The input resistance of a 74CO4 is 10^6 MΩ. We therefore choose a value of 1 MΩ for R_2. Since we must keep $2 R_2 \leq R_1$, we will let $R_1 = 2$ MΩ. The timing capacitor is next found from

$$C = \frac{T_o}{2.2 R_2} = \frac{1}{2.2 f_o R_2} = \frac{1}{2.2(3000 \times 10^6)} = 152 \text{ pF}$$

3.3 AND GATE

ALTERNATE NAMES Logical-product gate, coincidence gate.

PRINCIPLES OF OPERATION The AND gate provides a HIGH output if and only if all inputs are HIGH. Conversely, for negative logic it provides a LOW output if any input is LOW. This circuit therefore performs the OR function for negative logic.

Input configurations for the AND gate range from two inputs per gate (four gates per package) up to eight inputs (usually one to a package). More inputs are possible using input expansion microcircuits, which are available for most digital families. Figures 3.6 and 3.7 show the logic symbol and timing diagram for this device. Output-circuit options

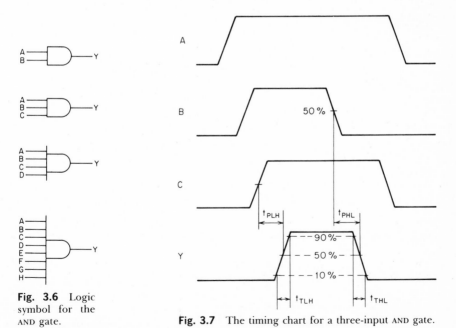

Fig. 3.6 Logic symbol for the AND gate.

Fig. 3.7 The timing chart for a three-input AND gate.

include the open-collector output and the tristate output (see Chap. 2). The truth table and boolean equation for this device are shown in Table 3.4.

TABLE 3.4   Truth Table and Boolean Equation for a Three-Input AND Gate

Truth table								
Inputs			Output	Inputs			Output	Boolean
A	B	C	Y	A	B	C	Y	equation
0	0	0	0	1	0	0	0	Positive logic:
0	0	1	0	1	0	1	0	$Y = ABC$
0	1	0	0	1	1	0	0	Negative logic:
0	1	1	0	1	1	1	1	$Y = A + B + C$

3.4   NAND GATE

ALTERNATE NAMES   AND-NOT gate, inverting AND gate, NOT-AND gate.

PRINCIPLES OF OPERATION   This circuit is identical to the AND gate except that the output is inverted. It outputs a LOW level if and only if all inputs are HIGH. Conversely, for negative logic the device output goes HIGH if any input is LOW. This circuit is therefore a NOR gate for negative logic. Figure 3.8 is the logic symbol for the NAND gate.

Input circuits for this device range from two inputs per gate (four to a package) up to eight inputs per gate (one per package). More inputs are possible using special expanders provided for some logic families.

The output-circuit options are open-collector, tristate, and special out-

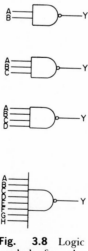

Fig.   3.8 Logic symbol for the NAND gate.

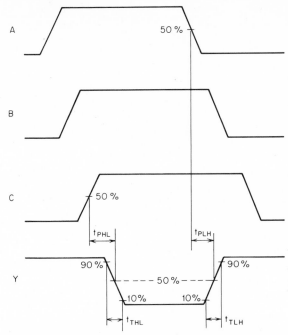

Fig. 3.9 The timing chart for a three-input NAND gate.

put-expander circuits to increase fan-out. The output waveform relative to the input waveforms is shown in Fig. 3.9.

The NAND gate also comes with a Schmitt-trigger option. These devices are similar to the analog Schmitt trigger in that, once triggered, the input must backtrack by a safe margin before the output changes back to the original state. This provides higher noise immunity than ordinary NAND gates. The NAND-gate truth table and boolean equation are shown in Table 3.5.

TABLE 3.5   Truth Table and Boolean Equation for a Three-Input NAND Gate

Truth table								
Inputs			Output	Inputs			Output	Boolean
A	B	C	Y	A	B	C	Y	equation
0	0	0	1	1	0	0	1	Positive logic:
0	0	1	1	1	0	1	1	$Y = \overline{ABC}$
0	1	0	1	1	1	0	1	Negative logic:
0	1	1	1	1	1	1	0	$Y = \overline{A} + \overline{B} + \overline{C}$

NAND-GATE APPLICATION; OSCILLOSCOPE LOGIC MULTIPLEXER [2]   A logic designer can always use a circuit which simultaneously displays four digital signals on a single-trace oscilloscope. The circuit shown in Fig. 3.10 performs this function efficiently. The resulting scope display is shown in Fig. 3.11.

This circuit is, in effect, a four-channel level shifter. The four logic inputs v_1, v_2, v_3, and v_4 are all assumed to have identical HIGH and LOW levels, which are compatible with G11 to G14. These input levels are shifted to new levels on the output line, as shown in Fig. 3.11. The LOW-level output for each waveform is determined by the sizes of R_2 to R_9. The HIGH-level output for each waveform likewise depends on R_6 to R_9. Resistor R_{10} affects all HIGH- and LOW-output levels.

A clock composed of G1 to G3 toggles a two-stage counter (FF1 and FF2). As the counter sequences through its four states, the decoder (G7 to G10) sequences a HIGH state through lines A, B, C, and D, as shown in the truth table. When A is HIGH, gates G15 and G19 are activated. If v_1 is LOW at this instant, G15 and G19 establish the v_{1b} level at the output, as shown in Fig. 3.11. If v_1 is HIGH while A is HIGH, G19 causes the v_{1a} level to appear at v_o. When B is HIGH, waveform v_2 appears at v_o as levels v_{2a} and v_{2b}. This is accomplished with G16

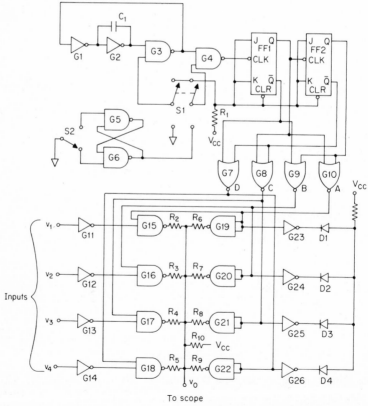

Fig. 3.10 A four-channel logic-signal multiplexer for an oscilloscope display.

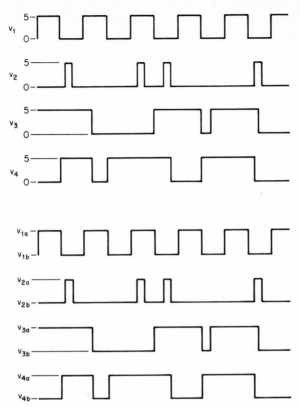

Fig. 3.11 Input and output waveforms of the four-channel oscilloscope multiplexer; 5-V TTL logic is assumed. The outputs will appear in the order shown if $R_5 < R_4 < R_3 < R_2$ and $R_9 < R_8 < R_7 < R_6$.

and G20. Likewise, v_3 and v_4 are translated to new levels at v_0 through use of G17, G18, G21, and G22.

The circuit in Fig. 3.10 shows six ways in which the NAND gate can be used:

1. NAND gate G3 is part of the feedback loop of the clock circuit. The second input to G3 is used to inhibit the clock using the DPDT switch S1.

2. G4 is a NAND gate used as a negative-logic NOR which transfers clock pulses or toggled pulses (from G5, G6 latch) to the two-stage counter.

3. G5 and G6 are NAND gates used to implement a latch circuit which is toggled using S2.

4. G7 through G10 are negative-logic NAND gates made from NOR gates.

5. G15 through G18 are used as NAND gates on the input but as a four-level (2-bit) digital-to-analog (D/A) converter on the output. These are open-collector NAND gates.

6. G19 through G22 are used as inverters, and their open-collector outputs operate as a four-level (2-bit) D/A converter.

The truth table for the G7 through G10 decoder is shown in Table 3.6.

TABLE 3.6   Truth Table for the Decoder Shown in Fig. 3.10 (G7 through G10)

Inputs				Outputs			
Q_2	Q_1	$\overline{Q_2}$	$\overline{Q_1}$	A	B	C	D
0	0	1	1	1	0	0	0
0	1	1	0	0	1	0	0
1	0	0	1	0	0	1	0
1	1	0	0	0	0	0	1

DESIGN PARAMETERS

Parameter	Description
A, B, C, D	Decoder output lines
C_1	Capacitor which determines oscillator frequency
D1–D4	Light-emitting diodes (LEDs) used to indicate channel being sampled while operating in manual sequencing mode
FF1 and FF2	Two-stage ripple-carry counter
G1 and G2	Main portion of free-running oscillator
G3	Part of free-running oscillator which can be used as a disable input
G4	Used to gate oscillator or manually toggled pulses into two-stage counter
G5 and G6	Manually toggled latch
G7–G10	Negative-logic NAND gates made from positive NOR gates
G11–G14	Inverting interface gates which can be used to translate from the logic levels of v_1 through v_4 to the logic levels used in this circuit
G15–G18	Open-collector NAND gates used to implement a four-level D/A converter; these four output levels combine with signals from G19 to G22 to establish the LOW output levels of v_0
G19–G22	Open-collector NAND gates (or inverters) used to implement a four-level D/A converter; these four output levels affect both the HIGH and LOW output levels of v_0
G23–G26	Inverters used to drive the LEDs which indicate channel being sampled
Q_1 and $\overline{Q_1}$	TRUE and FALSE outputs of FF1
Q_2 and $\overline{Q_2}$	TRUE and FALSE outputs of FF2
R_1	Pull-up resistor which holds JK inputs of FF1 and FF2 HIGH and enables G3 and G4 in clock mode
R_2–R_5	Resistors which determine the levels v_{1b}, v_{2b}, v_{3b}, and v_{4b}
R_6–R_9	Resistors which control both upper and lower output levels
R_{10}	Resistor which controls overall amplitude of all output levels
S1	Switch to select free-running clock or toggled-clock mode
S2	Switch used to toggle latch circuit
v_n	One of the input logic waveforms
v_{na}	The HIGH output level which represents the HIGH level of v_n
v_{nb}	The LOW output level which represents the LOW level of v_n
V_{CC}	Supply voltage if TTL logic used (called V_{DD} if CMOS logic used)
Vsat	Average output saturation voltage of G15 through G22 while in LOW state

DESIGN EQUATIONS

Description	Equation	
Oscillator frequency if standard TTL is used	$f_o \approx \dfrac{25\ \text{MHz}}{\{1 + [C_1/(3 \times 10^{-10})]^2\}^{1/2}}$	(1)
Oscillator frequency if low-power TTL is used	$f_o \approx \dfrac{5\ \text{MHz}}{\{1 + [C_1/(4 \times 10^{-10})]^2\}^{1/2}}$	(2)
Output levels:		
v_{1a}	$v_{1a} = \dfrac{R_6 V_{CC}}{R_6 + R_{10}} + V_\text{sat}$	(3)
v_{1b}	$v_{1b} = \dfrac{R_2 R_6 V_{CC}}{R_2 R_6 + R_{10}(R_2 + R_6)} + V_\text{sat}$	(4)
v_{2a}	$v_{2a} = \dfrac{R_7 V_{CC}}{R_7 + R_{10}} + V_\text{sat}$	(5)
v_{2b}	$v_{2b} = \dfrac{R_3 R_7 V_{CC}}{R_3 R_7 + R_{10}(R_3 + R_7)} + V_\text{sat}$	(6)
v_{3a}	$v_{3a} = \dfrac{R_8 V_{CC}}{R_8 + R_{10}} + V_\text{sat}$	(7)
v_{3b}	$v_{3b} = \dfrac{R_4 R_8 V_{CC}}{R_4 R_8 + R_{10}(R_4 + R_8)} + V_\text{sat}$	(8)
v_{4a}	$v_{4a} = \dfrac{R_9 V_{CC}}{R_9 + R_{10}} + V_\text{sat}$	(9)
v_{4b}	$v_{4b} = \dfrac{R_5 R_9 V_{CC}}{R_5 R_9 + R_{10}(R_5 + R_9)} + V_\text{sat}$	(10)

3.5 OR GATE

ALTERNATE NAMES Mixing gate, multi-input buffer gate, INCLU-SIVE-OR gate.

PRINCIPLES OF OPERATION The OR gate shown in Fig. 3.12 provides a HIGH output if any of its inputs is HIGH. For negative logic it provides a LOW output if and only if all inputs are LOW. This circuit performs the AND function for negative logic. OR devices are available with two inputs per gate (four gates per package) up to eight or more inputs per gate. More inputs are possible using expansion devices available for most digital families.

Output circuits provided for the OR gate in various digital families include open-collector outputs, both inverting and noninverting outputs, and tristate outputs. The timing chart of Fig. 3.13 shows the output signal relative to the input signals. The truth table and boolean equation for this device are shown in Table 3.7.

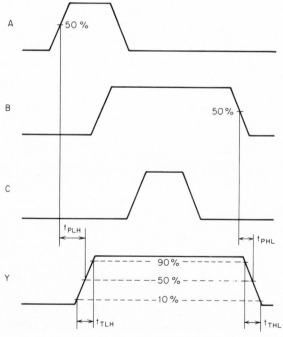

Fig. 3.12
Logic symbol for
the OR gate.

Fig. 3.13 The timing chart for a three-input OR gate.

TABLE 3.7 Truth Table and Boolean Equation for a Three-Input OR Gate

Truth table								
Inputs			Output	Inputs			Output	Boolean equation
A	B	C	Y	A	B	C	Y	
0	0	0	0	1	0	0	1	Positive logic:
0	0	1	1	1	0	1	1	$Y = A + B + C$
0	1	0	1	1	1	0	1	Negative logic:
0	1	1	1	1	1	1	1	$Y = ABC$

3.6 NOR GATE

ALTERNATE NAMES   OR-NOT gate, NOT-OR gate, inverting OR gate, mixing gate.

PRINCIPLES OF OPERATION   The NOR gate is identical to the OR gate except that the output is inverted. The logic symbols for various NOR gates are shown in Fig. 3.14. This circuit produces a LOW output if any of the inputs are HIGH. Likewise, for negative-logic systems it implements the NAND function. That is, if all inputs are LOW, the output is HIGH. The truth table and boolean equation are shown in Table 3.8.

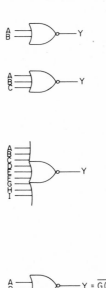

G = strobe

Fig. 3.14 Logic symbol for the NOR gate.

TABLE 3.8   Truth Table and Boolean Equation for a Three-Input NOR Gate

				Truth table				
Inputs			Output	Inputs			Output	Boolean
A	B	C	Y	A	B	C	Y	equation
0	0	0	1	1	0	0	0	Positive logic:
0	0	1	0	1	0	1	0	$Y = \overline{A + B + C}$
0	1	0	0	1	1	0	0	Negative logic:
0	1	1	0	1	1	1	0	$Y = \overline{ABC}$

This function is available with two inputs (four to a package) up to eight (one to a package). More than eight inputs are possible if input expanders are used.

Output-circuit options are open-collector, tristate, and output expanders for increased fan-out.

The NOR gate is also available with a strobe input. A LOW output will occur only if both the strobe and any input go HIGH. Figure 3.15 shows the timing relationship for input-output signals of the NOR gate.

3.7   XOR GATE

ALTERNATE NAMES Anticoincidence gate, inequality comparator, matching circuit.

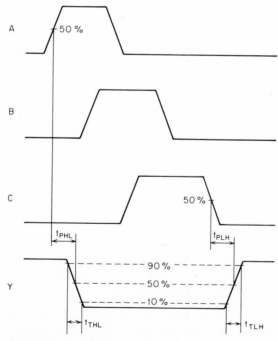

Fig. 3.15   The timing chart for a three-input NOR gate.

PRINCIPLES OF OPERATION The EXCLUSIVE-OR (XOR) gate shown in Fig. 3.16 provides a HIGH output if any single input is HIGH. If more than one input is HIGH or if all inputs are LOW, the output is LOW.

This circuit is available with two to five inputs per gate. Some digital families also provide an inverted output. Figure 3.17 shows some of the ways the XOR gate is implemented in different logic families. A simple application of De Morgan's laws shows that all these methods are equivalent.

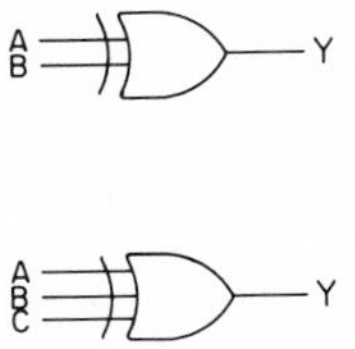

Fig. 3.16 Logic symbol for the XOR gate.

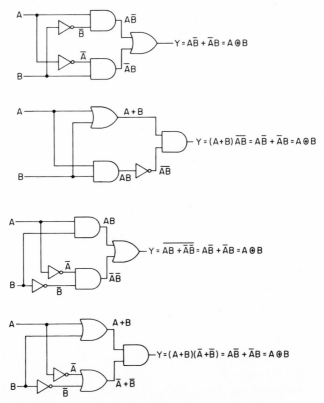

Fig. 3.17 Four ways to implement the two-input XOR gate.

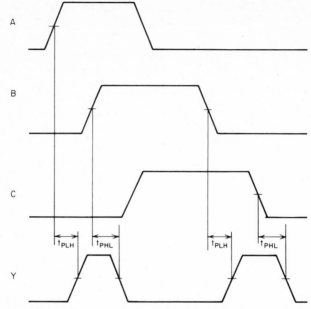

Fig. 3.18   The timing chart for a three-input XOR gate.

Figure 3.18 shows typical input-output waveforms for the XOR gate. The truth table and boolean equation are shown in Table 3.9.

TABLE 3.9.   Truth Table and Boolean Equation for a Three-Input XOR Gate

Truth table								
Inputs			Output	Inputs			Output	
A	B	C	Y	A	B	C	Y	Boolean equation
0	0	0	0	1	0	0	1	$Y = A\overline{B} + \overline{A}B$
0	0	1	1	1	0	1	0	or
0	1	0	1	1	1	0	0	$Y = A \oplus B$
0	1	1	0	1	1	1	0	

3.8   AND-OR-INVERT GATE

ALTERNATE NAMES M-wide N-input AND-OR-INVERT gate, AND-NOR gate, AND-OR-NOT gate.

PRINCIPLES OF OPERATION   As shown in Fig. 3.19, the AND-OR-INVERT gate is available in a number of configurations, all similar in that a group of two-, three-, or four-input AND gates converge into a single NOR gate. It is possible to put two of the smaller circuits on one chip, e.g., the two-wide two-input AND-OR-INVERT gate. Many of these circuits provide expansion terminals where additional AND gates can be connected. These sum into the NOR gate just like the other AND gates.   Figure

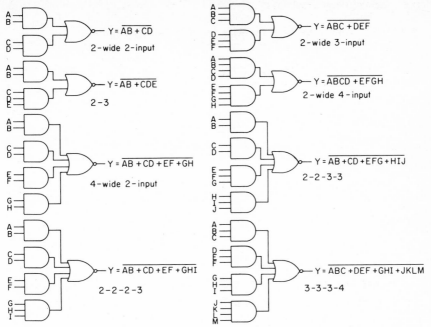

Fig. 3.19 Some of the available AND-OR-INVERT gates and their respective boolean equations.

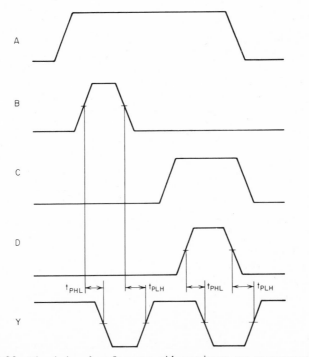

Fig. 3.20 The timing chart for a two-wide two-input AND-OR-INVERT gate.

TABLE 3.10   Truth Table for a Two-Wide Two-Input AND-OR-INVERT Gate

Inputs				Output	Inputs				Output
A	B	C	D	Y	A	B	C	D	Y
0	0	0	0	1	1	0	0	0	1
0	0	0	1	1	1	0	0	1	1
0	0	1	0	1	1	0	1	0	1
0	0	1	1	0	1	0	1	1	0
0	1	0	0	1	1	1	0	0	0
0	1	0	1	1	1	1	0	1	0
0	1	1	0	1	1	1	1	0	0
0	1	1	1	0	1	1	1	1	0

3.20 shows some typical input-output waveforms for this circuit. The truth table for this device is shown in Table 3.10.

3.9   DIGITAL MULTIPLEXER

ALTERNATIVE NAMES   Multiplexer, data selector, digital switch, MUX, N-line–to–1-line selector, N:1 multiplexer.

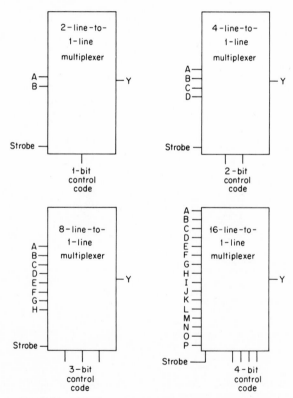

Fig. 3.21   The four most common digital multiplexers.

PRINCIPLES OF OPERATION   The multiplexer has two sets of input lines and one output line.   One set of input lines is for input data. The other set of input lines is for control.   The control lines determine which data line is selected to transfer its information over to the single output line.   If the multiplexer has 2^N data input lines, N control lines are required.

Figure 3.21 shows the four most common types of digital multiplexers, while Fig. 3.22 indicates the internal logic required for the 4:1 multiplexer.   For this particular multiplexer $N = 2$ control lines are sufficient to select one of $2^N = 2^2 = 4$ data lines.

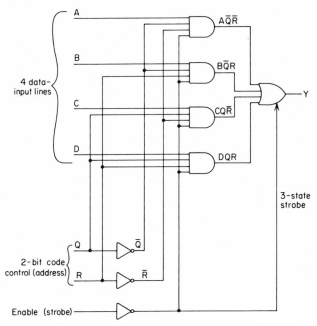

Fig. 3.22   One method used to implement a 4:1 digital multiplexer with Enable.

Most digital multiplexers have an additional control input called *enable* (or *strobe*).   This line controls all data input lines simultaneously. In the usual case if enable is LOW, all data inputs are controlled via the control inputs.   If enable is HIGH, the output terminal remains LOW regardless of the logic levels on the data input lines or the code control lines.   In microcircuits with three-state output circuits the strobe line is used to switch to the high-impedance state (see Chap. 2).

The timing diagram for a 2:1 digital multiplexer is shown in Fig. 3.23.   The truth table and boolean equation for the multiplexer in Fig. 3.22 are shown in Table 3.11.

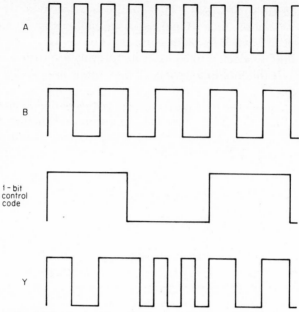

Fig. 3.23 The timing chart for a 2:1 digital multiplexer.

TABLE 3.11 Truth Table and Boolean Equation for a 4:1 Digital Multiplexer

Truth table*				
Inputs			Output	Boolean equation
Q	R	Enable	Y	
0	0	0	A	
0	1	0	B	$Y = \overline{\text{ENABLE}} \; (A\bar{Q}\bar{R} + B\bar{Q}R + CQ\bar{R} + DQR)$
1	0	0	C	
1	1	0	D	
X	X	1	0	

* $X = 0$ or 1.

3.10 DIGITAL DEMULTIPLEXER-DECODER

ALTERNATE NAMES Demultiplexer, decoder, M-line to 2^N-line decoder, M:N line decoder, 1-to-2^N demultiplexer.

PRINCIPLES OF OPERATION Although the decoder and demultiplexer names are often used interchangeably, there is a slight difference between these two circuits:

Demultiplexer. A single data line is channeled into one of 2^N output lines according to the code present on the N control lines. This is exactly opposite in function to a multiplexer.

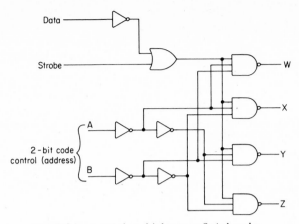

Fig. 3.24 A 1:4 demultiplexer or 2:4 decoder.

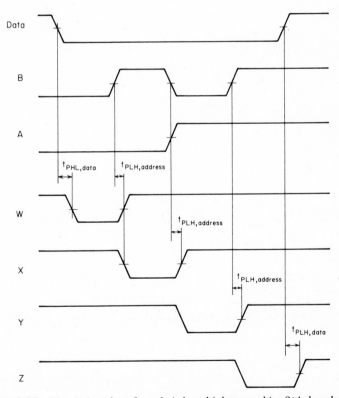

Fig. 3.25 The timing chart for a 1:4 demultiplexer and/or 2:4 decoder.

Decoder. One of the 2^N output lines is forced to a specified state depending on the code present at the N control lines.

If the demultiplexer data (or strobe) input line is fixed at a specified state, the circuit performs the decoder function. This is obvious from Fig. 3.24. If the data input is LOW or the strobe input is HIGH, the circuit is a 2:4 decoder. If the data line actually contains time-variable data, the circuit is a 1:4 demultiplexer.

This circuit is commonly available with a 4-bit BCD code input and 10 output lines, in which case it is a BCD-to-decimal decoder, or 4:10 decoder.

The timing chart for a 1:4 demultiplexer is shown in Fig. 3.25. The truth table and boolean equation for a 1:4 demultiplexer are shown in Table 3.12.

TABLE 3.12   Truth Table and Boolean Equations for a 1:4 Demultiplexer or 2:4 Decoder

			Truth table				
Inputs*				Outputs			
A	B	Data	Strobe	W	X	Y	Z
0	0	0	X	0	1	1	1
0	1	0	X	1	0	1	1
1	0	0	X	1	1	0	1
1	1	0	X	1	1	1	0
0	0	X	1	0	1	1	1
0	1	X	1	1	0	1	1
1	0	X	1	1	1	0	1
1	1	X	1	1	1	1	0
X	X	1	0	1	1	1	1

Boolean equations

$$W = \overline{(\text{DATA} + \text{STROBE})\ \overline{A}\overline{B}} \qquad X = \overline{(\text{DATA} + \text{STROBE})\ \overline{A}B}$$

$$Y = \overline{(\text{DATA} + \text{STROBE})\ A\overline{B}} \qquad Z = \overline{(\text{DATA} + \text{STROBE})\ AB}$$

* $X = 0$ or 1 (don't care).

3.11   FULL ADDER

ALTERNATE NAMES   Adder, gated full adder, N-bit adder, carry-save full adder, N-bit binary full adder, half adder with carry input, parallel-add–serial-carry circuit.

PRINCIPLES OF OPERATION   A 1-bit full adder has three inputs and two outputs, as shown in Fig. 3.26a. Logic inputs A and B are the input quantities to be added. Logic input C_i is the carry from the previous adder. Output S is the sum of A, B, and C_i. Output C_o is the carry resulting from the single-bit addition of A, B, and C_i. If this circuit had only the A and B inputs (no carry input), it would be called a *half adder*. Figure 3.26*b* shows how two or more 1-bit adders are connected to make adders for large words.

The timing chart for a typical 1-bit full adder is shown in Fig. 3.27.

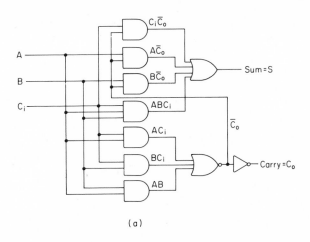

(a)

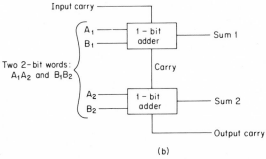

(b)

Fig. 3.26 *(a)* A 1-bit adder and *(b)* a 2-bit adder made from two 1-bit adders.

The truth table and boolean equation for 1-bit full adder are shown in Table 3.13.

An adder of 4 or more bits has substantial delay in the generation of the higher-order carry signals. This occurs since the carry signals are essentially generated in series while the sum outputs are formed in parallel. The speed of operation can be increased by using a *look-ahead carry generator*. Some manufacturers call this circuit a *fast parallel-carry circuit*. Newer microcircuits have this faster circuit on the same chip as the adder.

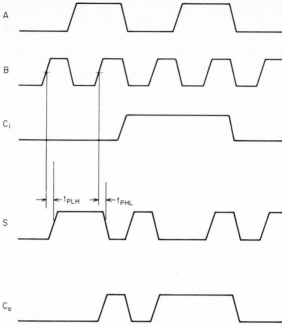

Fig. 3.27 The timing chart for a 1-bit full adder.

TABLE 3.13 Truth Table and Boolean Equations for a 1-Bit Full Adder

Truth table				
Inputs			Outputs	
A	B	C_i	S	C_o
0	0	0	0	0
0	1	0	1	0
1	0	0	1	0
1	1	0	0	1
0	0	1	1	0
0	1	1	0	1
1	0	1	0	1
1	1	1	1	1

Boolean Equations

$$\text{Sum} = S = C_i\,(AB + \overline{A}\overline{B}) + \overline{C}_i(A\overline{B} + \overline{A}B)$$
$$\text{Carry} = C_o = AB + C_i(A + B)$$

REFERENCES

1. Dean, J. A.: Astable and Monostable Oscillators Using RCA COS/MOS Digital Integrated Circuits, *RCA Appl. Note* ICAN-6267, 1970.
2. Corson, R.: Four Channel Multiplexer for Logic Experiments, *Radio-Electronics*, February 1974, p. 58.

Flip-Flops and Shift Registers

INTRODUCTION

The most common digital data-storage devices are the flip-flop and shift register. The simple RS flip-flop, shown in Fig. 4.1, is the basic internal circuit used in most of these storage devices. Several popular variations of flip-flops and shift registers will be outlined in this chapter. In most cases a truth table and timing chart will be included.

The random-access memory (RAM) can be fabricated from an array of flip-flops. This important device will be covered separately in the following chapter.

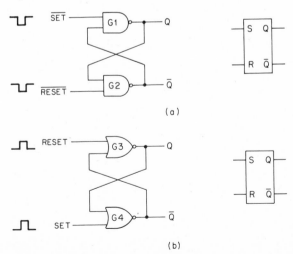

Fig. 4.1 Two versions of the RS flip-flop and their logic symbols: (a) using NAND gates and (b) using NOR gates.

4.1 *RS* FLIP-FLOP

ALTERNATE NAMES   *SR* flip-flop, latch, set-reset flip-flop, binary, bistable, Eccles-Jordan bistable element.

PRINCIPLES OF OPERATION   The *RS* flip-flop is the basic circuit from which most volatile data-storage devices are constructed.   Figure 4.1 shows two of the possible ways this circuit is implemented using NAND or NOR gates.   As shown in Fig. 4.1*a*, the inputs of the NAND gate *RS* flip-flop are normally maintained at a HIGH logic level.   Whenever a set or reset is required, the appropriate input is momentarily switched to a LOW level.   We must not allow both inputs to drop to the LOW level simultaneously, as this will cause both outputs to go HIGH.   When the input pulses disappear, the outputs will go to a state determined by the longest input pulse.   That is, if the set pulse is longer, the flip-flop will end up with the *Q* output HIGH.

The NOR-gate *RS* flip-flop shown in Fig. 4.1*b* requires positive input pulses for proper operation.   Both input lines are normally held in the LOW state.   A positive set pulse causes *Q* to go HIGH, and a positive reset pulse causes $\overline{Q}$ to go HIGH.

Figure 4.2 shows the timing chart for a NOR-gate *RS* flip-flop.   The truth tables for both types of *RS* flip-flops are shown in Table 4.1.

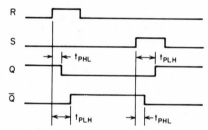

Fig. 4.2   Timing chart for a NOR-gate *RS* flip-flop.

TABLE 4.1   Truth Tables

NOR-gate *RS* flip-flop		Next state $n+1$	NAND-gate *RS* flip-flop		Next state $n+1$
Present state *n*			Present state *n*		
R	*S*	Q_{n+1}	*S*	*R*	Q_{n+1}
0	0	Q_n	⊔	⊔	*
0	⊓	1	⊔	1	1
⊓	0	0	1	⊔	0
⊓	⊓	*	1	1	Q_n

* Indeterminate.

Referring to Fig. 4.1*b*, we note that a reset pulse must propagate through one gate delay time *t$_{PHL}$* before it causes the *Q* output to change state. Two gate delay times are required before $\overline{Q}$ changes state. This is graphically shown in Fig. 4.2. A set pulse passes through the lower gate first, causing $\overline{Q}$ to change state before *Q.*

4.2 MASTER-SLAVE CLOCKED *RS* FLIP-FLOP

ALTERNATE NAMES   Clocked *RS* flip-flop, *SR* flip-flop, set-reset flip-flop, binary, bistable, latch.

PRINCIPLES OF OPERATION   A simple clocked *RS* flip-flop is shown in Fig. 4.3*a,* and the full master-slave version is presented in Fig. 4.3*b*. In Fig. 4.3*a* we note that the G3, G4 circuit is identical to the NAND-gate *RS* flip-flop shown in Fig. 4.1*a*. In normal operation the output line from G1 will have a HIGH quiescent state. This will be true if *S* or clock or both are LOW. Likewise, the G2 output line is normally in the HIGH state if *R* or clock or both are LOW.

Assume *Q* is initially LOW. If *S* changes to the HIGH state on the next positive clock pulse, the G3, G4 flip-flop will change states. As shown in Fig. 4.4, G1 changes states one gate delay time *t$_P$* after the clock pulse rises. Gates G3 and G4 require 2*t$_P$* and 3*t$_P$*, respectively, after the clock pulse. The flip-flop remains in this state regardless of any subsequent changes in *S* or clock. If *R* goes HIGH, however, G3 and G4 change states on the next clock pulse. As with the simple *RS*

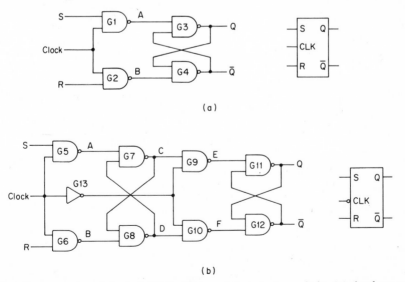

(a)

(b)

Fig. 4.3   Two types of clocked *RS* flip-flops and their logic symbols: *(a)* simple NAND version and *(b)* master-slave implementation.

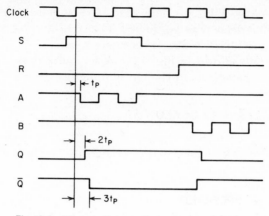

Fig. 4.4 Timing diagram for a clocked *RS* flip-flop.

flip-flop, we again note that the output waveform going from LOW to HIGH occurs one gate delay before the output which goes HIGH to LOW.

The master-slave clocked *RS* flip-flop shown in Fig. 4.3*b* is made from two identical clocked *RS* flip-flops. The second flip-flop, however, is driven with an inverted clock. This allows us to clock data (set or reset) into the first stage (the master flip-flop) when clock goes HIGH, and then these data are transferred to the second stage (the slave flip-flop) when clock goes LOW.

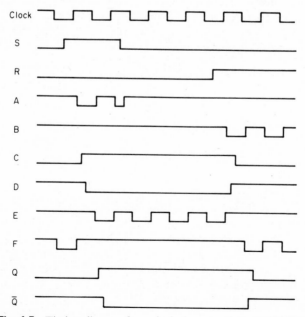

Fig. 4.5 Timing diagram for a clocked master-slave *RS* flip-flop.

A master-slave clocked *RS* flip-flop has two advantages over the simple clocked *RS* flip-flop: (1) the clock can be held LOW or HIGH, and the output stage will remain stable; and (2) the inputs and outputs of the flip-flop are isolated by one-half clock period. This minimizes the possibility of race conditions. Thus, the clock pulse width can be as long or short as desired within the speed capabilities of the logic family.

External measurements on a master-slave flip-flop lead one to believe that the device is sensitive to the negative-going edge of the clock since the slave flip-flop changes state soon after this falling edge.

As shown in Fig. 4.5, the *Q* output has a two-gate-time delay from the negative-going clock edge. Other circuit delays are clearly shown in the figure. The truth tables for these two clocked *RS* flip-flops are shown in Tables 4.2 and 4.3.

TABLE 4.2   Truth Table for a Clocked *RS* Flip-Flop

Present state n			Next state $n+1$	Present state n			Next state $n+1$
Clock	S	R	Q_{n+1}	Clock	S	R	Q_{n+1}
0	0	0	Q_n	1	0	0	Q_n
0	0	1	Q_n	1	0	1	0
0	1	0	Q_n	1	1	0	1
0	1	1	Q_n	1	1	1	*

* Indeterminate.

TABLE 4.3   Truth Table for a Clocked Master-Slave *RS* Flip-Flop

Present state n			Next state $n+1$	Present state n			Next state $n+1$
Clock	S	R	Q_{n+1}	Clock	S	R	Q_{n+1}
0	0	0	Q_n	1	1	0	Q_n
0	0	1	Q_n	1	1	1	Q_n
0	1	0	Q_n	⊓	0	0	Q_n
0	1	1	Q_n	⊓	0	1	0
1	0	0	Q_n	⊓	1	0	1
1	0	1	Q_n	⊓	1	1	*

* Indeterminate

The clocked *RS* flip-flop and the master-slave clocked *RS* flip-flop are not generally available as packaged microcircuits. Their unpopularity stems from the indeterminate output when both *R* and *S* inputs are HIGH. The other flip-flops in the following pages are widely used since they incorporate the best features of the *RS* flip-flops and also overcome the indeterminate-output problem. The RS flip-flops are included here to provide the reader with a smooth advance from the simple two-gate RS flip-flop up to the JK flip-flop.

4.3   DATA FLIP-FLOP

ALTERNATE NAMES   Delay flip-flop, D flip-flop.

PRINCIPLES OF OPERATION   As illustrated in Fig. 4.6, the data flip-flop can be constructed using a simple clocked RS flip-flop and an inverter. This inverter guarantees that HIGH logic levels do not simultaneously appear at the R and S inputs. Whatever data appear at the S

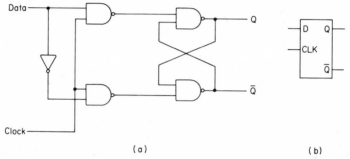

Fig. 4.6   The data flip-flop is made from a clocked RS flip-flop and an inverter: *(a)* one common implementation and *(b)* logic symbol.

input, i.e., the D input, just the opposite appears at the R input (output of inverter). When the clock input goes HIGH, the data logic level is transferred over to the Q output.

The D flip-flop is useful for many reasons. Several obvious applications follow:

1. The leading edge of data can be made synchronous with the clock by sending it through a D flip-flop.

TABLE 4.4   Truth Table for a Data Flip-Flop

Present state n		Next state $n+1$
Clock (strobe)	Data	Q_{n+1}
0	0	Q_n
0	1	Q_n
1	0	0
1	1	1
⊓	0	0
⊓	1	1

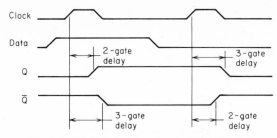

Fig. 4.7   Timing diagram for the data flip-flop.

2. Information which appears at irregular intervals along with other information on the data line can be strobed into the *D* flip-flop and saved. For example, in microprocessor systems the data bus may be connected to many loads, but at any given instant only one of these loads should be "listening" to the microprocessor. If the microprocessor sends out a strobe (clock) signal (a positive pulse) just as the correct data bit appears, a particular *D* flip-flop can receive and store it. The other unstrobed devices on the line do not store the information.

The truth table for a data flip-flop is shown in Table 4.4, and its timing diagram is presented in Fig. 4.7.

D FLIP-FLOP APPLICATION; SYNCHRONIZED ONE-SHOT The circuit shown in Fig. 4.8 is very popular with logic designers. It generates a single pulse without using an *RC* timing circuit. The output pulse width will be precisely one-half clock period wide if the gate input is synchronous with the falling edge of the clock. This requirement is easily achieved in synchronous systems.

Figure 4.9 helps clarify the operation of this one-shot. We first note that the output pulse occurs only when both *D* and $\overline{Q}$ are HIGH. Logic level $\overline{Q}$ is normally HIGH if we assume the *D* input is normally LOW. As the *D* input rises in synchronism with a clock falling edge, the AND gate is turned ON. At the next rising edge of the clock the *D* input is entered into the flip-flop and $\overline{Q}$ goes LOW. The AND-gate output likewise goes LOW and terminates the output pulse.

This particular circuit requires a positive-edge-triggered *D* flip-flop.

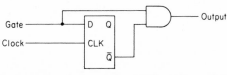

Fig. 4.8   A synchronized one-shot which utilizes a *D* flip-flop.

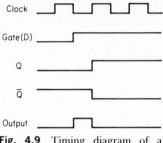

Fig. 4.9 Timing diagram of a synchronized one-shot.

A negative-edge-triggered flip-flop could be used if the circuit were modified slightly.

4.4 BASIC *JK* FLIP-FLOP

ALTERNATE NAMES Master *JK* flip-flop, preset-clear *JK* flip-flop, clocked *JK* flip-flop, binary, bistable.

PRINCIPLES OF OPERATION The basic *JK* flip-flop evolved from the *RS* flip-flop in an attempt to overcome the possibility of an indeterminate output state. The simple four-gate *JK* flip-flop shown in Fig. 4.10 overcomes the indeterminate output state through the use of feedback. The input gates of the *RS* flip-flop are changed to three-input gates, and these third inputs are tied to the Q and $\overline{Q}$ outputs. This circuit is

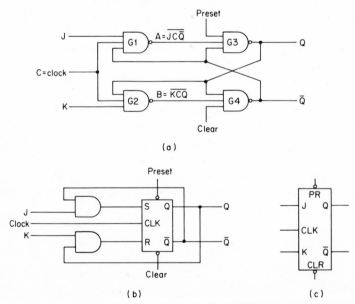

Fig. 4.10 The basic *JK* flip-flip: *(a)* one possible implementation using NAND gates; *(b)* abbreviated logic diagram using an *RS* flip-flop; *(c)* the common *JK* flip-flop logic symbol.

identical to the clocked *RS* flip-flop shown in Fig. 4.3*a* except for the new feedback connections.

The truth table for the basic *JK* flip-flop is shown in Table 4.5. Figure 4.11 shows the timing relationships of all waveforms in Fig. 4.10*a*.

TABLE 4.5 Truth Table for the Basic *JK* Flip-Flop

		Present state* n			Next state $n+1$
J	*K*	Clock	Preset	Clear	Q_{n+1}
X	X	0	0	0	†
X	X	0	0	1	1
X	X	0	1	0	0
X	X	0	1	1	Q_n
0	0	⊓	1	1	Q_n
0	1	⊓	1	1	0
1	0	⊓	1	1	1
1	1	⊓	1	1	$\overline{Q}_n$

* X = either 0 or 1.
† Indeterminate.

Gates G1 and G2 utilize the Q and $\overline{Q}$ feedback levels to prevent an indeterminate output when J and K are both high. If Q is high and $\overline{Q}$ is LOW, G1 is inhibited and logic signal A cannot change states. Therefore, if HIGH levels appear at both J and K, only G2 is able to pass a LOW logic level into the G3, G4 latch. A LOW level at B will change the state of the latch.

This basic flip-flop is not produced commercially since it has a potential race-around problem. If the clock pulse width T_p is wider than the delay of three gates, the circuit will oscillate. This circuit therefore operates satisfactorily only if $T_p < T_d$ (see Fig. 4.11). The explanation of this problem is as follows. Suppose Q is HIGH and $\overline{Q}$ is LOW. Also assume that high levels appear on both J and K. When a clock pulse goes HIGH, G2 changes states, causing B to go LOW. If clear and preset are both HIGH (as they should normally be), G4 changes states, causing $\overline{Q}$ to go HIGH. This causes Q to go LOW. However, the $\overline{Q}$ signal also allows G1 to change states if the clock is still high. This is undesirable since a LOW level at A will cause the latch to flip again. The oscillation will continue as long as the clock line is HIGH.

The potential oscillation problem described above is overcome in the master-slave *JK* flip-flop of Sec. 4.5.

The toggle flip-flop, often called the T flip-flop, is made from the *JK* flip-flop by simply connecting the J and K inputs to the positive supply voltage. Each time a clock pulse occurs, the Q and $\overline{Q}$ outputs will change states. The bottom line of the truth table in this section shows the toggle mode.

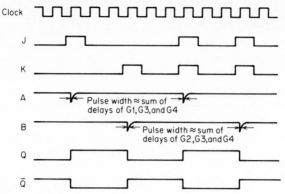

Fig. 4.11   Input, output, and internal waveforms in the basic *JK* flip-flop.   Note that preset = clear = **1**, where **1** stands for logic 1.

4.5   MASTER-SLAVE *JK* FLIP-FLOP

ALTERNATE NAMES   Preset-clear master-slave *JK* flip-flop, clocked master-slave *JK* flip-flop, binary, bistable.

PRINCIPLES OF OPERATION   All commercially available *JK* flip-flops are of the master-slave variety.   This device eliminates all the shortcomings of *RS* and simple *JK* flip-flops. There are no indeterminate conditions or race-around problems.

One common implementation of the master-slave *JK* flip-flop is shown in Fig. 4.12*a*.   By comparing Figs. 4.3*a*, 4.10*a*, and 4.12*a* we see that the master-slave *JK* flip-flop is a basic *JK* flip-flop followed by a clocked *RS* flip-flop.   In addition, the feedback signals for the basic *JK* flip-

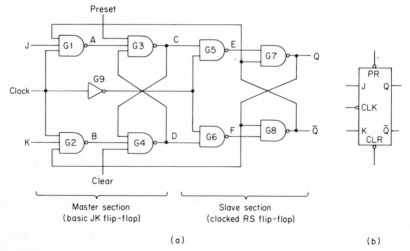

Fig. 4.12   (*a*) Simplified logic diagram of one common type of master-slave flip-flop; (*b*) logic symbol.

flop now comes from the output of the *RS* flip-flop. We will refer to these two flip-flops as the *master* and the *slave.*

Data are clocked into the master flip-flop when the clock line goes HIGH. Gate G9 inverts the clock so that data are clocked into the slave flip-flop when the clock goes LOW. Therefore, the slave flip-flop is inhibited while the master flip-flop is changing states. Since the feedback comes from the inhibited flip-flop, no race-around condition is possible.

The truth table for the master-slave flip-flop is shown in Table 4.6. The timing diagram for this device is given in Fig. 4.13.

TABLE 4.6   Truth Table for the Master-Slave *JK* Flip-Flop

		Present state* n			Next state $n+1$
J	K	Clock	Preset	Clear	Q_{n+1}
X	X	0	0	0	†
X	X	0	0	1	1
X	X	0	1	0	0
X	X	0	1	1	Q_n
0	0	⊓	1	1	Q_n
0	1	⊓	1	1	0
1	0	⊓	1	1	1
1	1	⊓	1	1	$\overline{Q}_n$

* $X=$ either 0 or 1.
† Indeterminate.

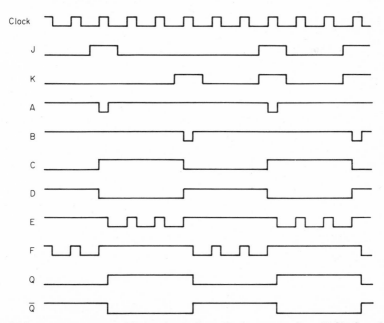

Fig. 4.13 Input, output, and internal waveforms in the master-slave *JK* flip-flop shown in Fig. 4.12. Note that preset = clear = **1.**

The *T* flip-flop is made from the master-slave *JK* flip-flop by applying HIGH levels to the *J* and *K* inputs. Each time a clock pulse appears, the flip-flop will toggle to the other state.

4.6 EDGE-TRIGGERED *JK* FLIP-FLOP

ALTERNATE NAMES   AC-coupled *JK* flip-flop, positive-edge-triggered *JK* flip-flop, binary, bistable.

PRINCIPLES OF OPERATION   This type of flip-flop is useful in applications where the *JK* information must be sensed and transferred to the output circuit—all on the positive edge of the clock pulse. This is the same operation performed by the basic *JK* flip-flop in Sec. 4.4, but the race-around condition is not present in the edge-triggered flip-flop discussed here.

The edge-triggered flip-flop shown in Fig. 4.14 is a substantial departure from the symmetrical flip-flops shown in earlier sections of this chapter. The main latch for this flip-flop is the G7, G8 gates. The clock is transferred over to the latch through gates G5 and G6. The feedback of the *Q* and $\overline{Q}$ signals into gates G1 and G2 makes this device perform as a true *JK* flip-flop. However, the *J* and *K* signals can be clocked into the latch through G1 and G2 only if signal point *F* is HIGH. By providing this extra gating (through G1, G2, and G3) a race-around condition during the clock pulse is prevented. The simple *JK* flip-flop shown in Fig. 4.10*a* is also positive-edged-sensitive, but it suffers from race-around.

The truth table for the positive-edge-triggered *JK* flip-flop is shown in Table 4.7. Figure 4.15 shows the timing diagram for this device for all combinations of the *J* and *K* input lines.

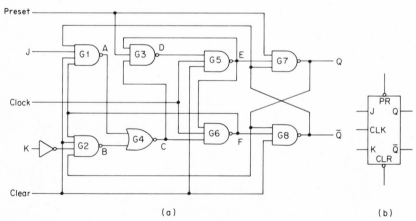

(a)   (b)

Fig. 4.14   (*a*) The edge-triggered flip-flop and (*b*) its logic symbol.

TABLE 4.7   Truth Table for the Edge-triggered *JK* Flip-Flop

		Present state* n			Next state $n+1$
J	K	Clock	Preset	Clear	Q_{n+1}
X	X	X	0	0	†
X	X	X	0	1	1
X	X	X	1	0	0
X	X	0	1	1	Q_n
0	0	⎍	1	1	Q_n
0	1	⎍	1	1	0
1	0	⎍	1	1	1
1	1	⎍	1	1	$\overline{Q}_n$
X	X	⎁	1	1	Q_n

* $X=$ either 0 or 1.
† Indeterminate.

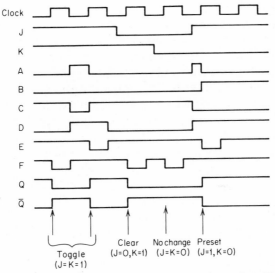

Fig. 4.15   Timing diagram of the positive-edge-triggered *JK* flip-flop.

4.7   SERIAL-IN–SERIAL-OUT SHIFT REGISTER

ALTERNATE NAMES   Serial shift register, serial memory, N-bit shift register.

PRINCIPLES OF OPERATION   For small values of N this device is normally constructed with a cascade of *RS* flip-flops, as shown in Fig. 4.16. These are often clocked master-slave *RS* flip-flops of the type shown in Fig. 4.3*b*.   Nearly all the logic families discussed in Chap. 2 contain

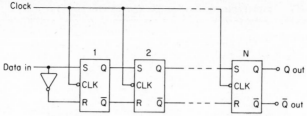

Fig. 4.16 A serial shift register using *RS* flip-flops.

serial shift registers of low *N*. For large *N* the MOS and CCD families dominate the market. From the circuit designer's viewpoint we are most interested in the external characteristics of a shift register. The important questions are:

1. Does the shift register use true static memory cells which do not need refreshing?

2. If refreshing is required, is it mostly performed on the chip or is complex external circuitry required?

3. How many clocks are required for the shift register? Can they be overlapping clocks, or do the two, three, or four clock phases require carefully spaced nonoverlapped pulses?

4. Could the shift register be replaced with a random-access-memory (RAM) chip in conjunction with a counter chip?

Static *RS* flip-flop serial shift register Static shift registers using *RS* flip-flops were the natural first step in evolving from discrete circuits to microcircuits. As shown in Fig. 4.16, the *Q* output of each stage is connected to the *S* inputs of the following stage. Likewise, each $\overline{Q}$ output goes to the following *R* input. This guarantees that the *S* and *R* inputs of each stage will always be at different states. The indeterminate output state mentioned in Secs. 4.1 and 4.2 is thereby avoided.

As the clock pulse rises in Fig. 4.16, the master section of each flip-flop reads in the data from the previous stage. When the clock pulse falls, the data in each master flip-flop are transferred over to their corresponding slave flip-flop (see Sec. 4.2). This master-slave clocking procedure is necessary so that each stage can solidly read data from a previous stage before the output of that previous stage begins to change to its new state. The master-slave *RS* flip-flop is a natural for this application, but because it requires many active devices (about 20) per flip-flop, large-*N* shift-register microcircuits never use this type of memory cell.

Static CMOS serial shift register Figure 4.17 shows the logic diagram of a typical CMOS static shift register which uses *D* flip-flops. Operation of this circuit is best understood by referring to its timing diagram in Fig. 4.18. CMOS logic circuits are usually built from transmission gates and inverters. Switches are used to represent the transmission gates. We have shown a separate diagram for each phase of the clock.

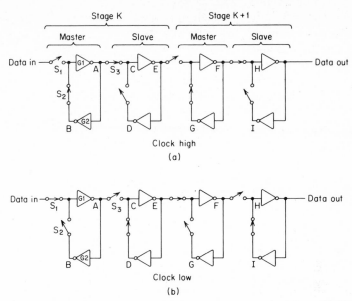

Clock high
(a)

Clock low
(b)

Fig. 4.17 Two stages of a serial static CMOS shift register which uses D flip-flops: *(a)* clock HIGH; *(b)* clock LOW.

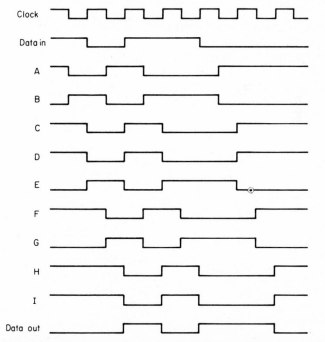

Fig. 4.18 Timing diagram for two stages of a CMOS static shift register which uses D flip-flops.

When the clock is HIGH, switch S_1 is open. The input data are allowed to change states during this interval. When S_1 closes (clock LOW), the input data drive A and $B,$ as shown in Fig. 4.18. S_3 prevents data from being transferred to the slave portion of stage 1. When the clock goes HIGH and S_1 opens, the stored charge on the gates of G1 and G2 maintains the same logic levels until S_2 closes. Now that the loop is closed, the positive feedback will maintain the logic state which existed before the opening of S_1. At the same time the loop is closed, S_3 closes and the slave portion of stage 1 receives the shifted data. The slave flip-flop is identical to the master flip-flop except that it becomes a closed-loop circuit when the clock is LOW whereas G1 and G2 form a closed loop when the clock is HIGH.

The data are shifted at the rate of one flip-flop (such as G1 and G2) per half clock pulse. Thus, as shown in Fig. 4.18, two complete clock pulses are required to shift the input data through two stages.

We note in passing that standard MOS serial static shift registers are also fabricated using two-phase D flip-flops. These circuits require only half the parts of the CMOS devices, but their power consumption is an order of magnitude higher.

Dynamic two-phase serial shift register Dynamic circuits are widely used because they require less than half as many active devices as static circuits. This automatically means that more than twice the number of stages can be packed into a given chip size. In the majority of shift-register applications the dynamic nature of these devices is of no concern since many systems require continual data recirculation.

A typical four-transistor two-phase dynamic storage cell is shown in Fig. 4.19. The master section is Q_1, Q_2, C_1, and C_2 while the slave section is Q_3 Q_4, C_3, and C_4. All stages are isolated from each other

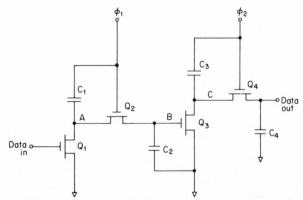

Fig. 4.19 A typical four-transistor two-phase dynamic memory cell from a serial shift register.

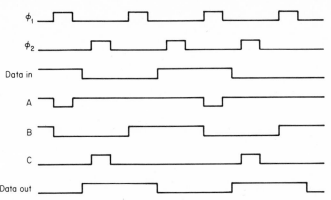

Fig. 4.20   Input, internal, and output waveforms of a typical stage in a two-phase dynamic shift register (n-channel MOSFETs assumed).

with transmission gates (Q_2, Q_4, etc.).   Logic levels are stored on the capacitors which go to ground (C_2, C_4, etc.)   The timing diagram in Fig. 4.20 shows how logic signals are passed through this single stage. We first note that the drain supply voltage for Q_1 is available only when ϕ_1 is HIGH.   It is capacitively coupled to Q_1 through C_1.   During the positive ϕ_1 pulse the inverse of the input data appears at A.   Since ϕ_1 also turns Q_2 ON, $A = B$ during this pulse.   After the ϕ_1 pulse terminates, this inverted logic level stays on C_2.   When the ϕ_2 pulse goes HIGH, the inverse of the logic level at B appears at C and data out. When ϕ_2 goes low, C_4 momentarily holds the logic signal at data out until it is shifted into the next stage.

Dynamic four-phase serial shift register   Several types of four-phase shift-register circuits are used in the industry.   The type shown in Fig. 4.21 requires nonoverlapping clocks, as shown in Fig. 4.22.   Some devices allow overlapping of the clocks at the expense of larger geometry. This is all insignificant to the average circuit designer since most manufacturers place the clock-phase generating circuits on the chip.   These types require only a single input clock, thus eliminating this design problem for the device user.

Figure 4.22 shows the timing diagram for one memory cell of a four-phase shift register.   We assume 10100110 is being shifted through this master-slave memory cell.   As ϕ_1 goes HIGH, node B (C_2) is precharged HIGH.   When ϕ_2 goes HIGH and turns Q_2 ON, nodes B and A have two possible states.   If data in is LOW, Q_3 remains OFF during ϕ_2 and B (and A) remains HIGH.   However, if data in is HIGH, Q_3 turns ON during ϕ_2 and B and A are pulled down to a LOW state.   At the end of ϕ_2 the master section Q_1, Q_2, and Q_3, will have caused the logical inverse of data in to be stored on C_2.

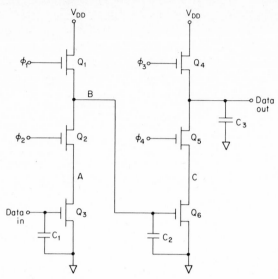

Fig. 4.21 A nonoverlapping-clock four-phase dynamic-shift-register memory cell.

The slave section Q_4, Q_5, and Q_6 is identical in operation to that of the master section. Therefore, at the end of ϕ_4 the inverse of B will be stored on C_3. Thus, data in has been moved from C_1 to C_3 in four clock cycles (ϕ_1, ϕ_2, ϕ_3, and ϕ_4).

The four-phase dynamic shift register can operate at higher clock rates than the two-phase dynamic shift register. Dissipation is very low

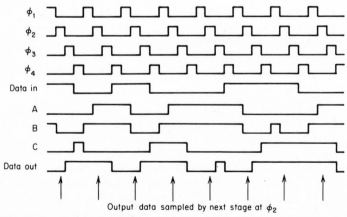

Fig. 4.22 Timing diagram for the nonoverlapping-clock four-phase dynamic shift register (n-channel MOSFETs assumed).

since the three transistors in each section are never ON at the same time. This prevents any dc path from V_{DD} to ground during any phase of the clock.

Charge-coupled-device dynamic serial shift register The longest serial shift registers use charge-coupled devices (CCD). Registers with 100,000 or more stages on a single chip are commercially available. As shown in the conceptual cutaway drawing in Fig. 4.23, the CCD shift register is essentially a row of capacitors. The negative electrodes of these capacitors are made from deposited metal over a layer of silicon dioxide insulation. The silicon dioxide is deposited on a uniformly doped semiconductor substrate. The substrate is grounded, and the metal electrodes are connected to three-level three-phase clocks. Some CCD devices are designed to operate with two- or four-phase clocks.

When a negative voltage is applied to an electrode, it causes a potential well to be created on the surface of the semiconductor substrate immediately below the electrode. This potential well has the ability to attract and store positive charge, if any, within 0.3 μm or less. If an adjacent electrode is given a more negative voltage, it will remove the charge from the first. A more negative charge on an electrode produces a deeper potential well. The charge behaves like water; i.e., it runs into the deepest well in the immediate vicinity.

Figure 4.24 shows a typical set of clock waveforms for a three-phase CCD device. The figure shows that charge is transferred whenever a −15-V electrode appears between −5- and −10-V electrodes. In this particular clocking scheme the −15-V electrode receives the charge. This same electrode then drops to −10 V while the −10-V electrode, which initially supplied the charge, returns to the base value of −5 V.

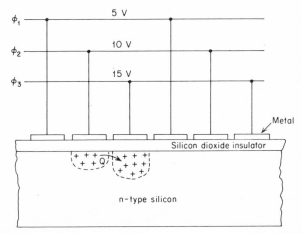

Fig. 4.23 A cutaway of a section of a three-phase CCD shift register.

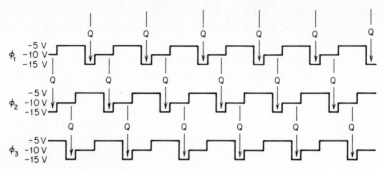

Fig. 4.24 Typical clock waveforms for a three-phase CCD shift register. Q represents the time at which a bundle of charge is transferred.

The next electrode down the shift register then drops to -15 V and removes the charge from the cell which just changed to -10 V. The charge traveling from cell to cell is large if it represents a HIGH logic level. Likewise, the bundle of charge is very small if it represents a LOW logic level.

4.8 PARALLEL-IN–SERIAL-OUT SHIFT REGISTER

ALTERNATE NAMES Parallel-to-serial converter, N-bit shift register.

PRINCIPLES OF OPERATION Microprocessor systems typically operate with 4-, 8-, 12-, or 16-bit parallel words. Many input-output devices

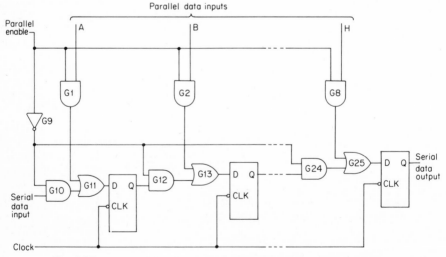

Fig. 4.25 A common type of 8-bit parallel-in–serial-out shift register.

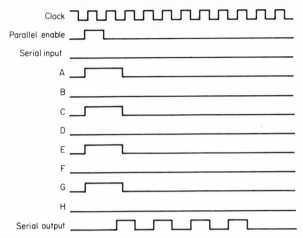

Fig. 4.26   Timing diagram of an 8-bit parallel-in–serial-out shift register.

used in microprocessor systems, however, are designed to communicate with the rest of the system using 8-bit serial words.   Parallel-in–serial-out shift registers are ideally suited for the required parallel-to-serial conversion when the microprocessor talks to the input-output device.

Figure 4.25 portrays a common logic configuration for an 8-bit parallel-in–serial-out shift register.   Individual stages use the standard D flip-flop discussed in Sec. 4.3.   The timing diagram for an 8-bit version of this device is shown in Fig. 4.26.   Parallel data are synchronously entered into all stages using AND gates G1 through G8.   When parallel enable is HIGH and parallel data are present, the parallel data are entered on the falling edge of the clock.   After the clock line settles on the LOW state, parallel enable is allowed to return to the LOW state.   Data are now shifted 1 bit to the right on each falling edge of the clock.   After eight clock pulses all the original parallel data will be shifted through the device and out of the serial output terminal.

Most parallel-in–serial-out shift registers also allow data to be entered serially.   In Fig. 4.25 this is done by using the serial-data input while the parallel-enable line is LOW.   Under these conditions the G9 output goes HIGH and the G10 output is under control of the serial-data input. As before, data are shifted to the right on each negative transition of the clock.

SHIFT-REGISTER APPLICATION: MICROPROCESSOR-TELETYPEWRITER INTERFACE

Many commercial microcircuits are available for the parallel-to-serial conversion required between microprocessors and teletypewriters.

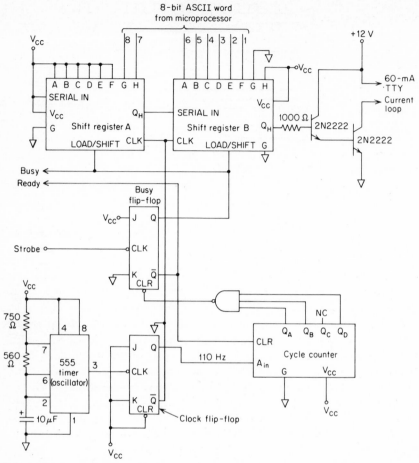

Fig. 4.27 A parallel-to-serial converter for driving a teletypewriter from an 8-bit microprocessor.

The circuit shown in Fig. 4.27 is included here to clarify some of the basic principles of these circuits.

As shown in the timing diagram in Fig. 4.28, the strobe from the microprocessor initiates the sequence of events. Before the strobe pulse the circuit is in the ready mode. The two shift registers are simultaneously in the load mode, and the cycle counter is locked in its reset state. A HIGH state is present at the output of the shift register since the first parallel input bit is connected to V_{CC}.

The negative strobe pulse causes the busy flip-flop to put out HIGH busy and LOW ready signals. The LOW ready signal allows the cycle counter to begin counting the 110-Hz square wave from the clock flip-flop. The HIGH busy logic signal puts the shift registers into the shift

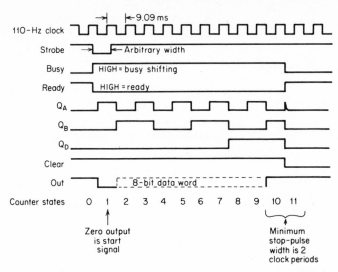

Fig. 4.28 Timing diagram for the 8-bit microprocessor-teletype parallel-to-serial converter.

mode. The shift registers and cycle counters operate synchronously. For each shift pulse the cycle counter advances one state. On state 1 a LOW logic level is presented to the 2N2222 since the second bit in shift register B is hard-wired to ground. This corresponds to the teletypewriter start pulse. The 8-bit microprocessor word is next shifted out as the cycle counter advances through states 2 to 9. When state 10 is reached, the shift register transmits a HIGH level, corresponding to the teletypewriter stop pulse.

The stop pulse must be at least 1½ clock periods wide for most teletypewriters. This circuit produces a stop pulse 2 clock periods wide. As the cycle counter reaches state 11, the NAND-gate output goes LOW. This clears the busy flip-flop, which in turn clears the cycle counter. The shift registers are now in the load mode. The serial output terminal remains at a HIGH level during state 11 since the shift register will not begin shifting the next microprocessor word until the next negative clock transition. Thus, the shift-register output continuously remains HIGH during states 10 and 11. If no strobe signal is received, the ready signal and output data will both remain in HIGH states indefinitely.

4.9 SERIAL-IN–PARALLEL-OUT SHIFT REGISTER

ALTERNATE NAMES Serial-to-parallel converter, N-bit shift register.

PRINCIPLES OF OPERATION Shift registers of this type have many applications, e.g., converting serial alphanumeric keyboard data to the parallel data format required in microprocessor systems. Quite often this

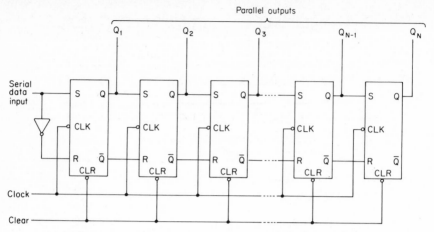

Fig. 4.29   A serial-in–parallel-out shift register which uses RS flip-flops.

function is on the same chip as the parallel-in–serial-out shift register mentioned in Sec. 4.8.

The internal devices for this register can be either RS or D flip-flops. As observed from outside the package, performance of these two flip-flop types is identical.   The RS flip-flop used here was introduced in Sec. 4.2.

The logic diagram and timing diagram for this device are shown in Figs. 4.29 and 4.30, respectively.

Serial data are clocked into the slave section of the first RS flip-flop on the negative transition of the clock.   Each successive clock pulse will move the data toward the right by one stage.   New data must appear at the serial data input before the negative transition of the clock.

Parallel output data are available at all Q output terminals.   Some shift registers also provide $\overline{Q}$ output data.

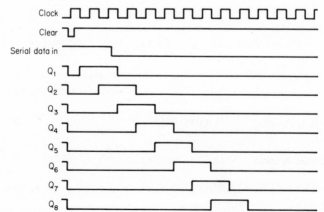

Fig. 4.30   Timing diagram of an 8-bit serial-in–parallel-out shift register.

All flip-flops are set to zero ($Q = \mathbf{0}$)[1] when the clear line is driven to the LOW state. The clear line must be held in the HIGH state during normal data input and shifting operations.

4.10 PARALLEL-IN–PARALLEL-OUT SHIFT REGISTER

ALTERNATE NAMES   Shift-right–shift-left register, right-shift–left-shift register, parallel-to-serial converter, serial-to-parallel converter, serial shift register, parallel shift register, bidirectional universal shift register.

PRINCIPLES OF OPERATION   The parallel-in–parallel-out shift register is the most general type and can be used in place of serial, serial-to-parallel, or parallel-to-serial shift registers. However, since the parallel-to-parallel shift register usually has so many extra features, such as shift right or left, it seldom is available in lengths of more than 8 bits. A typical implementation of a 4-bit bidirectional universal shift register is presented in Fig. 4.31.

[1] For clarity logic $\mathbf{0}$ and $\mathbf{1}$ are set in boldface type in the text.

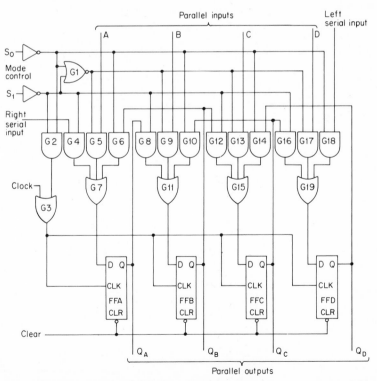

Fig. 4.31   A 4-bit bidirectional universal shift register featuring parallel or serial inputs, parallel or serial outputs, and right-shift–left-shift capability.

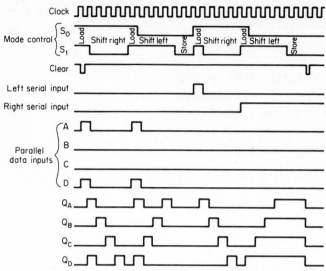

Fig. 4.32 Timing diagram of the bidirectional universal shift register shown in Fig. 4.31.

The operation of this device is best understood by referring to the timing diagram in Fig. 4.32 and the truth table in Table 4.8.

Parallel data are accepted on the A, B, C, and D inputs when both mode-control lines are HIGH. The mode-control lines must change states only when the clock line is HIGH. The parallel data are entered into the register at the next positive clock transition. If both control lines are HIGH, the G1 output is HIGH. This allows G5, G7, G9, G11, G13, G15, G17, and G19 to pass the parallel input data to the D inputs of the flip-flops.

If data are to be shifted right, the S_0 line remains HIGH and S_1 is

TABLE 4.8 **Truth Table for a Parallel-in–Parallel-out Shift Register***

	Inputs (present state)									Outputs (next state)			
	Mode control			Serial		Parallel							
Clear	S_1	S_0	Clock	Left	Right	A	B	C	D	Q_A	Q_B	Q_C	Q_D
H	X	X	0	X	X	X	X	X	X	nc	nc	nc	nc
H	1	1	⌐	X	X	a	b	c	d	a	b	c	d
H	0	1	⌐	X	1	X	X	X	X	1	a	b	c
H	0	1	⌐	X	0	X	X	X	X	0	a	b	c
H	1	0	⌐	1	X	X	X	X	X	b	c	d	1
H	1	0	⌐	0	X	X	X	X	X	b	c	d	0
H	0	0	X	X	X	X	X	X	X	nc	nc	nc	nc
L	X	X	X	X	X	X	X	X	X	0	0	0	0

* a, b, c, d = data on A, B, C, D at positive transition of clock; nc = no change.

brought to the LOW state. The data present on the right-shift serial input move through G4 and G7 to flip-flop A. They are then moved into the slave portion of flip-flop A at each positive clock transition. Likewise, for left shifting the S_0 and S_1 states are reversed. The serial data from the outside arrive via G18 and G19 and are transferred into the slave portion of flip-flop D at each positive clock transition.

Internal shifting to the right utilizes gates G8, G11, G12, G15, G16, and G19. Internal shifting to the left uses gates G14, G15, G10, G11, G6, and G7.

We note that if both S_0 and S_1 are LOW, the G2 output goes HIGH. This passes through the G3 NOR gate and locks all flip-flop clocks to the HIGH state. Thus, whenever $S_0 = S_1 = 0$, the register is idle and holds the state that existed before this mode-control command.

4.11 SHIFT-REGISTER APPLICATION; PULSE-SEQUENCE GENERATOR

ALTERNATE NAMES Pulse-burst generator, programmable pulse generator, finite-number pulse source, pulse-train generator, gated clock.

PRINCIPLES OF OPERATION The serial-in–parallel-out shift register provides an economical means of generating short pulse sequences.

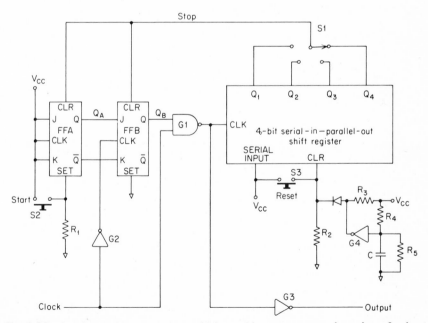

Fig. 4.33 A pulse-sequence generator which provides a programmed number of pulses according to the position of S1.

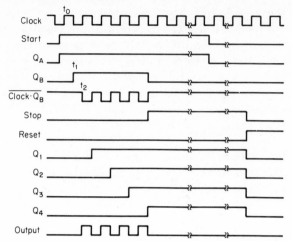

Fig. 4.34 Voltage waveforms at various locations in the pulse-sequence generator in Fig. 4.33.

Long pulse sequences are best generated using counters and digital comparators. In Fig. 4.33 we find that a pulse-sequence generator capable of producing 1 to 4 pulses requires only three packages. This can be extended to 8 pulses by replacing the 4-bit shift register with an 8-bit shift register. For sequences up to 16 pulses we could use two 8-bit shift registers in a cascade connection.

As illustrated by the timing chart in Fig. 4.34, the duration of the output pulse sequence is determined by the state of flip-flop B. The output pulse train occurs only while Q_B is HIGH.

Before operation of the start push button we must clear the shift register using the reset button. This puts the stop line in the LOW state, thus making the two flip-flops ready for a new pulse sequence. The start push button sets flip-flop A, making the J input of flip-flop B go HIGH. This can occur at any arbitrary point in the clock cycle. At the next negative transition of the clock flip-flop B will be set. The flip-flop output, however, does not change states until after the propagation of this device t_P. Therefore, gate G1 cannot perform the NAND function and transfer the clock to the output until the next positive transition of the clock. Since the serial input of the shift register is maintained at a HIGH logic level, Q_1 will rise at the first positive transition of $\overline{CLOCK \cdot Q_B}$. Output Q_2 will rise with the second positive transition of $\overline{CLOCK \cdot Q_B}$, Q_3 rises with the third, etc. When Q_4 rises, switch S1 routes this stop signal to the clear inputs of both flip-flops, thus causing these devices to be cleared.

It is most likely that the pulse sequence will end before the operator's finger is removed from the start button. The set and clear inputs of flip-flop A will both be HIGH simultaneously. This will cause both the Q and $\overline{Q}$ outputs of this flip-flop to be HIGH until the start button is disengaged. The situation is different with flip-flop B. The clear input totally overides any data on the J, K, or clock inputs. This flip-flop will therefore be cleared, and gate G1 will cease to pass the pulse train.

DESIGN PARAMETERS

Parameter	Description
C	Capacitor which determines power-on reset delay
CLOCK	Input clock signal
$\overline{\text{CLOCK}} \cdot Q_B$	Gated clock which drives the shift register
f_{max}	Maximum clock frequency
FFA	Flip-flop used to debounce the start signal from S2
FFB	Timing flip-flop which synchronizes the start signal with falling edge of the clock
G1	NAND gate which passes the clock signal to the shift register only when Q_B is HIGH
G2	Gate which inverts the clock so that FFB will change states on the negative transition of the clock
G3	Inverter gate used only to provide positive output pulses
G4	Open-collector output gate used to provide power-on reset signal
$I_{in,0}$	Logical 0 input current for any flip-flop or shift register input terminal
Output	Pulse-sequence output signal
Q_A	Latched start signal from FFA
Q_B	Synchronized start signal from FFB
$Q_1–Q_n$	Shift-register parallel outputs
R_1	Pull-down resistor to hold FFA set input at a nominal LOW logic level
R_2	Pull-down resistor to hold shift-register clear input at a nominal LOW logic level
R_3	Pull-up resistor for open collector of G4
R_4	Resistor which (along with C) determines delay of power-on reset circuit
R_5	Large resistor used to discharge C slowly when power is turned off
Reset	A signal line which goes HIGH when S3 is depressed or power-on reset circuit is activated; it clears all registers in the shift register
S1	Switch used to select number of pulses in pulse train
S2	Switch (push button) used to provide start signal
S3	Switch (push button) which resets shift register
SR	Shift register
Start	Signal at set input of FFA which initiates pulse sequence
Stop	Signal passing through S1 which terminates pulse sequence
t_0	Time at which start signal goes HIGH
t_1	First negative transition of clock after start command is given
t_2	Leading edge of first clock signal to pass through G1
$t_{P,G1}$	Propagation delay time of G1
$t_{P,FFB}$	Propagation delay time of FFB
T_d	Power-on reset delay time
T_r	Power-off recovery time for power-on reset circuit
V_{CC} or V_{DD}	Power-supply voltage
V_D	Diode forward drop ≈ 0.7 V
$V_{in,0}$	Logical 0 input voltage for any flip-flop or shift-register input terminal

DESIGN EQUATIONS

Description	Equation
Maximum allowed clock frequency	$f_{max,clock} < f_{max,FFB}$ (1*a*)
	$f_{max,clock} < f_{max,SR}$ (1*b*)
	$f_{max,clock} < \dfrac{1}{t_{P,G2} + t_{P,FFB}}$ (1*c*)
Resistance values:	
R_1	$R_1 < \dfrac{V_{in,0}}{I_{in,0}}$ (2) (the maximum values of these parameters at the clear input of FFA are required)
R_2	$R_2 < \dfrac{V_{in,0}}{I_{in,0}}$ (3) (the maximum values of these parameters at the clear input of the shift register are required)
R_3	$R_3 < \dfrac{R_2(V_{CC} - V_D - V_{in,1})}{V_{in,1}}$ (4) ($V_{in,1}$ refers to the minimum value of this parameter for the clear input of the shift register; use V_{DD} for V_{CC} if CMOS)
R_4	$R_4 \approx \dfrac{T_d}{C}$ (5)
R_5	$R_5 \approx 10 R_4$ (6)
Capacitor value	$C \approx \dfrac{T_d}{R_4}$ (7)

DESIGN STEPS

Step 1. What is the maximum required length of the pulse sequence? If it is 16 or less, the shift-register approach described in this section is quite economical. If the pulse sequence is longer than 16, we advise using a counter-comparator circuit.

Step 2. Choose a logic family which is compatible with the clock input and the pulse-sequence output interfaces. If compatible interfaces are not possible, interface gates will be needed to shift logic levels and/or increase fan-in and fan-out.

Step 3. Compute the allowable clock frequency for the logic type used. This maximum must be smaller than both (1) the worst case minimum clock frequency of flip-flop *B* or the shift register and (2) the reciprocal of the sum of G1 (maximum) propagation delay and FFB (maximum) propagation delay; see Eqs. (1).

Step 4. If the start command is to be initiated by a mechanical switch, as shown in Fig. 4.33, the FFA latch is required. This latch will be set (Q_A HIGH) on the very first positive transition of the set input. Most switches have contact bounce, which results in many positive and negative transitions at the moment of closure before a total connection is made.

If a synchronous start command is possible, it can be in the form of a positive pulse on the J input of FFB. This pulse should begin at approximately t_0 and end at approximately t_2.

Step 5. Use Eq. (2) to determine a nominal resistance value for R1.

Step 6. The possible requirement of a power-on reset should be considered. If damage or other problems can possibly be incurred by the inadvertent generation of a pulse train when power is turned on, a power-on reset is mandatory. The power-on reset circuit shown in Fig. 4.33 is a typical implementation. The gate is the open-collector type. Resistors R_2 through R_5 are determined from Eqs. (3) to (6). Capacitor C is assumed to be some practical value such as 0.01 μF.

Step 7. Are the flip-flops and shift register positive- or negative-edge-triggered devices? Figure 4.33 assumes the former. If negative-edge-triggered devices are used, G2 and G3 must be deleted and G1 changed to an AND gate.

Step 8. Can MSI or LSI be used anywhere in this circuit to cut the component count? These larger digital building blocks are usually more economical even if they are only partially utilized.

EXAMPLE USING DESIGN STEPS Suppose we need a manual pulser capable of producing 1 to 4 pulses at a 100-kHz rate. The unit is to be hand-held and will use the CMOS clock described in Sec. 3.2. Low power consumption dictates the use of CMOS. A power-on reset time of 10 ms should be adequate. A 6-V battery is assumed for V_{DD}.

Device Parameters

$$f_{\text{max},FFB} = 1.5 \text{ MHz} \qquad f_{\text{max},SR} = 1.5 \text{ MHz}$$
$$I_{\text{in},0,\text{max}} = 10 \text{ pA} \qquad t_{P,FFB} = 300 \text{ ns}$$
$$t_{P,G1} = 75 \text{ ns} \qquad V_D = 0.7 \text{ V} \qquad V_{\text{in},0,\text{max}} = 1.5 \text{ V}$$

Step 1. Shift-register approach assumed.

Step 2. CMOS output assumed.

Step 3. The 4000 series CMOS family will be used.

a. The 4027 dual *JK* flip-flop has a minimum f_{max} of 1.5 MHz. The 4035 shift register also has a minimum f_{max} of 1.5 MHz.

b. The sum of the maximum propagation delays of G1 (a 4011 assumed) and FFB (a 4027) is 75 + 300 ns. This corresponds to a frequency of $1/(375 \text{ ns}) = 2.67$ MHz. We must therefore use the 1.5-MHz rate as the upper frequency for this circuit.

Step 4. Mechanical start and reset are assumed.

Step 5. Resistance R_1 is found from

$$R_1 < \frac{V_{\text{in},0}}{I_{\text{in},0}} < \frac{1.5}{10^{-11}} = 1.5 \times 10^{11} \ \Omega$$

A 100-kΩ resistor is satisfactory.

Step 6. The need for a power-on reset circuit is assumed. The resistor values are computed as follows:

$$R_2 < \frac{V_{\text{in},0}}{I_{\text{in},0}} < \frac{1.5}{10^{-11}} = 1.5 \times 10^{11} \ \Omega$$

A 100-kΩ resistor is satisfactory. If we assume $C = 0.001 \ \mu\text{F}$,

$$R_3 < \frac{R_2(V_{DD} - V_D - V_{\text{in},1})}{V_{\text{in},1}} = \frac{10^5(6 - 0.7 - 2.5)}{2.5} = 112 \ \text{k}\Omega$$

$$R_4 \approx \frac{T_d}{C} = \frac{10^{-2}}{10^{-8}} = 1 \ \text{M}\Omega$$

$$R_5 \approx 10 R_4 = 10 \ \text{M}\Omega$$

REFERENCES

1. Terman, L. M.: MOSFET Memory Circuits, *IEEE Proc.*, vol. 59, no. 7, pp. 1050–1055, July 1971.
2. Karp, J., and E. deAtley: Use Four-Phase MOS IC Logic, *Electron. Des.*, Apr. 1, 1967, p. 62.
3. Kroeger, J., and B. Threewitt: Review the Basics of MOS Logic, *Electron. Des.*, Mar. 15, 1974, p. 98.
4. Boyle, W. S., and G. E. Smith: Charge-coupled Devices: A New Approach to MIS Device Structures, *IEEE Spectrum*, July 1971, p. 18.
5. Digital Integrated Circuits Catalog, National Semiconductor Corp., Santa Clara, Calif., 1974.
6. Tarui, M.: A Simplified Variable-Width Pulse-decoding Circuit, *EDN*, Dec. 5, 1973, p. 90.
7. Pasco, R. C.: Converter Lets Processor Drive Teletypewriter, *Electronics*, Oct. 30, 1975, p. 97.
8. Millman, J., and C. C. Halkias: "Integrated Electronics: Analog and Digital Circuits and Systems," McGraw-Hill, New York, 1972, p. 628.
9. Blakeslee, T. R.: "Digital Design with Standard MSI and LSI," Wiley, New York, 1975, p. 143.

Chapter 5

Read-Write Random-Access Memories

INTRODUCTION

The data-storage circuits introduced in Chap. 4 are available in arrays containing thousands of identical circuits on a single chip of silicon. These arrays are usually contained in dual-in-line packages with 16 to 24 pins. The packaged devices are commonly called RAMs (random-access memories) although a more accurate terminology is read-write RAMs. The ROMs (read-only memories) and PROMs (programmable read-only memories) discussed in the next chapter are also random-access memories. By convention, these devices are always called ROMs and PROMs, so that read-write RAMs can keep the shorter name RAM. In this handbook RAM will always refer to a read-write random-access memory.

RAMs are grouped into two major categories, static and dynamic. Static RAMs contain true flip-flops for each storage element. Once data are stored in these circuits, they are maintained in the same state by dc power supplies on appropriate terminals. No clock or other dynamic signal is required to maintain the states of the flip-flops. Dynamic RAMs, however, require a refreshing signal on appropriate terminals every few milliseconds. This is usually done by performing a read operation while all combinations of one-half of the address lines are exercised. The data-output line (or lines) is disengaged from the rest of the system during this operation. In a microprocessor system the microprocessor is temporarily placed in the halt mode during the refresh period.

We will find in the following sections that designing a memory system utilizing dynamic RAMs is not much more difficult than designing a system with static RAMs. A dynamic RAM operates almost identically to a static RAM except for the short period of time devoted to refresh. The refresh period might typically require 100 μs out of each 2000 μs. During this 100 μs the system data buses are all placed in the high-impedance state and the system comes to a standstill. For the other 1900 μs, however, the system behaves as if refresh did not exist.

5.1 RAM NOMENCLATURE

The large number of RAM devices on the market and their complexity mean that a large nomenclature has developed. Symbols used by one company may resemble symbols used by another, but the designer must be cautious when comparing data sheets. Until a uniform nomenclature is developed, both word descriptions and timing diagrams must be consulted if one wants to judge one RAM against another honestly. Table 5.1 has been prepared to assist the designer in this task.

TABLE 5.1   Nomenclature for RAMs (*continued*)

Symbol	Alternate symbols	Word description
A0–AM	A_0–A_M	Address buses
C_{AD}		Address-bus input capacitance (one-line per device)
C_{CE}		CE-line input capacitance (one device)
C_{CS}		$\overline{CS}$-line input capacitance (one device)
C_I	C_{in}	Input-terminal capacitance
C_O	C_{out}	Output-terminal capacitance
$\overline{CAS}$		Column address strobe
CE	$\overline{CE}$	Chip enable, usually MOS level logic, that is, +12 V
$\overline{CS}$	CS	Chip select, usually TTL logic level, which enables input-output gates
CSW	t_{CSR}, t_{CSW}, t_{CRW}	Chip-select pulse width
DI	D_{in}	Data input bus (if one per device)
DO	$\overline{DO}$, D_{out}	Data output bus (if one per device)
D0–DN	D_0–D_N	Data buses (input and/or output)
I_{BB}		V_{BB} supply current (one device)
I_{CC}		V_{CC} supply current (one device)
I_{CC1}		V_{CC} supply current if CE OFF (one device)
I_{CH}	I_{HC}	$\overline{CS}$ or CE input current in HIGH state (one device)
I_{CL}	I_{LC}	$\overline{CS}$ or CE input current in LOW state (one device)
I_{CP}		$\overline{CS}$ or CE peak input current in HIGH state (one device)
I_{DD}	I_{DD2}, I_{SX}	V_{DD} supply current, selected chip
I_{DDS}		V_{DD} supply current, unselected chip with reduced V_{DD}
I_{DDU}	I_{DD1}, I_{SXU}	V_{DD} supply current, unselected chip
$I_{DD,av1}$		Average I_{DD} if $t_c = 400$ ns and $t_{CE} = 230$ ns
$I_{DD,av2}$		Average I_{DD} if $t_c = 1000$ ns and $t_{CE} = 230$ ns
I_{DD1}, I_{DD2}		See I_{DDU} and I_{DD}
I_{DO}	$|I_{LO}|$, I_{OZH}, I_{OZL}, I_{LO}	Output terminal leakage current, high-Z state
I_{HC}		See I_{CH}

RAM devices from various manufacturers have many similarities and in some cases can be directly plugged into the sockets of a board designed for another device. Sometimes a minor alteration, e.g., changing V_{BB} from -5 to -3, is all that is required. Other situations may be more involved. A few devices use an active HIGH chip select while most devices require an active LOW for this signal. Many devices invert the output data while others leave it unchanged. A good designer must study all the common data sheets and become familiar with the similarities and differences of RAM devices. If management dictates which RAM device must be used, the task is simplified, but designers who must make a trade-off between all devices of a certain class have their work cut out for them.

The most critical design task in a memory system is implementing the correct timing. The input-output timing waveforms provided by a RAM manufacturer must be adhered to if successful operation is to be achieved. Table 5.2 has been provided to acquaint the reader with some of the more common timing parameters. Many other parameters

TABLE 5.1 Nomenclature for RAMs

Symbol	Alternate symbols	Word description
I_{HI}	I_H, I_I	Logic-terminal input current, HIGH state
I_{IL}	I_{LI}	Logic-terminal input current, LOW state
I_{LC}		See I_{CL}
I_{LO}		See I_{DO}
I_{OH}		Output-terminal source current in HIGH state
I_{OL}		Output-terminal sink current in LOW state
I_{OZH}, I_{OZL}		See I_{DO}
I_{SX}, I_{SXU}		See I_{DD}, I_{DDU}
ME	$\overline{ME}$	Memory enable, similar to CE or $\overline{CE}$
P_D		Power dissipation of RAM device
$\overline{RAS}$		Row address strobe
$R/\overline{W}$	$\overline{R}/W$, WE, $\overline{WE}$	Read-write control terminal, often called write enable (WE or $\overline{WE}$)
RMW		Read-modify-write
V_{BB}	V_{SX}	Negative supply voltage in n-channel RAMs, positive in p-channel RAMs
V_{CC}		Positive supply voltage, always 5 ± 0.25 V
V_{CH}	V_{IHC}	$\overline{CS}$ maximum voltage requirement
V_{CL}	V_{ILC}	$\overline{CS}$ minimum voltage requirement
V_{DD}		Drain-supply voltage, usually $+12$ V for n-channel RAMs and -12 to -15 V for p-channel RAMs
V_{IH}		Logic-input HIGH voltage
V_{IHC}		CE-input HIGH voltage
V_{IL}		Logic-input LOW voltage
V_{ILC}		CE-input LOW voltage
V_{OH}		Logic-output HIGH voltage
V_{OL}		Logic-output LOW voltage
V_{SS}		Usually ground in n-channel RAMs
WE	$\overline{WE}$	Write enable, same as $\overline{R}/W$ ($\overline{WE}$ is same as $R/\overline{W}$)

TABLE 5.2 Timing Relations between RAM Parameters

Symbol	Name	When used*	Definition	Limits specified	Diagram
t_A, $t_{A,AD}$, $t_{A,CS}$, t_{CAC}, t_{PLH}, t_{PHL}, t_{RAC}	Access time	Read, RMW	Time from start of CS = 1 to start of valid D_{out}	Max	CS; t_A; D_{out} Output data not valid / Output data valid
t_{AC}, t_{AS}, t_P	Address to CE set-up time	Read, write, RMW	Time address must be stable before start of CE = 1	Min	Address can change / Address stable; t_{AC}; CE
t_{ACC}	Address to output access	Read, RMW	Time address must be stable before valid D_{out}	Max	Address can change / Address stable; t_{ACC}; D_{out} Data not valid / Data valid; Note: $t_{ACC} = t_{AC} + t_{CO} + t_T$
t_{AH}, t_H, t_{HA}	Address hold time	Read, write, RMW	Time address must be stable while CE = 1	Min	Address stable / Address can change; CE (or CS); t_{AH}; ($\overline{RAS}$ or $\overline{CAS}$); Alternate t_{AH} definition (16-pin 4-K RAMs)
t_C, t_{CY}, t_{RC}	Cycle time	Read, write	One period of CE for complete read or write cycle	Min	t_C; CE

Symbol	Type	Description	Min/Max
t_{CC}	Read, write, RMW	CE OFF time	Min
t_{CE}	Read, write	CE ON time	Min to max
t_{CR}; t_{CF}	Read, write, RMW	Chip-select rise and fall times	Rise and fall times of CS — Max
t_{CF}†	Read, write, RMW	CE OFF to output high-Z state	Time from start of CE = 0 to D_{out} high-Z state — Min
t_{CO}	Read, RMW	CE output delay	Time from start of CE = 1 to start of valid D_{out} — Max
t_{CRR}	Read	Chip-select read recovery time	Minimum allowed CS OFF time — Min
t_{CRW}	RMW	CE width during RMW	Minimum allowed and maximum recommended CE on time — Min to max
t_{CSR}	Read	Chip-select read pulse width	Time CS pulse is HIGH — Min

TABLE 5.2 Timing Relations between RAM Parameters (continued)

Symbol	Name	When used*	Definition	Limits specified	
t_{CSW}	Chip-select write pulse width	Write	Time CS pulse is HIGH	Min	
t_{CW}	CE to $\overline{\text{WE}}$	Write	Time from start of CE = 1 to end of $\overline{\text{WE}}$ = 0	Min	
t_{CWR}	Chip-select write recovery time	Write	Minimum allowed CS off time before next write cycle	Min	
t_D	D_{in} to CE setup	Write	Minimum time valid D_{in} must be present before start of CE = 0	Min	
t_{DH}, t_{HD}, t_H	D_{in} hold time	Write	Minimum time after start of CE = 0 before D_{in} can become invalid	Min	
t_{DR}, t_{EN}, t_{PHZ}, t_{PLZ}	Data recovery time	Read	Time for D_{out} to become nonvalid after start of CS = 0	Min	

Symbol	Parameter	Applies to	Type	Description
t_{DW}, t_{SU}	D_{in} to $\overline{WE}$ setup	Write	Min	Time valid D_{in} must precede start of $\overline{WE} = 0$ 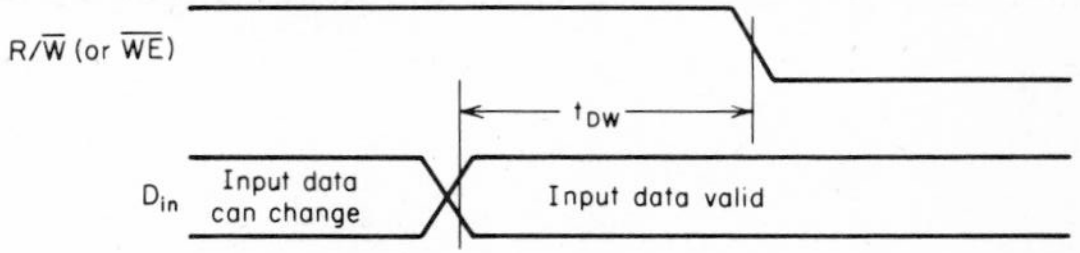
t_{ref}, t_{RFSH}	Time between refresh	Refresh	Max	Maximum time between refresh cycles 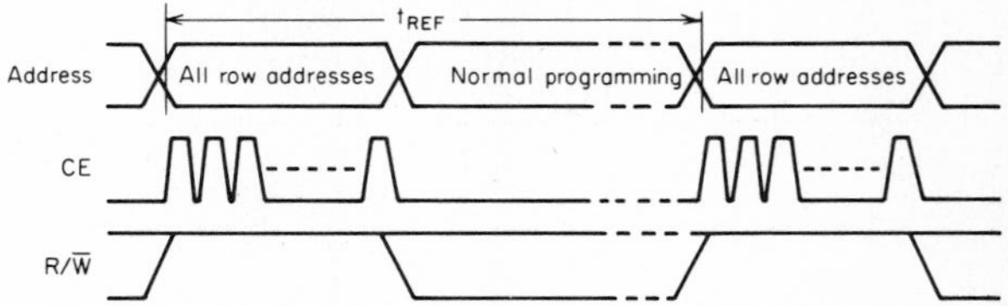
t_{RWC}	RMW cycle time	RMW	Min	Minimum time allowed for complete RMW cycle 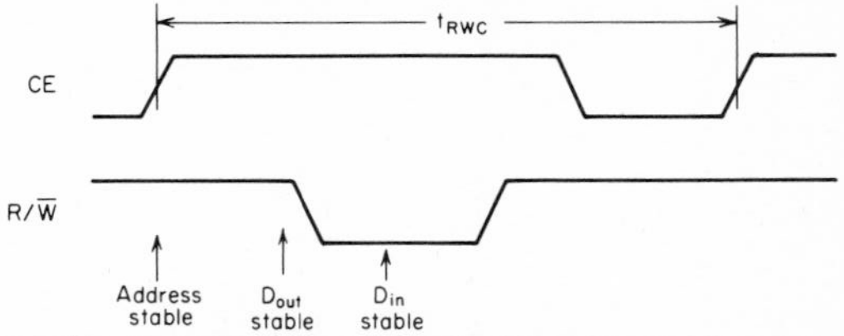
t_T	CE transition time	Read, write, RMW	Min and max	Rise or fall time of CE pulse 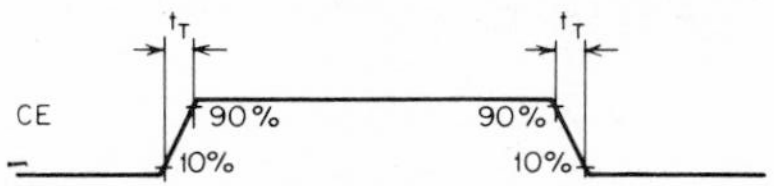
t_W	$\overline{WE}$ to CE off	Write	Min	Time from start of $\overline{WE} = 0$ to end of CE = 1 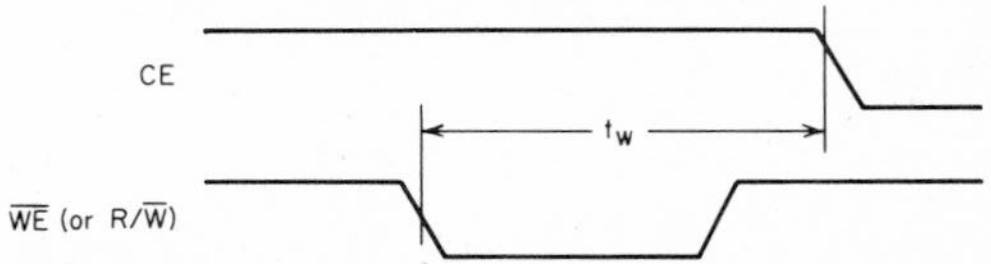

Diagram labels:

$R/\overline{W}$ (or $\overline{WE}$) — t_{DW} — Input data can change / Input data valid — D_{in}

t_{REF} — Address — All row addresses / Normal programming / All row addresses — CE — $R/\overline{W}$

t_{RWC} — CE — $R/\overline{W}$ — Address stable — D_{out} stable — D_{in} stable

t_T — CE — 90% — 10% — $R/\overline{W}$

t_W — CE — $\overline{WE}$ (or $R/\overline{W}$)

TABLE 5.2 Timing Relations between RAM Parameters (*continued*)

Symbol	Name	When used*	Definition	Limits specified	
t_{WC}	$\overline{WE}$ to CE on	Read, RMW	Time from start of $\overline{WE}=1$ to end of $\overline{CE}=0$	Min	
t_{WL}	CE to $\overline{WE}$	Read	Time from end of $\overline{CE}=1$ to end of $\overline{WE}=1$	Min	
t_{WP}, $t_{W,WR}$, $t_{SU,WR}$	$\overline{WE}$ pulse width	Write	Time $\overline{WE}$ is LOW	Min	

* RMW = read, modify, write. † Second definition.

can be found on data sheets since each manufacturer and each chip technology has developed its own set of parameters. A common set of parameters ought to emerge in a few years.

5.2 $(N \times 1)$-BIT RANDOM-ACCESS MEMORIES

ALTERNATE NAMES $1 \times N$ RAM, 256×1 RAM, 512×1 RAM, 1024×1 RAM, 2048×1 RAM, 4096×1 RAM, 8192×1 RAM, $16{,}384 \times 1$ RAM.

PRINCIPLES OF OPERATION As shown in Fig. 5.1, this device usually has one data input line D_{in} and one data output line D_{out}. The data output line is inverted $(\overline{D_{out}})$ in many of the more popular 4096×1 devices. Some RAM chips have input and output on one line. This offers quite an advantage since in most microprocessor systems the data bus is bidirectional; i.e., each line may have data flowing in either direction.

The required number of address lines is

$$M = 1.4427 \ln N \tag{5.1}$$

A particular M-bit address selects one memory cell in the RAM device. We can transfer the contents of that cell to the $\overline{D_{out}}$ terminal if R/$\overline{W}$ is HIGH. The cell contents are unchanged by this operation. If R/$\overline{W}$ is LOW, the logic level present on the D_{in} line is stored in the cell.

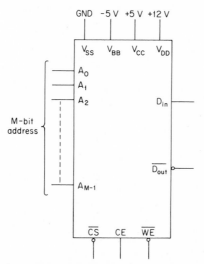

Fig. 5.1 Pin configuration for a typical $N \times 1$ RAM. $M = 1.4427 \ln N$.

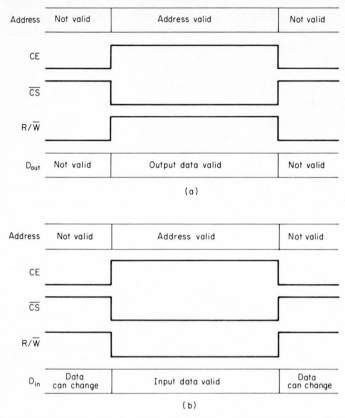

Fig. 5.2 Idealized read and write timing charts for a RAM: *(a)* read cycle; *(b)* write cycle.

When the chip-enable (CE) line goes HIGH, it starts internal clocks, latches the address into an internal register, and switches V_{DD} power ON to the internal MOS circuits. CE is applied after the correct address is available since its leading edge performs the latching operation.

The chip-select ($\overline{\text{CS}}$) line turns on the RAM's input-output buffers. If $\overline{\text{CS}}$ is LOW, the D_{in} and D_{out} lines are connected to the RAM internal circuitry. When $\overline{\text{CS}}$ is HIGH, the D_{in} line is disconnected and the D_{out} line is placed in the high-impedance state. $\overline{\text{CS}}$ can remain HIGH during a refresh cycle.

The multitude of timing signals listed in Table 5.2 is confusing to someone who has not designed a memory system. To assist the reader in understanding these parameters it is helpful to first carefully study the ideal read and write timing charts shown in Fig. 5.2. A timing chart for a real RAM circuit requires only minor changes in the ideal charts.

An ideal read cycle starts when the following simultaneous events take place:

1. A valid address appears on A_0 to A_{M-1}.

2. CE goes HIGH to gate power to the MOS circuitry.

3. $\overline{CS}$ goes LOW to power the input-output buffers.

4. R/$\overline{W}$ goes HIGH so that RAM cell data are transferred to the output buffer.

5. The RAM cell data appear at D_{out}.

A real device, as shown in Fig. 5.3, requires the following modifications to the five ideal events noted above:

1. All signals must be shown with finite rise and fall times.

2. The address and $\overline{CS}$ are allowed to reach the RAM first. All other inputs must be delayed.

3. A HIGH CE is provided to the RAM after a delay of t_{AC} (see Table 5.2).

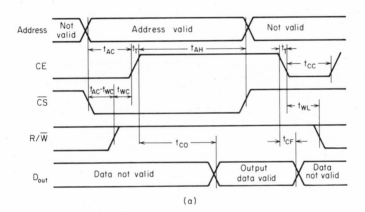

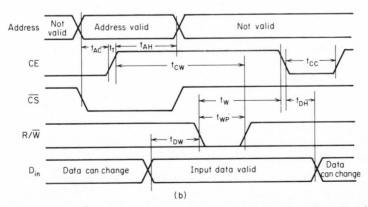

Fig. 5.3 A typical timing chart for a real RAM (compare with Fig. 5.2): *(a)* read cycle; *(b)* write cycle.

4. $R/\overline{W}$ must go HIGH at least t_{WC} before CE goes HIGH. Thus, $R/\overline{W}$ occurs $t_{AC} - t_{WC}$ after appearance of A_0 to A_{M-1} and $\overline{CS}$.

5. When CE goes HIGH, valid D_{out} will not appear until after a delay t_{CO}. Thus D_{out} appears $t_{CO} + t_{AC}$ after the addresses and $\overline{CS}$ arrive at the RAM chip.

6. The rise time of t_T of CE must be added to the above delays. The total read access time is therefore $t_{ACC} = t_{AC} + t_{CO} + t_T$.

7. The addresses and $\overline{CS}$ do not have to be stable the entire time CS is HIGH. These signals need be present only for the hold time t_{AH} after the start of CE = 1.

8. The output data become nonvalid t_{CF} after the end of CE = 1.

9. $R/\overline{W}$ must stay HIGH for a period of t_{WL} after the start of CE = 0.

10. The next read or write cycle cannot start until CE has been kept LOW for a time t_{CC}.

Referring back to Fig. 5.2, we note that an ideal write cycle starts when the following simultaneous events take place:

1. A valid address appears at A_0 to A_{M-1}.

2. CE switches HIGH to gate power to the internal MOS circuitry.

3. $\overline{CS}$ goes LOW to activate the input-output buffers.

4. $R/\overline{W}$ is LOW, so that D_{in} will be latched into the selected RAM cell.

5. Input data are present on the D_{in} terminal.

As with the read cycle, a real write cycle requires only a few minor changes to the ideal timing chart. Comparison of Fig. 5.3b with Fig. 5.2b shows the following differences:

1. All signals are shown with finite rise and fall times in the real timing chart.

2. The address and $\overline{CS}$ lines are allowed to arrive at the RAM first.

3. CE goes to the HIGH state $t_{AC} + t_T$ after the addresses and $\overline{CS}$ have stabilized.

4. D_{in} must be stable at least t_{DW} before $R/\overline{W}$ goes LOW.

5. The $R/\overline{W}$ must stay LOW for at least t_{WP}.

6. The $R/\overline{W}$ line must not return to the HIGH state until at least t_{CW} after CE changes to the HIGH state.

7. The start of the LOW state for $R/\overline{W}$ must not be closer than t_W to the end of the HIGH CE.

8. The addresses and $\overline{CS}$ must remain valid for at least t_{AH} after the start of a HIGH CE.

9. Input data must remain valid at least until t_{DH} after CE returns to the LOW state.

10. The CE LOW time must be at least t_{CC} long.

5.3 $N \times 1$ RAM APPLICATION: 4-KWORD MEMORY

ALTERNATE NAMES 4k × 8 memory, 4096-word by 8-bit memory, 32,768-bit memory.

PRINCIPLES OF OPERATION A 4-kword by 8-bit memory circuit is ideal for one page of memory in an 8-bit microprocessor system. As shown in Fig. 5.4, this circuit is easily implemented using eight 4096 × 1 RAM devices. All 12 address buses are tied together. These buses are driven by high-power buffers, which are in turn driven by the microprocessor address bus.

The address bus in most 8-bit microprocessor systems is 16 bits wide.

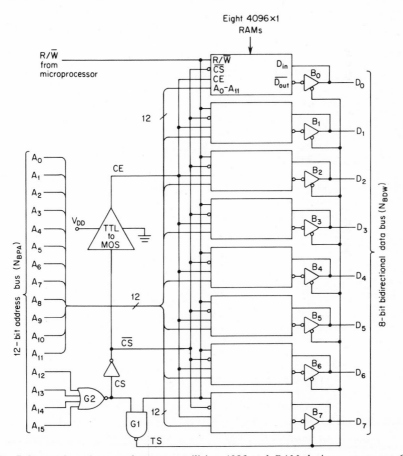

Fig. 5.4 A 4-kword page of memory utilizing 4096 × 1 RAM devices. NOR gate G2 assumes that this is the first page of memory.

We therefore have 4 bits left over to create chip-select signals (CS and $\overline{\text{CS}}$) for this page of memory. If the page of memory shown in Fig. 5.4 is the first page of a memory system, the task is simple. One need merely decode A_{12} to A_{15} using the NOR gate G2 in Fig. 5.4. We want CS to go HIGH when A_{12} to A_{15} are all LOW. This could be implemented with a negative-logic NAND or a positive-logic NOR as follows:

$$\text{CS} = \overline{A}_{12}\,\overline{A}_{13}\,\overline{A}_{14}\,\overline{A}_{15} = \overline{A_{12} + A_{13} + A_{14} + A_{15}}$$

This is inverted to obtain the $\overline{\text{CS}}$ required by most RAMs.

The CS signal is passed through a TTL-to-MOS converter to provide a MOS level CE waveform. The CE signal is usually a 12-V pulse, which switches ON the V_{DD} supply to most of the internal MOS circuits. In many systems CE can be merely the inverse of $\overline{\text{CS}}$ with a slight delay T_{AC} (see Fig. 5.2). Some RAM devices allow t_{AC} to go to zero. If the CE waveform is substantially different from the inverse of $\overline{\text{CS}}$, gating with other waveforms may be required.

The output data are inverted and buffered using B0 to B7. These are three-state devices which have a high-impedance output at all times except when a read command is received for this page of memory. Thus B0 to B7 are enabled when both CS and R/$\overline{\text{W}}$ are HIGH.

DESIGN PARAMETERS FOR A 4-K MEMORY*

Parameter	Description
N_{BAW}	Bits per address word
N_{BC}	Bits per chip
N_{BDW}	Bits per data word
N_{BMA}	Bits per module address word
N_{BPA}	Bits per page address word
N_{WS}	Words per system
TS	Tristate signal (high-Z state if TS = HIGH)

* In addition to those in Tables 5.1 and 5.2.

DESIGN STEPS FOR A 4-K MEMORY

Step 1. Draw a block diagram of one page of the memory system similar to that shown in Fig. 5.4. If $N \times 1$ RAM devices are used, each page of memory will require N_{BDW} RAMs. The page of memory shown in Fig. 5.4 is therefore a matrix of memory cells having dimensions of $N_{BDW} = 8$ bits wide by $N_{BC} = 4096$ bits deep. This is a 4096-word page, i.e., a $8 \times 4096 = 32{,}768$-bit page.

TABLE 5.3 *N* × 1 RAM Comparison Chart

RAM parameter	Design goal for parameter	Device 1	Device 2	Device 3
1. Chip size, square mils	Minimum			
2. Second sources	Maximum			
3. Cost	Minimum			
4. Number of power supply voltages	Minimum			
5. Operating power	Minimum			
6. Standby power	Minimum			
7. Input-voltage range	TTL			
8. Output-voltage range	TTL			
9. Output data inverted?	No			
10. Input capacitances	Minimum			
11. Output capacitances	Minimum			
12. Number of clocks	Minimum			
13. Number of refresh cycles per interval	Minimum			
14. Maximum access time	Minimum			
15. Maximum cycle time	Minimum			

Step 2. Prepare a memory-chip comparison table as shown in Table 5.3. Since many devices can be eliminated by inspection, only the final candidates need be placed in the table. Do not use vendors' preliminary data sheets, as these data are often given as typical. True worst-case data are required for a good memory design. A top designer never uses typical data in any part of a design.

Use the table to make a final decision on the RAM. If we later find this RAM choice has too many incompatibilities with the microprocessor system, the second or third candidate may be used.

Step 3. If the RAM output data are inverted, inverting gates as shown in Fig. 5.4 are required on all $\overline{D_{out}}$ terminals. They must be tristate buffers since this memory system must not have any effect on the bus except when being read. A LOW TS level places these buffers in the active mode (read). TS is LOW only when both R/$\overline{W}$ and CS are HIGH.

Each data buffer must be capable of driving all devices attached to that data line. Both current and capacitive loading must be considered.

Step 4. Make timing diagrams of required read and write waveforms for the chosen RAM. Make a list showing levels required, load capacitances, and input currents for each RAM terminal.

Step 5. Make a timing diagram of all microprocessor system waveforms which are recommended by the microprocessor vendor to be used for memory interfaces. These may include:

1. Address bus
2. Data bus
3. Read/$\overline{write}$ (R/$\overline{W}$) line

4. Enable, select, or valid-memory-address (VMA) lines
5. Data-bus-enable (DBE) line
6. Clock phases
7. Three-state-control (TSC) line

Again, tabulate levels and drive capability (current and capacitance).

Step 6. Compare the sets of data obtained in steps 4 and 5. If possible, connect some of these signals directly to the page of memory shown in Fig. 5.4. Note, however, that in most microprocessor systems buffers should be used on all microprocessor lines which drive more than one low-power TTL load.

Some signals will need to be combined with clock phases, inverted, or NAND-NOR-modified to provide the correct timing required by the memory. Carefully consider fan-in and fan-out requirements, especially where one line drives many RAMs in parallel.

Step 7. Design the page-address decoder. If this page of memory is the lowest page in the system, the NOR gate (G2) shown in Fig. 5.4 is a sufficient decoder. If one of the upper 15 pages is to be accessed, a circuit like Fig. 5.5 can be used. This circuit contains a standard 16-pin dip switch, which allows easy selection of any combination of A_{12}, $\overline{A}_{12}$, A_{13}, $\overline{A}_{13}$, A_{14}, $\overline{A}_{14}$, A_{15}, and $\overline{A}_{15}$.

For example, if we wanted to make a particular memory board respond only to addresses A000 to AFFF (hex addresses) we would close switches

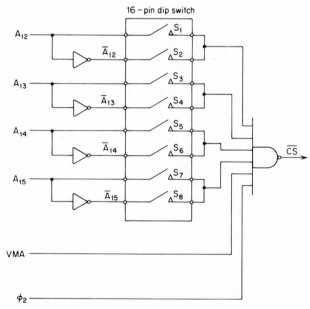

Fig. 5.5 A typical switching scheme which allows decoding 16 pages of memory from address lines A_{12} to A_{15}.

S2, S3, S6, and S7. This provides decoding of $A_{15}A_{14}A_{13}A_{12} = 1010$ = A (hex). The 4-bit word $A_{15}A_{14}A_{13}A_{12}$ has 16 combinations, of which we have selected combination 10 (decimal) or A (hex). Since each page contains 4096 (decimal) or 1000 (hex) words, page A contains words A000 through AFFF (0000 through 0FFF is the hex equivalent of decimal numbers 0000 through 4095).

In 6800 microprocessor systems a signal called *valid memory address* (VMA) is available to combine with the high-order addresses, as shown in Fig. 5.5. VMA indicates to the memory that the 6800 microprocessor is in a read or write cycle. $\overline{\text{CS}}$ goes to the LOW state only when VMA is HIGH and the proper $A_{15}A_{14}A_{13}A_{12}$ address is present.

The NAND gate must be capable of driving the $\overline{\text{CS}}$ input of many RAM devices in parallel. Both input current and input capacitance of the memory chips should be considered.

Step 8.   Design the chip-enable (CE) circuit. The CE input on most memory chips requires MOS levels, e.g., a pulse voltage of $V_{DD} = 12$ V. Many RAM devices allow CE to be synchronous or slightly delayed with respect to CS. In reality, if a TTL-to-MOS-level shifter is used (see Fig. 5.4), CE will be delayed several nanoseconds.

The TTL-to-MOS-level shifter must be capable of supplying CE current into the entire page of RAM devices. The capacitance of all loads must also be compared with the capacitive-load-driving capability of the level shifter.

Step 9.   Check the available power-supply voltages to make sure they fall within the recommended RAM voltage limits.

Step 10.   If the RAM chips are dynamic, the refresh circuitry must be designed. This circuitry can be designed to handle all pages of memory with minimal hardware. Refresh is accomplished in most types of RAM devices by performing a read operation approximately every 2 ms on all row addresses. For example, consider a 4096-bit RAM which has 12 address lines A0 through A11. A0 through A5 are row addresses, and A6 through A11 are column addresses. If we had to read all RAM addresses, 4096 separate 12-bit addresses would have to be sent sequentially to the memory while the $R/\overline{W}$ line was HIGH. However, the 6-bit row address word A_0 through A_5 has only 64 combinations. Thus, to refresh the RAM chips we run through 64 addresses while $R/\overline{W}$ is HIGH.

During refresh we do not use the data appearing at the D_{out} terminals of each RAM. Consequently, in most microprocessor systems the data output buffers (B_0 through B_7 in Fig. 5.4) remain in the high-Z state during refresh. The microprocessor is placed in the halt mode during this operation so that no instructions will be erroneously executed.

In a typical microprocessor system all the row addresses can be refreshed is less than 5 percent of the time between refreshes. Thus, if

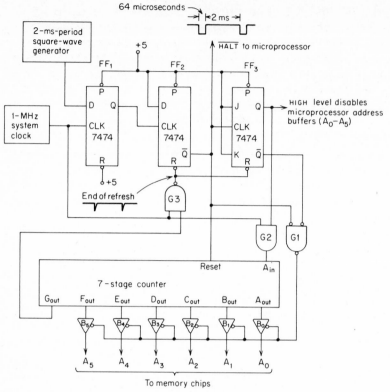

Fig. 5.6 One possible method of refreshing 64 row addresses every 2 ms in a 4096-word memory page using 4096×1 RAM devices.

the time between refresh t_{ref} is 2 ms, a typical time for exercising a read on all row addresses is less than 100 μs.

Figure 5.6 indicates one possible circuit to refresh a microprocessor memory system of any size. The only constraint is that B_0 to B_5 must be capable of driving every RAM chip using A_0 to A_5. This system uses a 2-ms-period clock which is edge-synchronized with the 1-MHz system clock using D flip-flop FF_1. For simplicity of discussion assume a starting state of $\overline{Q} = 1$ in both FF_2 and FF_3. This causes the G1 output to be HIGH, which forces B_0 through B_5 into the high-Z state. The seven-stage counter is also initialized in the reset state. Since the Q output from FF_3 is initially LOW, the microprocessor buffers are in the active state. Likewise, since $\overline{Q} = 1$ from FF_2, the microprocessor $\overline{HALT}$ signal is inactive.

A refresh cycle begins with a negative edge from the Q output of FF_1. This causes $\overline{Q}$ of FF_2 to change to the LOW state. A number of events immediately take place:

1. The seven-stage counter reset line is released, so that counting can begin.

2. A HIGH state is clocked into the Q output of FF_3.

3. With two LOW inputs into G1 its output goes LOW, causing B_0 through B_5 to go active.

4. Gate G2 is turned ON, so that 1-MHz clock pulses are allowed to toggle the seven-stage counter.

5. A $\overline{HALT}$ signal is sent to the microprocessor, and the microprocessor output address buffers are driven to the high-Z state.

The above five conditions remain in effect until all 64 combinations of A_0 through A_5 have been exercised. When the sixty-fifth clock pulse tries to pass through the seven-stage counter, the G_{out} line goes HIGH. This causes the G3 output to go LOW, resulting in reset commands for FF_2 and FF_3. This action closes G1 and G2 so that the counter and its six buffer gates are stopped. The sixty-fifth clock pulse passing into A_{in} is therefore truncated into a sliver pulse whose width is equal to the delay times of the seven-stage counter, G3, FF_3, and G2. The reset command in FF_2 releases the $\overline{HALT}$ line to the microprocessor. Likewise, the Q output of FF_3 allows the microprocessor buffers to return to the active state.

REFERENCES

1. M6800 Application Manual, Motorola Semiconductor Products, Inc., 1975.
2. Springer, J.: Making Sense Out of Delay Specs in Semiconductor Memories, *Electronics,* Oct. 25, 1971, p. 82.
3. Frankenberg, R. J., Designer's Guide to Semiconductor Memories, Pts. 6–8, *EDN,* Oct. 20, 1975, p. 44; Nov. 5, 1975, p. 59; and Nov. 20, 1975, p. 127.

Chapter 6

Read-only Memory Devices

INTRODUCTION

Fast and permanent data storage is an absolute necessity in a useful microcomputer system. This storage medium allows one to use the microcomputer system for applications immediately after the ON switch is thrown. No one wants to be bothered with loading a monitor program every time one sits down to use the system. Likewise, if one has BASIC, FORTRAN, APL, or COBOL stored permanently in the microcomputer, one can also use these programs immediately after power turn-on. The storage medium most commonly used for this permanent storage in microcomputer systems is read-only memory (ROM).

The name ROM is usually given to a class of microcircuits which are programmed at the factory by changing one mask pattern. These devices contain a permanent program and cannot be changed by the user. A memory chip which can be programmed by the user is called a *programmable read-only memory* (PROM). These user-programmable devices come in three general categories: (1) those which can be programmed once by the user with no possibility of changes thereafter, (2) those which can be erased by exposure to ultraviolet light and reprogrammed, and (3) those which are electrically erasable and reprogrammable. In this chapter we discuss all three types of PROM in addition to the mask-programmable ROM.

6.1 MASK-PROGRAMMABLE ROMs

ALTERNATE NAMES ROM, MOS-ROM, bipolar ROM, fixed-program ROM.

PRINCIPLES OF OPERATION A block diagram of a typical ROM is shown in Fig. 6.1. This device is almost identical to a random-access memory (RAM). In fact, ROMs and RAMs are often found side by side on a circuit board, tied to the address bus and data bus in an identical manner except that the ROM does not use the write and data-input lines.

To simplify the description, Fig. 6.1 is shown for a 512-word by 8-bit ROM. Since we have 512 words, 9 address lines are required. The buffer decoder separates the address input into row and column drive signals. For a particular 9-bit address the row and column addresses will intersect at a particular group of 8 bits and send these bits over to the 8-bit latch. If strobe is HIGH, the 8-bit word is transferred directly to the tristate drivers. If $\overline{CE_1}$ is LOW and CE_2 is HIGH, the tristate drivers will transfer the 8-bit word to the outside world.

When the strobe line goes LOW, the 8-bit latch will retain the 8-bit word indefinitely. If the $\overline{CE_1}$ and CE_2 lines still enable the D flip-flop, the tristate drivers will also output the 8-bit word indefinitely. If $\overline{CE_1}$ or CE_2 disables the chip, the tristate buffers go to the high-Z state.

Mask-programmable ROMs are purchased with fixed programs such as character generators, function-look-up tables, keyboard encoders, etc. The program is established by one mask pattern at the end of the wafer-fabrication process. Many users also have custom ROMs developed

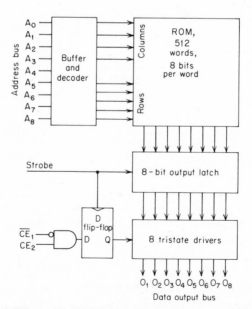

Fig. 6.1 A typical 512-word by 8-bit ROM with output latches and tristate output lines (*Signetics 82S215*).

for a high-volume product. In this case, however, one must be willing to wait 6 to 8 weeks for design and fabrication of the new ROM. The price tag for custom ROMs ranges from $10,000 to $50,000, but for a high-volume application this cost is only a few cents per ROM.

6.2 MASK-PROGRAMMABLE ROM APPLICATION: CRT CHARACTER GENERATION

ALTERNATE NAMES   Video display, alphanumeric generator, CRT display, MOS character generator, video RAM.

PRINCIPLES OF OPERATION   The character generator outlined in Fig. 6.2a is quite typical of those used in most video-output microprocessor systems. The basic requirements are a random-access memory (RAM), a read-only memory (ROM), and a parallel-to-serial converter (often called a dot shifter). Not shown in this simplified block diagram is the sequencing logic which drives the RAM, ROM, and shift register.

The RAM block shown in Fig. 6.2a must be large enough to store

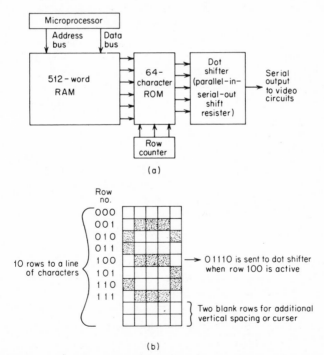

Fig. 6.2   (a) A generalized block diagram of a character-generator circuit; (b) one typical character showing code for each row.

one page of video information. That is, each word (6 to 8 bits) stored in the RAM is reserved for one character location on the cathode-ray tube (CRT). If a 16-line by 32-character display is required, the RAM must hold $16 \times 32 = 512$ words.

The ROM stores the information necessary to construct a different character on the CRT display for each code appearing on the ROM inputs. If 128 different types of characters are desired, the RAM must send a 7-bit word to the ROM. The ROM output is 5 parallel lines in a standard 5×7 character-generator circuit. A 7×9 character font would require 7 parallel output lines.

Each time a ROM output is required, the RAM data and the three inputs from the row counter must be stable. As shown in Fig. 6.2b, if the character selected is S and the row counter is 100, the ROM output will be 01110.

The outputs from the ROM are parallel-loaded into the shift register. The shift register is then switched to the shift-right mode, and the ROM outputs sequentially appear at the shift-register output. These serial data are combined with blanking information and sent to the CRT display. We note that only one row of a character is sent to the CRT at a given time. The same row of the next character to the right is sent out next, and so on until the end of that horizontal scan row. On the next horizontal trace the next row of all these characters is sent out. This is repeated 10 times until a line of characters and 3 blank rows are complete. All other lines of characters utilize 10 rows, where each row is always 1 scan line of the CRT. After all lines of characters are laid out on the CRT, the electron beam returns to the top of the display and the process is repeated. Most systems repeat the entire process, i.e., refresh the display, 30 or 60 times per second.

Figure 6.3 shows a character-generator circuit in more detail. This particular ROM (the 2513) is probably the most widely used character-generator chip. Some of the newer character generators contain all the logic shown in Fig. 6.3 on one chip. However, for educational purposes we will discuss this older workhorse circuit.

In Fig. 6.3 we note that an 8-bit shift register is used. This register is called the dot shifter in some systems. The second and eighth inputs are loaded with 0s, and the first input is loaded with a 1. As the dot oscillator drives the clock input of the dot shifter, the video output starts from 0, passes a 5-bit dot code, then returns to 0. Since the serial input is also tied HIGH, when the LOW input from B_{in} arrives at H_{out}, we have HIGH outputs from B_{out} to G_{out}. This causes the output of G4 to go HIGH, thus placing the dot shifter back in the load mode so that the next code can be received from the ROM.

The ROM has two internal decoders. The 6:64 decoder selects the proper character, and the 3:8 decoder selects the proper row for the

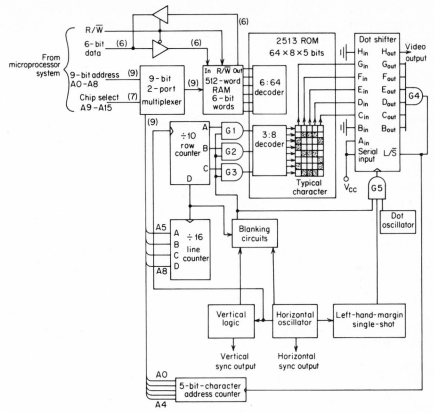

Fig. 6.3 Block diagram of a character generator for a video display using the 2513 ROM.

chosen character. Both these inputs must simultaneously arrive at the ROM with a timing accuracy of better than 50 ns in good-quality video systems. The 6-bit code is obtained from data stored in the RAM. The 3-bit row address comes from a modulo-10 counter, which is incremented at the end of each horizontal scan row. Three gates between the row counter and ROM are used for blanking. If the outputs of these gates are all **0**, row 000 in the ROM is selected. This always transfers **0**s to the dot shifter no matter what character is selected. The same blanking line can also inhibit the dot oscillator to further reduce the chances of data appearing on the CRT during retrace periods.

The 9-bit RAM address is normally obtained from the character address counter (A0 to A4) and the line counter (A5 to A8). The address counter is incremented at the completion of each row of each character when the output of G4 goes HIGH. The line counter is incremented whenever the row counter overflows (after the last row of a character string is completed).

The three counters (row, line, and character) are continually operating and lay down a new screen of characters every $\frac{1}{30}$ (or $\frac{1}{60}$) s. If the microcomputer wishes to insert a new character on the CRT, the proper address must appear on the two-port multiplexer. Otherwise the MUX continually transfers the line and character counter addresses to the RAM. New character information is transferred into the RAM from the microprocessor data bus only when $R/\overline{W}$ is low and A9 to A15 have a specific code. The operation is so fast that the CRT display does not appear disturbed except for the one changed character. The microcomputer address bits A0 to A8 tell the RAM where to place the new character.

The horizontal oscillator is normally operated at 15,750 Hz. The output from this oscillator triggers the horizontal sweep circuits in the CRT system. The vertical circuits are merely dividers and miscellaneous logic needed to generate a vertical trigger output.

The left-hand-margin single-shot allows the horizontal scan to start before characters appear. Otherwise, characters could be slightly distorted if they were at the extreme edge of the CRT.

6.3   FIELD-PROGRAMMABLE ROMs

ALTERNATE NAMES   PROM, EPROM, EAROM, EEROM.

PRINCIPLES OF OPERATION   This type of ROM is programmable by the user and is appropriately called the PROM. These devices come in two types, those which provide permanent data storage once programmed and those which can be erased electrically or with ultraviolet light and reprogrammed.

Permanent-data-storage PROMS   The most popular permanent-storage PROMS are a group of devices which are programmed by blowing out selected Nichrome fusible links. Some companies also make devices with polycrystalline silicon fuses instead of Nichrome. Each fuse represents 1 bit in the memory. The output data from the PROM may be normally LOW or HIGH; i.e., some devices require the programmer to blow out links where LOW data are to be established and some require the blown out links to represent HIGH states. The fuses, however, never connect directly with the PROM output terminals. Several stages of buffering are utilized, so that TTL, open-collector, or tristate outputs can be provided.

The fusible links must be blown out by carefully following the manufacturer's recommendations. Since large voltages and currents are required, the programmer may inadvertently burn out other parts of the PROM. For example, the Schottky TTL PROM series by Signetics (82S series) has a programming procedure as follows:

1. Terminate all outputs with 10 kΩ to V_{CC}.
2. Select the address to be programmed and raise V_{CC} to 8.75 V.
3. After 10 μs delay apply 17 V to the output to be programmed.
4. After another 10 μs, pulse the $\overline{CE}$ input to a logic LOW for 1 or 2 ms.
5. After another 10 μs, remove the 17 V from the output.
6. Verify programming by lowering V_{CC} to 5.5 V, applying a logic LOW to $\overline{CE}$, and observing to see if the programmed output stays in the HIGH state. Lower V_{CC} to 4.5 V and see if the output still remains in the LOW state.

Ultraviolet-erasable PROMS   This type of PROM is more expensive than those previously discussed in this chapter, but for prototyping or small production runs, where the program must be changed fairly often, the overall cost is lower. The ultraviolet erasable PROM can be erased and reprogrammed almost endlessly. It will hold a program for 10 years with little degradation. The only disadvantage of this device is that all bits must be erased before it can be reprogrammed.

The most popular ultraviolet PROMS are those with eight parallel tristate outputs. These types can be tied directly to a microcomputer data bus with no interface circuits. Some of the newer ultraviolet PROMS require only +5 V for operation whereas older devices needed up to three voltages. Likewise, newer devices are designed so that programming can be performed on the same circuit board. When the prototyping is done and the program is well established, many PROMs have direct-replacement mask-programmable ROMs which can be plugged into the same socket with no circuit changes.

The only disadvantage with in-circuit programming of ultraviolet PROMs is that there is no convenient way of erasing the devices. Erasure is accomplished by exposing the window of the PROM to ultraviolet light with a wavelength shorter than 4000 Å. The required integrated dose for the 2716 PROM, for example, is 15 W·s/cm². Thus, if a typical lamp with 0.012 W/cm² is used, the exposure time must be $15/0.012 = 1250$ s $= 21$ min. The lamp is normally placed about 1 in above the PROM to obtain the recommended dosage.

Electrically alterable PROMs   This class of PROM comes in two varieties, the electrically erasable ROM (EEROM) and the electrically alterable ROM (EAROM). The first version requires that all bit locations be erased when the proper erasure voltages are applied. The second type allows one to select specific words (usually 4- or 8-bit words) to erase and reprogram. One can immediately see the advantages of the second type. In a typical application for an EAROM the other data stored in the device must not be disturbed while a correction is made to one word. If one had to erase the entire EAROM each time a single word change was made, a separate RAM circuit would be required just to retain these data while the erase–program-correction process took place. This would make the EAROM memory much too expensive.

The EAROM opens up a host of new applications. Probably the two largest applications which will eventually reach into every home are TV channel memories and telephone-number memories. The EAROM is ideal for these memories since the TV and telephone (in most homes) are not in use for long periods of time.

REFERENCE

1. Cayton, B.: Designing with Nitride-Type EAROMs, *Electronics,* Sept. 15, 1977, p. 107.

Chapter 7

Data-Word Sorting
and Checking

INTRODUCTION

Many electronic systems process digital words organized in parallel groups of 4, 8, 12, 16, or more bits. All digital computers convert information into this format before computations are made. This chapter will discuss several widely used word-processing microcircuits and indicate some of their possible applications. All the functions performed by these circuits can be easily performed by a microprocessor, but it is often advantageous to free the microprocessor from some of these more mundane tasks so that it will not be slowed down. In a microcomputer system where the microprocessor has much free time the functions mentioned in this chapter should probably be included in the microprocessor software. Speed is often a factor in word processing. Most MOS-type microcomputers are over an order of magnitude slower than the fastest TTL circuits. If some of the word processing must be performed at a high speed, special circuits outside the microprocessor can be used.

This chapter first discusses word comparators. This class of device takes two words and decides whether they are equal. If they are not equal, it tells the outside world which word is larger in magnitude.

The larger part of this chapter is devoted to error-checking and error-correcting microcircuits. When data words are transferred in or out of storage media (RAM, tape, disk) or through communication channels, errors are occasionally introduced. In many cases a lost bit can be the cause of disaster in a system. We discuss the parity-generation and parity-checking circuits used to detect data errors. The chapter will close with a discussion of real-time and non-real-time error correction.

7.1   DIGITAL MAGNITUDE COMPARATOR

ALTERNATE NAMES N-bit comparator, digital comparator, word comparator.

PRINCIPLES OF OPERATION   A 4-bit digital comparator is shown in Fig. 7.1. The concepts discussed here for a 4-bit comparator can be extended to digital words of any length. We note that this circuit has inputs for two 4-bit words $A = A_3 A_2 A_1 A_0$ and $B = B_3 B_2 B_1 B_0$. The output labeled $A > B$ will go HIGH whenever the A inputs contain a larger binary number than the B inputs. Obviously, the $A < B$ output is just the opposite. The $A = B$ output means that every bit in A exactly matches every bit in B.

The digital comparator is quite useful when looking through a serial stream of data for a specific code word. Suppose the serial data are shifted into a serial-in–parallel-out shift register. We would then tie

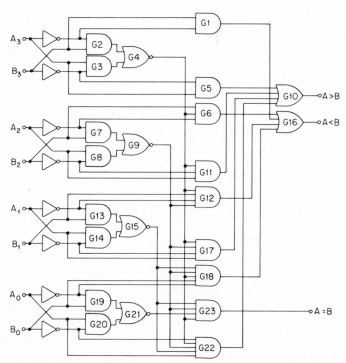

Fig. 7.1 Internal-logic diagram of a 4-bit magnitude comparator.

the shift-register outputs to the A inputs of the digital comparator. The B inputs are tied HIGH or LOW according to the code being sought. When a match occurs, the $A = B$ line will go HIGH. A device like the 74L85 will make this comparison in 55 ns.

We will now describe how the circuit of Fig. 7.1 operates. A bit-by-bit comparison is made on the two input words. For example, bits A_3 and B_3 are tested in G2, G3, and G4. The output of G4 will be HIGH only if A_3 and B_3 are identical. These three gates plus the two inverters feeding G2 and G3 are called a *coincidence gate,* represented by $A_3 \odot B_3$. We also note that the output of G9 is $A_2 \odot B_2$, the output of G15 is $A_1 \odot B_1$, and the output of G21 is $A_0 \odot B_0$. Now if all four of these coincidences occur together, we must have $A = B$. This coincidence of all bits is determined with the AND gate G23.

The logic for computing $A > B$ is a little more difficult than that required to compute $A = B$. The output from OR gate G10 goes HIGH if any of its four input lines goes HIGH. The first line from G5 simply compares the most significant bits of each word. If $A_3 = 1$ and $B_3 = 0$, we must have $A > B$. However, if $A_3 \odot B_3$, that is, $A_3 = B_3 = 1$ or $A_3 = B_3 = 0$, we must go to G11 for a comparison. This gate requires $A_3 \odot B_3$, $A_2 = 1$, and $B_2 = 0$ to put out a HIGH level. What we are doing here is disregarding the most significant bits (since they are equal) and making the comparison at the next bit level.

If $A_2 \odot B_2$ and $A_3 \odot B_3$, we must go down to G17 and test for $A_1 = 1$ and $B_1 = 0$. If all these conditions are met, we have again determined $A > B$. Gate G22 is the same story once again. Its output will be true only if $A_3 \odot B_3$, $A_2 \odot B_2$, $A_1 \odot B_1$, and $A_0 \overline{B}_0 = 1$.

Calculation of $B > A$ is identical in concept to the calculation of $A > B$. In this case, however, we test to see if $\overline{A}_3 B_3 = 1$, $\overline{A}_2 B_2 = 1$, etc. The output of G16 will go HIGH if any of G1, G6, G12, or G18 detects a B bit larger than an A bit while all higher-order bits are coincident.

7.2 THE PARITY TREE

ALTERNATE NAMES Parity generator, parity checker, odd-even parity generator-checker.

PRINCIPLES OF OPERATION Parity trees of any size can be made from an array of XOR gates. The parity tree shown in Fig. 7.2 will generate or detect parity for an 8-bit word using one additional bit for parity. One can obtain parity-tree microcircuits with capabilities up to 13 bits.

Let us consider how the circuit of Fig. 7.2 generates a parity bit. Even parity means we must attach an extra bit to the 8 data bits to produce an even number of 1s. For example, if the data word is

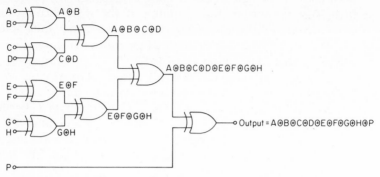

Fig. 7.2   A parity tree for an 8-bit word plus an additional parity bit.

10101011, we must add a **1** to make a 9-bit word 101010111.   The 9-bit word now has six **1**s.

Each 8-bit word is first tested with the parity tree to see if it has an odd or even number of **1**s.   The rules of operation for Fig. 7.2 are as follows:

 P input HIGH

  Output HIGH for even number of **1**s

  Output LOW for odd number of **1**s

 P input LOW:

  Output HIGH for odd number of **1**s

  Output LOW for even number of **1**s

We would probably use this device with the *P* input LOW if even-parity generation is required.   Thus, whenever an odd number of **1**s is detected in the 8-bit data word, the parity-tree output goes HIGH.   This HIGH level can be transferred over to the register holding the data word and added to one end of the data word.   The data word can then be shifted out to its destination.

Figure 7.3 indicates how an 8-bit word on a microcomputer data bus can have a parity bit attached (generated) before the word is sent into a serial channel.   To simplify the discussion we will not show the sequencing logic which generates the control signals EN, PEN, DEN, and L/$\overline{\text{S}}$.   These control signals perform the following functions:

 EN: When HIGH, this enable signal allows clock pulses to enter the 8-bit latch-shift register.

 L/$\overline{\text{S}}$:   If this control signal is HIGH, each positive transition of the clock will latch the microcomputer data bus into the eight internal registers.   If L/$\overline{\text{S}}$ is LOW, each positive transition of the clock will shift the 8-bit register contents to the right by one stage.

 DEN: This is the data-enable signal which allows data to be shifted out of the 8-bit register into the serial output.

 PEN: This is the parity-enable signal which allows the parity bit to be transferred into the serial output.

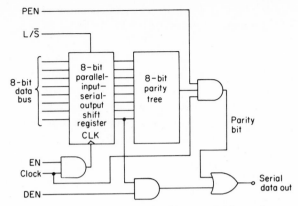

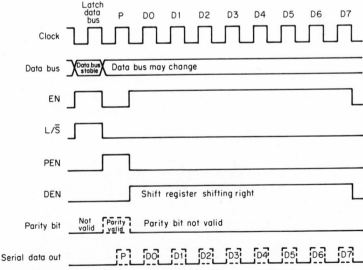

Fig. 7.3 One way to generate an even parity bit as an 8-bit word on a parallel bus is converted into a serial bit stream.

The control signals must be synchronized with the clock signal, as shown in Fig. 7.4. We note that all control-signal transitions occur on the trailing edge of the clock. The positive edge of the clock performs the load and shift functions after the control signals have stabilized for one-half clock period.

As shown in Fig. 7.4, the data bus is first latched into the 8-bit register when EN and L/$\overline{\text{S}}$ are both HIGH. The 8-bit register outputs immediately cause the parity tree to generate a HIGH or LOW output signal, depending on the number of 1s present. The parity bit, however, is not placed in the serial data stream until the next clock pulse, when PEN is HIGH.

Fig. 7.4 Timing relationships of control signals and clock with the serial output data of Fig. 7.3.

As PEN falls, DEN and EN both go HIGH, causing the 8 data bits to be shifted into the serial data stream. After 8 bits are shifted out, the EN and DEN control signals go LOW to close out the cycle. The next word is then placed on the data bus for another conversion.

The parity of a word is checked in a manner similar to parity generation. In this case, however, we generate a parity bit and compare it with the parity bit arriving with the word. If a match of parity bits is not made, a signal is sent out to the system to initiate some corrective action. In most cases the system is shut down, and the operator must repeat some process, e.g., reading a tape.

7.3   ERROR-CHECKING AND ERROR-CORRECTING CODES

Parity checking is a welcome feature in a digital system, but unfortunately a parity circuit merely detects the presence of a single-bit error somewhere in a data word. Many new digital systems contain circuits which not only detect bit errors but also are capable of correcting most types of bit errors. A simple check for parity errors causes more problems than it solves because a parity error merely raises a warning flag. Since the equipment operator has no way of adjusting the system's reaction to match the seriousness of the error, drastic action must be taken every time the warning flag is raised. The operator must then determine the nature of the problem, where it happened in the system, and what to do about it.

Error detection can be implemented with a number of techniques other than simple parity. Some of these methods allow real-time error correction (within a few microseconds), and others require a second reading of the tape (or other storage or communication device).

REAL-TIME ERROR CORRECTION   A real-time error checking and correction (ERCC) circuit has been developed by Data General Corporation for use in their 16-bit memory boards. When each 16-bit data word is written into memory, the ERCC logic adds a 5-bit check word. Each stored word is therefore 21 bits long. The 5-bit check word contains the results of five separate parity calculations. As illustrated in Fig. 7.5, the five parity calculations use the following rules:

Check bit 0: Data bits 2, 5, 10 to 15, and check bit 0 must satisfy odd parity.

Check bit 1: Data bits 4 to 10, 15, and check bit 1 must satisfy even parity.

Check bit 2: Data bits 1 to 3, 7 to 9, 14, 15, and check bit 2 must satisfy odd parity.

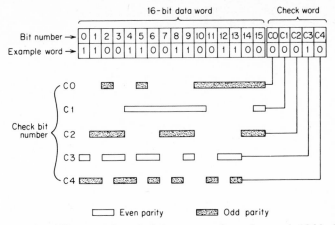

Fig. 7.5 How five different parity calculations are performed on each 16-bit data word. A 5-bit check word is then appended to the data word before it is written into memory.

Check bit 3: Data bits 0, 2, 3, 5, 6, 9, 12, 13, and check bit 3 must satisfy even parity.

Check bit 4: Data bits 0, 1, 3, 4, 6, 8, 11, 13, and check bit 4 must satisfy odd parity.

When each 21-bit word is read from memory, the same set of five parity calculations is performed. As shown in Fig. 7.6, a new 5-bit word, called the *error word,* is generated from the results of these calculations. If no error has occurred, the error word is all **0**s. Otherwise, this error word is compared with the data shown in Table 7.1, which

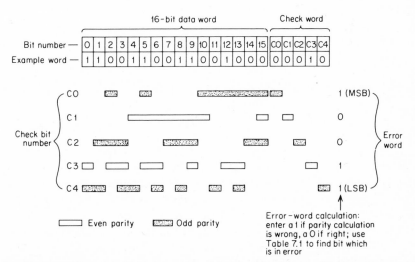

Fig. 7.6 Procedure by which an error word is constructed from a 21-bit memory word. We discover by using Table 7.1 that bit 13 is in error.

TABLE 7.1   Interpretation of Each Generated Error Word

Error word					Meaning
0	0	0	0	0	No error
0	0	0	0	1	Check bit 4 in error
0	0	0	1	0	Check bit 3 in error
0	0	0	1	1	Data bit 0 in error
0	0	1	0	0	Check bit 2 in error
0	0	1	0	1	Data bit 1 in error
0	0	1	1	0	Multiple-bit error
0	0	1	1	1	Data bit 3 in error
0	1	0	0	0	Check bit 1 in error
0	1	0	0	1	Data bit 4 in error
0	1	0	1	0	All 21 bits are **1**
0	1	0	1	1	Data bit 6 in error
0	1	1	0	0	Data bit 7 in error
0	1	1	0	1	Data bit 8 in error
0	1	1	1	0	Data bit 9 in error
0	1	1	1	1	Multiple-bit error
1	0	0	0	0	Check bit 0 in error
1	0	0	0	1	Data bit 11 in error
1	0	0	1	0	Data bit 12 in error
1	0	0	1	1	Data bit 13 in error
1	0	1	0	0	Data bit 14 in error
1	0	1	0	1	All 21 bits are **0**
1	0	1	1	0	Data bit 0 in error
1	0	1	1	1	Multiple-bit error
1	1	0	0	0	Data bit 10 in error
1	1	0	0	1	Multiple-bit error
1	1	0	1	0	Data bit 5 in error
1	1	0	1	1	Multiple-bit error
1	1	1	0	0	Data bit 15 in error
1	1	1	0	1	Multiple-bit error
1	1	1	1	0	Multiple-bit error
1	1	1	1	1	Multiple-bit error

is stored in ROM on the memory board. The appropriate correction is made to the 16-bit data word before it leaves the memory board. If a multiple error occurred, the ERCC circuit cannot correct the data word. In this case a flag can be sent to the host computer for some corrective action.

As an example, suppose an error occurred in bit 13 of the data word 1100110011001100 shown in Fig. 7.5. The computed error word shown in Fig. 7.6 is 10011. This error word is compared with ROM data (Table 7.1) to determine the next course of action. The ROM says bit 13 is in error, and since bit 13 appears to be **0,** we must change it to a **1** before the data word is sent from the memory board.

In a hardware implementation of the above ideas all five parity checks are performed simultaneously as the 21-bit data word comes from the RAM chips. If the error word is 00000, the 16-bit data word is sent to the outside world with no delay.   If an error is detected, a slight delay occurs as the bit in error is inverted.

NON-REAL-TIME ERROR CORRECTION   Several methods are available for correcting errors found while reading a magnetic-tape unit.   These methods often require a second reading of the tape before sufficient data are received to make the error correction.   Non-real-time error correction is not as fast as the method outlined in the previous sections, but it can be automated so that the computer operator is barely aware that the tape unit reads a portion of tape twice.

One of the leading codes used for non-real-time error correction is the cyclic redundancy check (CRC).   This code can be generated or checked using a single 14-pin IC.   It is becoming a common fixture in floppy and hard disk systems, digital cassette and cartridge systems, and various other types of serial communication equipment.   CRC generating or checking can also be easily performed with software in a microcomputer system.   Different ways of using CRC codes are discussed in Refs. 3 and 4.

REFERENCES

1. Jensen, R. K.: Error Detecting Codes Are Simple, *Electron. Des.* 19, Sept. 13, 1967, p. 90.
2. Madren, F. S.: Memory Error Correction: More than Just a Little Better than Parity, *Control Eng.,* May 1975, p. 36.
3. Chien, R. T.: Memory Error Control: Beyond Parity, *IEEE Spectrum,* July 1973, p. 18.
4. Swanson, R.: Understanding Cyclic Redundancy Codes, *Comput. Des.,* November 1975, p. 93.

Chapter 8

Sequential Circuits

INTRODUCTION

Systems requiring various events to occur at different times utilize a sequential circuit to control the incidence of each event. Nearly every digital system requires some type of sequential operation. Simple sequencers are implemented with counters and gates. More complex sequencers use ROM chips, programmable logic-array chips, or microprocessors. We cover all these types except the microprocessor in this chapter. Several other chapters in this handbook are devoted to the microprocessor and its applications. The microprocessor is, by its very nature, a sequential device. It executes instructions one at a time and can be utilized to control thousands of events in any required sequence. We therefore cover only small sequencers in this chapter because microprocessors are an optimum choice whenever 10 or more ICs are required for a sequential circuit.

8.1 COUNTER

ALTERNATE NAMES Divider, frequency divider, binary counter, decade counter, pulse counter, ripple counter, ripple-carry counter, clocked counter, presettable counter, programmable counter, up-down counter, digitally trimmed frequency divider, synchronous counter.

PRINCIPLES OF OPERATION Most counter chips are merely arrays of flip-flops interconnected with gates. Each stage is usually made from toggle or D flip-flops. We briefly present each basic type of counter and then show how they can be utilized in a sequential circuit.

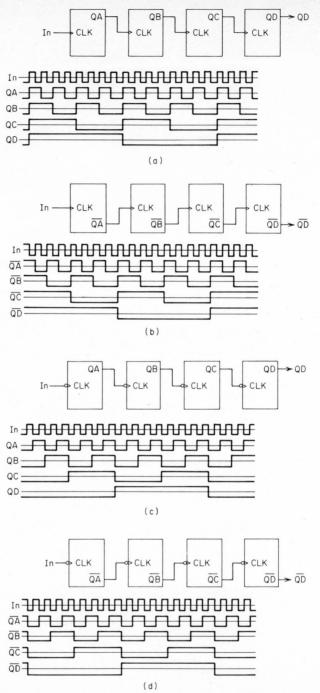

Fig. 8.1 Four ways of assembling a ripple-carry counter using toggle flip-flops: *(a)* positive clock, positive output, *(b)* positive clock, negative output, *(c)* negative clock, positive output, and *(d)* negative clock, negative output.

Ripple-carry counter   The most basic counters are made from a cascaded series of toggle flip-flops. Two types of toggle flip-flops are available, those which change states on the rising clock edge and those which change state on the falling clock edge. As shown in Fig. 8.1, a ripple-carry counter can be implemented with four basic methods. In two of the methods positive *(Q)* outputs are used for coupling, and in the other two methods negative *(Q̄)* coupling is utilized. Each of these methods can be implemented with positive-edge-triggered clocks or with negative-edge-triggered clocks. Of the four methods, we observe that only those shown in Fig. 8.1*b* and *c* behave the way we would expect a counter to operate. Starting from an all-cleared condition, only the circuits in Fig. 8.1*b* and *c* provide a true sequential operation. In the two other circuits, all four stages change state on the first clock pulse. This would not be a drawback in systems where the counter is running continuously. In systems where the count must start from an all-cleared state Fig. 8.1*a* and *d* would be unacceptable.

A ripple-carry counter sequencer   What is the feasibility of using a four-stage ripple-carry counter in a simple sequencer? As shown in Fig. 8.2, the *Q* outputs from all four stages are the inputs for a 1-of-16 line decoder. This provides outputs which are programmable for any count from 0000 to 1111 (15 decimal). Internally, this decoder is 16 four-input NAND gates. The timing diagram for this circuit is shown in Fig. 8.3.

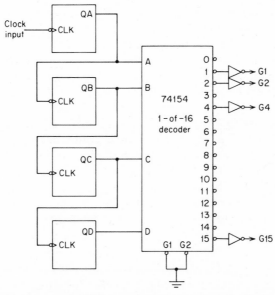

Fig. 8.2   A simple sequencer made from a ripple-carry counter and gates. Each output can be programmed to provide a pulse at any count from 0000 to 1111.

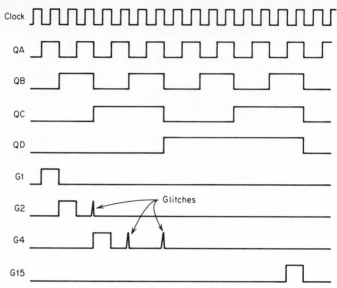

Fig. 8.3 Timing diagram for the circuit in Fig. 8.2.

Clocked counter Sequencers built from ripple-carry counters have a major problem which is solved using the clocked counter. This problem is caused by the delays in each stage which result in a *race condition;* i.e., the outputs of the flip-flops race each other to the four-input gates, causing some erroneous short-term outputs to occur. Since the flip-flops operate sequentially, the delays through the flip-flop stages accumulate. If each flip-flop has a clock-to-output delay of 50 ns, the output from the last flip-flop will change 200 ns after an input clock transition.

Figure 8.3 shows three possible race conditions which generate *spikes* or *glitches* in the otherwise clean output waveforms. The glitch on output 2 occurs when QA has dropped to the LOW state before QB drops to the LOW state or QC rises to the HIGH state. If each flip-flop has a 50-ns clock-to-output delay, this glitch could be up to 50 ns long. Output 4 has two possible race conditions. Since QB rises after QA falls, a glitch is generated. Also, QA and QB fall before QC falls and QD rises, thus creating another glitch.

The clocked counter shown in Fig. 8.4 contains flip-flops with the *JK* inputs tied together, thereby making *T* flip-flops. The common *JK* terminal is called the *T* input. When the *T* input is LOW, the flip-flop disregards the clock input. If *T* is HIGH, the stage toggles on each positive transition of the clock. Since all clock inputs are tied together, the outputs of all flip-flops having $T = 1$ should change states simultaneously. However, slight differences in the clock-to-output delay of individual flip-flops will cause small errors in this simultaneous output tim-

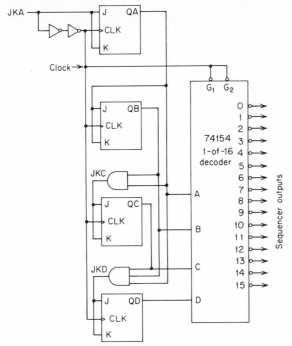

Fig. 8.4 A three-stage clocked binary counter with parallel carry. Sequencer gate connections are also shown.

ing. The race condition for clocked counters depends on the differences in delays of the various stages, whereas for a ripple-carry counter the delays accumulate, causing a race condition at least an order of magnitude worse.

The timing diagram in Fig. 8.5 portrays the counting technique of this synchronous counter. This circuit uses parallel look-ahead carry; i.e., the T input for each stage comes from an AND gate which is excited by every preceding flip-flop. The last flip-flop therefore receives its T input after the delay of only the first flip-flop and one gate. The last stage of a ripple-carry flip-flop must wait for the delay of all preceding stages before it can change states. Thus, synchronous counters will correctly operate in much faster sequencers. For example, in a four-stage counter the ripple-carry sequencer has four device delays and the synchronous circuit has only two device delays. The maximum frequency of operation for the synchronous circuit in this case would be twice that of the ripple-carry circuit.

An additional measure of safety to prevent glitches is provided by enabling the decoder only while the clock is LOW using the G_1 and G_2 inputs. This overcomes the problem of differences in delays between the various stages of the counter. Careful analysis of Fig. 8.5 shows

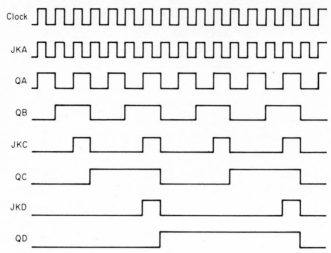

Fig. 8.5 Timing diagram of the clocked 4-bit counter in Fig. 8.4.

that no glitches are possible since all flip-flops are static while clock is LOW. No flip-flop will change states until clock returns to the HIGH state.

Programmable counter Many sequencer applications require counters which divide by any arbitrary number within a given range. The programmable counter (presettable counter) provides a very simple method of achieving division by any number. For example, division by any number under 16 requires only an inverter in addition to a four-stage programmable counter. Division by any number under 256 requires two of these counters and a NAND gate. A programmable counter implemented with the 161 counter is shown in Fig. 8.6 (74161, 74L161, 74LS161, 74C161, etc.). Since the 161 is a synchronous counter, we tie the clock inputs of both devices together, thereby making an eight-stage synchronous counter.

Programming is performed using the data inputs D0 to D7. These input levels can be established by a set of switches, or they can be active inputs from other circuitry. Whenever the $\overline{\text{LOAD}}$ line from G1 is LOW, the binary number on D0 to D7 is clocked into the eight registers on the positive-going edge of the clock. This $\overline{\text{LOAD}}$ signal occurs when G1 detects HIGH levels from the carry outputs of both 161 devices (see Fig. 8.7). CO1 is HIGH only when all output stages in the first device are HIGH. Likewise, CO2 detects an all HIGH state for the second device. The $\overline{\text{LOAD}}$ line immediately returns to the HIGH state since the loaded data appear at all eight outputs, causing CO1 and/or CO2 to fall.

Counting resumes on the next rising clock edge after $\overline{\text{LOAD}}$ has returned to the HIGH state. When all outputs reach the HIGH state, the

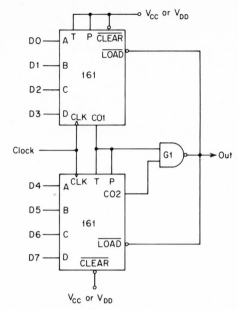

Fig. 8.6 A presettable (programmable) counter capable of any division from 1 to 255. The 161 device is utilized (74161, 74L161, 74LS161, 74C161, etc.).

$\overline{\text{LOAD}}$ line stops counting and starts the sequence over again. The programmable counter division ratio is $N - M$, where N is the binary value of an all-1s condition and M is the binary value of the preloaded data word D0 to D7. As shown in Fig. 8.7, if $N = 11111111 = 255$ (decimal) and $M = 11110110 = 246$ (decimal), the division ratio is $255 - 246 = 9$.

Examination of Fig. 8.7 reveals that all not Q outputs have clean

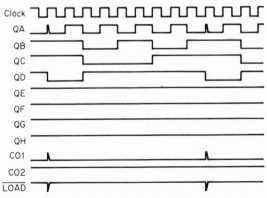

Fig. 8.7 Waveforms at important locations in Fig. 8.6. The data word loaded is D7 to D0 = 11110110.

rectangular waveforms. Stages which must be preloaded to a state different from the state which caused $\overline{\text{LOAD}}$ to go LOW will output a short pulse. For this example only output QA has this spike problem. If each output of the programmable counter is to be used as a trigger pulse, the spike can be used as is. If a clean rectangular output waveform is required, a flip-flop may be needed on that particular output. This, of course, lowers the output frequency by 2, so that the entire counter may be required to run at a higher clock frequency.

8.2   ROM-CONTROLLED SEQUENCER

ALTERNATE NAMES   ROM programmable sequencer, PROM-controlled sequencer, time-duration programmable sequencer.

PRINCIPLES OF OPERATION   A highly flexible sequencer can be made from a PROM and two counter chips. A sequencer of this type allows independent adjustment of the number of output states, the time of occurrence, and the duration of each state. Figure 8.8 indicates how two 74193 up-down counters are used to implement a programmable sequencer having a maximum capability of 256 clock periods. That is, each state can begin at any point in the 256-clock-period master cycle. Also, each state can remain ON for 1 to 256 clock periods.

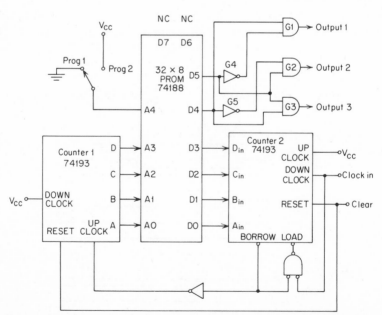

Fig. 8.8   A PROM programmable sequencer having a total count capability of 256. Both the time of occurrence and the duration of each output are controlled by the PROM.

However, the total number of periods and starting positions must add up to 256 or less. If the time of occurrence or duration of any (or all) states must be changed, one merely installs a new PROM. If the number of total available clock periods must be increased, one or both counters must be enlarged.

Figure 8.8 will be utilized to describe the sequence of operation for this circuit. Assume a clear command has just set both counters to state 0000. The borrow output of counter 2 will stay LOW while in state 0000. If the switch is on Prog1, the ROM address selected is 00000. As shown in Table 8.1, the 8-bit output word for address 00000 is 00010111. The 4 most significant bits make output 1 (G1) go HIGH. After clear returns to the HIGH state, the 4 least significant bits of the PROM are parallel-loaded into counter 2 on the next downward transition of the input clock. As this parallel load takes place, the borrow output of counter 2 returns to the HIGH state. This locks out any further parallel loading of counter 2 until it has counted down to zero. As counter 2 counts down from 0111 the PROM holds output 1 HIGH. When the countdown is complete, the borrow line goes down, causing counter 1 to advance to address 00001. Since output 1 is to remain HIGH for another countdown sequence of counter 1, the upper 4 bits of the PROM must remain at 0001. The lower 4 bits of the PROM are loaded into counter 2 and the countdown sequence starts again, this time from a different preloaded number.

We note from Table 8.1 that the duration of each output state is controlled by the 4 least significant bits in each PROM word. If an output state must last more than 15 clock cycles, 2 or more adjacent

TABLE 8.1   Typical Data Stored in the PROM of Fig. 8.8

PROM address		State number bits				State duration bits			
Switch	Counter 1	D7	D6	D5	D4	D3	D2	D1	D0
0	0000	0	0	0	1	0	1	1	1
0	0001	0	0	0	1	1	1	0	1
0	0010	0	0	0	1	1	1	1	0
0	0011	0	0	1	0	1	1	1	1
0	0100	0	0	1	0	1	1	1	1
0	0101	0	0	1	0	1	1	1	1
0	0110	0	0	1	0	1	1	1	1
0	0111	0	0	1	0	1	0	0	0
0	1000	0	0	1	0	0	0	1	0
0	1001	0	0	1	1	1	0	1	0
0	1010	0	0	1	1	1	0	1	0
0	1011	0	0	1	1	1	0	1	0
0	1100	0	0	1	1	1	0	1	0
0	1101	0	0	1	1	1	0	1	0
0	1110	0	0	0	0	0	0	1	1
0	1111	0	0	0	0	0	0	1	1

PROM words are used to define an output state. For example, output state 1 lasts $7 + 13 + 14 = 34$ clock cycles. The second output is HIGH for $4 \times 15 + 8 + 2 = 70$ clock cycles. Likewise, output state 3 is HIGH for $5 \times 10 = 50$ clock cycles. All outputs are then inactive for $2 \times 3 = 6$ clock cycles, as calculated from the last two PROM addresses.

Suppose this circuit is put into production to control a sequence for a washing machine. After several thousand units are produced, housewives complain that one of the outputs (say a wash cycle) should last 40 percent longer. All the manufacturer has to do is change the code stored in each PROM. Since no hardware change is required, the manufacturer can make this customer-requested change at almost no cost.

8.3 PROGRAMMABLE LOGIC ARRAY

ALTERNATE NAMES PLA, field-programmable logic array, FPLA, ROM logic array.

PRINCIPLES OF OPERATION The PLA is an ideal device for sequencers requiring many multiple-input AND gates (see Figs. 8.4 and 8.8). As indicated in Fig. 8.9, a typical PLA (or FPLA) has an array of 48 AND

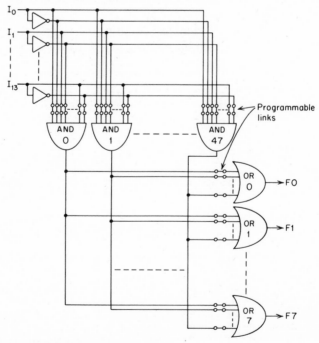

Fig. 8.9 A typical field-programmable logic-array device (FPLA) (*Intersil IM5200*).

gates whose inputs are selectable from 14 input lines (and their internally generated inverses). The outputs of the 48 AND gates drive the inputs of 8 OR gates. Any combination of the 48 AND-gate outputs can be selected for each of the OR gates. The 8 OR-gate outputs become the device outputs. For example, each OR output can be a logical equation such as

$$F_2 = I_1 \overline{I_3} I_7 + \overline{I_1} I_2 \overline{I_5} + I_7 \overline{I_{11}} \overline{I_{12}} I_{13}$$

or
$$F_6 = I_1 I_2 I_3 \overline{I_4} I_5 \overline{I_6} I_8 I_{10} \overline{I_{13}} + \overline{I_3} \overline{I_5} \overline{I_{12}}$$

If the OR gates are not needed, the device is programmed such that one AND gate drives each OR gate. The IM5200 FPLA also has the capability of optionally reversing the polarity of any output.

A fairly large sequencer can be implemented by connecting the IM5200 inputs to the outputs of a 14-stage counter. Eight highly versatile output waveforms are possible because of the 48 internal AND gates. The 14 inverters on the inputs save at least three hex inverters if all logic states are to be made available. This feature is also advantageous since many counters (10 or more stages) have only Q outputs (no $\overline{Q}$ outputs).

The FPLA does not eliminate the glitch problem if it is used as shown in Fig. 8.2 (see also Fig. 8.3). If the counter is synchronously clocked and the input clock is also tied to one of the FPLA inputs, glitches will not occur (see Figs. 8.4 and 8.5).

An FPLA can also be used in place of the PROM in Fig. 8.8. As in that earlier example, the FPLA provides both system outputs and next-state outputs. In this case, however, all gating is done inside the FPLA. Output gates G1 to G5 (in Fig. 8.8) will not be required. An FPLA also requires much less stored information than a ROM since it stores logic equations whereas the ROM stores bytes or bits at given addresses. In an address field of 14 bits, for example, one FPLA could be utilized to pick out any eight addresses. These addresses could be widely separated in the address space. A ROM requires data to be stored in compact groups.

In a typical application [4] a tradeoff was made for ROM vs. PLA memory for a laser encoder-tester. One PLA having 11 input variables and 15 outputs provides the permanent memory. If ROM storage were used, an array of 61,400 bits would be required. The PLA represented a reduction in memory size by a factor of 23.

REFERENCES

1. Bentley, J. H.: The Foolproof Way to Sequencer Design, *Electron. Des.*, May 10, 1973, p. 76.
2. Richards, C. L.: An Easy Way to Design Complex Program Controllers, *Electronics*, Feb. 1, 1973, p. 107.

3. Ettinger, M. A., and G. W. Jacob: An Algorithm for Sequential Circuit Design, *Comput. Des.,* May 1968, p. 46.
4. Maggiore, J.: PLA: A Universal Logic Element, *Electron. Prod.,* Apr. 15, 1974, p. 67.
5. Hemel, A.: The PLA: A Different Kind of ROM, *Electron. Des.,* Jan. 5, 1976, p. 78.
6. Tenny, R.: Build a Programmable Sequencer with a Broad Operating Range, *EDN,* May 20, 1976, p. 90.

Chapter 9

Digital Signal Generation

INTRODUCTION

Every microcomputer system needs a clock with one or more phases for synchronizing the flow of data and addresses. This chapter discusses some common clock circuits. Early microcomputers typically used a multiphase clock requiring three or four ICs. Newer systems utilize one clock generated inside the microprocessor chip. One need merely add a crystal or a capacitor to obtain the clock frequency of one's choice.

9.1 TWO-PHASE NONOVERLAPPING CLOCK

ALTERNATE NAMES Dual-phase clock, 2ϕ clock, biphase clock.

PRINCIPLES OF OPERATION

Single-shot approach Many systems utilizing two-phase clocks require that there be no overlap of the alternating clock pulses. Figure 9.1 shows typical output waveforms from a nonoverlapping positive-pulse two-phase clock. A straightforward circuit which guarantees no overlapping of phases is shown in Fig. 9.2. This type of circuit was used by some early microcomputer systems.

Four single-shot devices are required for the circuit in Fig. 9.2. The two output phases come from the Q outputs of SS1 and SS4. The spacing between the phases is set by SS2 and SS3. We note that SS1, SS3, and SS4 generate positive pulses and SS2 generates a negative pulse.

Assume SS2 is just turning OFF, i.e., returning to the positive state.

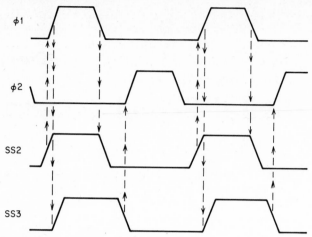

Fig. 9.1 The two upper traces are nonoverlapping pulses from a two-phase clock. Each trace is the output from one of the single-shots shown in Fig. 9.2.

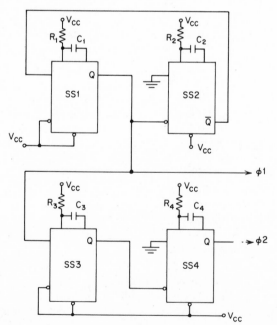

Fig. 9.2 A circuit which generates two clock phases with nonoverlapping positive pulses.

When it reaches 50 percent amplitude, it triggers SS1. After SS1 ($\phi1$) reaches 50 percent amplitude, SS3 is triggered. SS3 must generate a longer output pulse than SS1 ($\phi1$) since the falling edge of SS3 initiates SS4 ($\phi2$). Likewise, the falling edge of SS1 ($\phi1$) initiates the next negative SS2 pulse. To guarantee nonoverlapping of the $\phi1$ and $\phi2$ pulses we must have

$$t_{ss1} < t_{ss3} \quad \text{and} \quad t_{ss3} + t_{ss4} < t_{ss1} + t_{ss2}$$

The first of these inequalities guarantees a space between the falling edge of $\phi1$ and the rising edge of $\phi2$. The second inequality guarantees a space between the falling edge of $\phi2$ and the rising edge of $\phi1$.

Toggle flip-flop approach   A method to generate two nonoverlapping clock phases utilizing only digital microcircuits is shown in Fig. 9.3a. The waveforms in Fig. 9.3b indicate how each phase is generated by an AND gate. The AND-gate output is HIGH only when the corresponding toggle flip-flop output *and* the clock input are HIGH. This method assures that the separation of pulses will be one-half clock period.

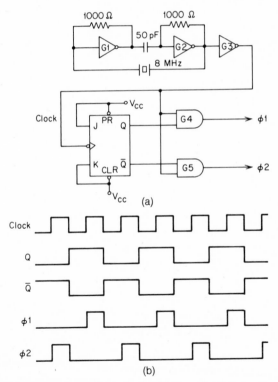

Fig. 9.3 A nonoverlapping two-phase clock utilizing only digital microcircuits: *(a)* the circuit and *(b)* selected waveforms.

The oscillator portion requires three inverters. The inverter driving the flip-flop isolates the semilinear oscillator from the flip-flop and squares up the waveform. Gates G1 and G2 have negative feedback via the 1000-Ω resistors. This keeps the gates in the active region longer and thereby allows the crystal to have better control over frequency. A crystal is a linear device and operates better in a linear circuit.

The component values (R and C) in the oscillator circuit are valid only for standard TTL gates. For gates operating at different voltage and/or current levels, other RC values will be needed.

9.2   SINGLE-CHIP TIMER

ALTERNATE NAMES   Clock generator, monolithic clock, timing circuit, oscillator, pulse generator.

PRINCIPLES OF OPERATION   The timer IC originally gained wide acceptance when the 555 device was introduced. A simplified block diagram of the 555 is given in Fig. 9.4. The main elements of this chip are two comparators and an SR flip-flop. The comparators are configured as level detectors using three resistors to establish trip levels. The upper-level detector A1 sends a set output to the flip-flop whenever the threshold input rises above $2V_{CC}/3$. Similarly, the lower-level detector A2 transfers a reset to the flip-flop when the trigger input falls below $V_{CC}/3$.

Since the flip-flop in the 555 is a dc-coupled device, we must be careful not to allow both set and reset inputs to be HIGH at the same time. Externally, this means that the threshold input cannot be above $2V_{CC}/$

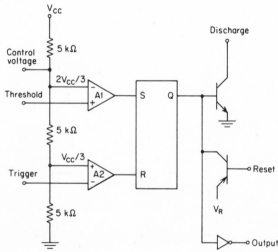

Fig. 9.4   Simplified block diagram of the 555 timer microcircuit.

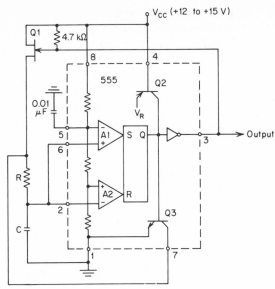

Fig. 9.5 Typical connections to the 555 timer for astable operation.

3 while the trigger input is below $V_{CC}/3$. If this happens, the flip-flop will have an indeterminate output.

The Q output from the flip-flop goes three places: (1) A totem pole buffer provides a TTL output with a source or sink capability of more than 100 mA. This stage inverts the Q waveform. (2) An open-collector output is available for a variety of applications. Its primary use is to discharge a timing capacitor attached to the threshold input (see following paragraphs). (3) The collector of a reset transistor is available to invert the output state independent of the flip-flop.

The timer used as a clock The 555 is useful as a clock, oscillator, pulse generator, or modulator. Figure 9.5 portrays a typical use for this device as a clock using its astable mode of operation. Many clocks require a 50 percent duty-cycle output waveform. A symmetrical waveform is assured by using a FET switch (Q1) to charge the timing capacitor C. Transistor Q3 inside the 555 discharges this capacitor. The ON resistance of Q1 must be less than 1 percent of R if symmetry error is to be less than 1 percent. Since the ON resistance of the discharging transistor (Q3) is very small compared with the FET, its error contribution is negligible.

The square wave is obtained at the 555 output terminal. When this voltage is HIGH, it turns Q1 ON, thereby charging C through R. During this period Q3 is OFF. When the capacitor voltage reaches $2V_{CC}/3$, the upper comparator A1 trips. This action sets the flip-flop, causing the 555 output terminal to go to the LOW state. A LOW output turns transistor Q3 ON and Q1 OFF. Transistor Q3 discharges C through R. When the capacitor voltage is reduced to $V_{CC}/3$, the lower comparator A2 trips. The flip-flop is cleared as a consequence of this action. Transistor Q3 now turns OFF and Q1 turns ON, causing the cycle to repeat.

The circuit of Fig. 9.5 operates best with V_{CC} in the +12- to +15-V range. A lower V_{CC} begins to crowd the performance of Q1. The frequency of oscillation is

$$f_0 = \frac{0.72}{RC}$$

and the period is

$$T_0 = 1.39\,RC$$

The comparator trip voltages vary directly with V_{CC}. However, the charging voltage through Q1 also varies directly with V_{CC}. These two actions cancel each other out, so that T_0 and f_0 are almost immune to V_{CC} changes. Typical errors of only 0.01 percent per volt change in V_{CC} can be expected.

9.3 MICROCOMPUTER-CONTROLLED CLOCK

ALTERNATE NAMES   Programmable frequency sources, remote-controlled clock, digital word-to-frequency controller.

PRINCIPLES OF OPERATION   Microcomputers can be used to control any parameter of a clock. Since the principal parameter of a clock is its frequency, we will here describe an astable pulse generator whose frequency is proportional to an 8-bit data word.

As shown in Fig. 9.6, the data bus must first be captured in a latch and held stationary since the clock circuit requires a steady input. The latched bus drives an 8-bit digital-to-analog converter (DAC) with a current output. This device generates an output current proportional to the binary magnitude of the input word. This current is integrated in the A1 active-integrator circuit. A current flowing into the integrator causes its output to slew downward. When the output reaches approximately 0.6 V, transistor Q3 turns OFF. This action causes the collector of Q3 to swing up to V_{CC} thereby triggering the single-shot. The 74121 single-shot is not retriggerable; i.e., once triggered, it will complete its pulse generation independent of its inputs.

The single-shot output pulse turns Q2 fully ON for a period determined by the pulse width, R_2, R_3, and C_2. The current through Q2 completely

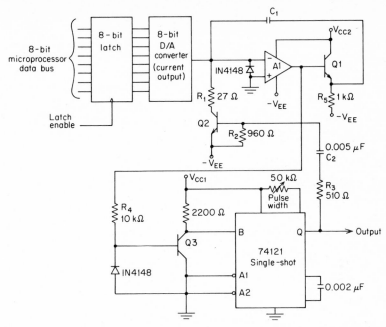

Fig. 9.6   A pulse generator whose frequency is controlled by a microcomputer data bus.

overwhelms the DAC current and quickly drives the integrator output to the positive rail.   Q2 is ON for a period of time

$$T_2 = \frac{-\ln\,(1 - V_{BE}/KV_{CC1})}{RC_2}$$

where $V_{BE} = 0.6\ \text{V}$ $K = \dfrac{R_2 h_{ie}}{h_{ie}(R_2 + R_3) + R_2 R_3}$

$h_{ie} =$ input resistance of Q2 $R = R_3 + \dfrac{R_2 h_{ie}}{R_2 + h_{ie}}$

The Q2 input resistance h_{ie} is an important factor only if R_2 is more than several hundred ohms.   Transistor Q2 must remain ON long enough to discharge C1 completely.

The time between pulses is equal to the sum of the integration time T_1 and the discharge time T_2.   The integration time is

$$T_1 = \frac{V_{CC2}C_2}{I_{DAC}}$$

In practice we make $T_2 \ll T_1$, so that the frequency of oscillation is

$$f_o = \frac{1}{T_1 + T_2} \approx \frac{1}{T_1}$$

or
$$f_o \approx \frac{I_{DAC}}{V_{CC2}C_2}$$

We see that the frequency of oscillation is linearly proportional to the DAC output current. The DAC output current is also linearly proportional to the binary magnitude of its 8-bit input. The frequency of oscillation is therefore linearly proportional to the 8-bit word from the microprocessor.

9.4 MONOSTABLE MULTIVIBRATOR

ALTERNATE NAMES   Single-shot, one-shot, pulse generator, retriggerable monostable multivibrator, resettable single-shot.

PRINCIPLES OF OPERATION   Single-shot microcircuits are available in all technologies, i.e., TTL, CMOS, ECL, etc. Most of these devices have several trigger inputs, where one or more inputs are sensitive to rising waveforms and one or more inputs are sensitive to falling waveforms. A reset input is also normally included so that the output pulse can be prematurely terminated if necessary.

Single-shots come in two general classes, retriggerable and nonretriggerable. This feature is normally specified in the title block of the specification sheet since it is important. A nonretriggerable single-shot produces only one output pulse per input trigger. If a second trigger pulse arrives while the output pulse is ON, it is ignored. In other words, these devices lock out all inputs during the pulse. A retriggerable single-shot does just the opposite. Each input trigger starts the pulse-generation circuitry over again. If a series of input trigger pulses spaced closer than the nominal output pulse width arrives, the output will remain continuously ON.

Figure 9.7 shows the logic diagram of the 4528 CMOS retriggerable and resettable monostable multivibrator. The pulse width is determined by the external RC network shown. If $C \geq 0.01$ μF, the output pulse width is

$$T_p = 0.2RC \ln (V_{DD} - V_{SS})$$

If V_{SS} is ground potential and V_{DD} is 5 V, the pulse width is

$$T_p = 0.32RC$$

For values of $C < 0.01$ μF the 0.32 constant increases as C decreases. The curves in the 4528 data sheet should be consulted.

As indicated in Fig. 9.7, this single-shot is basically two latches and a CMOS switch. The first latch made up of G2 and G3 stores the input triggers A or $\overline{B}$. If the A trigger is used, the B-trigger input

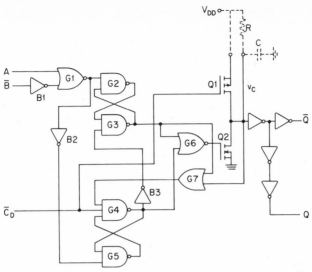

Fig. 9.7 Internal logic diagram of the 4528 CMOS retriggerable monostable multivibrator.

must be held in the HIGH state. If the B trigger is used, the A-trigger input must be held in the LOW state. The second latch, G4 and G5, stores the dc clear $\overline{C}_D$ input. When $\overline{C}_D$ goes LOW, this latch is set, causing the G2, G3 latch to be cleared.

In the standby state we require $\overline{B} = \overline{C}_D = 1$ and $A = 0$. The standby output from the G2, G3 latch holds Q2 OFF (through G6), and it also keeps the G4 output from the G4, G5 latch cleared (through G7). Since $\overline{C}_D$ is HIGH, Q1 is also OFF. With both transistors OFF the capacitor C charges to V_{DD} through R. The Q output terminal is LOW when v_C is charged to V_{DD}. During operation the v_C waveform is exponential. Several buffers in series provide sufficient gain to produce clean rectangular waveforms at Q and $\overline{Q}$.

A trigger edge on A or $\overline{B}$ forces the G2, G3 latch to change states. This action reverses the output of G6, thus turning transistor Q2 ON. The capacitor quickly discharges through Q2 resulting in two LOW signals into G7, one from Q2 and one from G3. The output of G7 goes to the LOW state, thereby changing the state of the G4, G5 latch. However, the output from G4 now forces the G2, G3 latch to change states again, causing G6 and Q2 to reverse. The capacitor therefore discharges only during the delays of Q2, G7, G4, B3, G3, and G6.

With Q2 OFF the capacitor C charges back toward V_{DD} through R. The Q output is HIGH for approximately $0.32\,RC$. Since the G2, G3 latch immediately resets itself (after the Q2, G7, G4, B3, G3, G6 delay), another trigger edge could be accepted to start the charge cycle on C over again.   Each time a new trigger edge appears, transistor Q2 discharges C.

REFERENCES

1. Jung, W. G.: Take a Fresh Look at New IC Timer Applications, *EDN*, Mar. 20, 1977, p. 127.
2. Shah, M. J.: Binary Input Determines Pulse Generator Frequency, *Electronics*, Aug. 16, 1973, p. 98.

Chapter 10

Serial Communications Microcircuits

INTRODUCTION

Data are most economically transferred over long distances in a serial format. One would never expect to perform a parallel transfer of an 8-bit microcomputer data bus to a peripheral device 200 mi away, as it would require eight telephone lines or eight separate radio frequencies operating simultaneously. Even for peripheral devices a few feet from the computer, serial communication is quite common. The most widely used microcircuit for interfacing the parallel microcomputer data bus to a serial device is called the *universal asynchronous receiver-transmitter* (UART). The transmitter section of this device performs a parallel-to-serial conversion, and the receiver section does the reverse.

Some of the typical serial interfaces with microcomputers are video displays, tape recorders, telephone lines, and radio links. Peripheral devices requiring high data-transmission rates are usually interfaced with parallel devices, discussed in the next chapter.

The UART is an asynchronous device because no clock reference is separately transmitted to synchronize the transmitter to the receiver. At the receiver end the required clock reference is extracted from the single data line. The clock at both ends typically operates at 16 times the data rate. Data rate is measured in bauds, which is the number of signal changes per second. A low-speed Teletype operates at 110 bauds. Thus, the internal clock in the UART for a Teletype must operate at $16 \times 110 = 1760$ Hz.

If synchronous serial communication with a computer is required, the universal synchronous receiver-transmitter (USRT) may be utilized. This chip contains all the logic necessary to interface a word-parallel system, such as a microcomputer, with a bit-serial synchronous communication network.

Many LSI microcircuits are available which will handle either synchronous or asynchronous serial communications. This class of device is called the universal synchronous-asynchronous receiver-transmitter (USART).

As shown in Fig. 10.1*a*, an asynchronous serial data word is organized as follows:

1. The serial line is normally HIGH between words. This is called a *mark*, and a LOW state is called a *space*.

2. A start bit is always LOW.

3. The data bits (5 to 8 in number) follow the start bit.

4. The parity bit (odd or even) is computed from the number of data bits used.

5. Either 1 or 2 stop bits (HIGH level) are utilized.

In many asynchronous applications there will be long periods of time between data words. During this interval the serial line remains in the HIGH state while the receiver is watching for a transition to a LOW state. When this transition occurs, the receiver initiates a checkout procedure to see if the transition is really data or just a noise spike. If real data are detected, the receiver begins to accept the data word.

Figure 10.1*b* shows the format for a serial bit stream using synchronous communication. The sync characters are transmitted only between blocks of data (i.e., during idle periods). At the end of every sync word the receiver readies itself for another block of data. Some devices allow for the use of two sync words.

If the serial communication link is purely digital, the USART merely

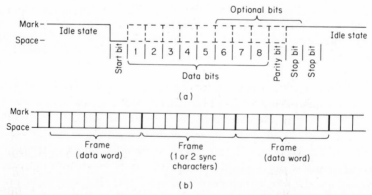

(a)

(b)

Fig. 10.1   *(a)* Serial format of each word used in asynchronous communication devices; *(b)* format for synchronous words.

needs buffering and/or level translation at the communication-line–computer interface. However, most serial communication is done through analog lines. This requires a system called a modulator-demodulator (modem) to convert digital levels to some type of analog levels. Frequency-shift-keying (FSK) circuits are most commonly used. This is an FM circuit where one frequency represents a logical HIGH and another discrete frequency represents a logical LOW. Phase or amplitude modulation can also be utilized for this serial analog interface. However, phase shift keying (PSK) is more complex and expensive than FSK, and amplitude modulation is too susceptible to noise degradation.

10.1 UNIVERSAL ASYNCHRONOUS RECEIVER-TRANSMITTER

ALTERNATE NAMES UART, USART, parallel-serial-parallel converter, serial communications interface, programmable communication interface, asynchronous communication interface adapter (ACIA).

PRINCIPLES OF OPERATION It is helpful to visualize the UART as two separate systems on a chip, i.e., the receiver system and the transmitter system. We have shown these two parts of a typical UART in two abbreviated block-diagram forms in Figs. 10.2 and 10.3. In a real chip the ×16 clock input, the five control lines, and the 8-bit data bus are common to the receiver and transmitter. We have shown them separately here for clarity. In most UARTs the five control lines time-share the data-bus inputs.

Referring to Fig. 10.2, we note that the five control inputs are first stored in flip-flops when the mode-control line is triggered. The input data are then applied to the parallel data bus. The input-data strobe rises to gate the parallel word into the transmit-data register (eight D flip-flops). A number of other things take place while the input data strobe is HIGH:

1. The word-length logic uses its 2 input bits to decide whether the word length is 5, 6, 7, or 8 bits. This utilizes gates G17 to G19 to transfer variable-length words from the transmit data register to the transmit shift register. Bypass logic must also be programmed so that the parity and stop bits will bypass data bit 8 if a 7-bit word is chosen, bypass data bits 7 and 8 if a 6-bit word is chosen, or bypass data bits 6, 7, and 8 if a 5-bit word is chosen.

2. The parity–no-parity logic is activated. This logic also has bypass circuitry so that the stop bits will be routed around the register used for parity if no parity is chosen. The odd-even parity logic computes a value for the parity bit using the transmit-data-register output lines.

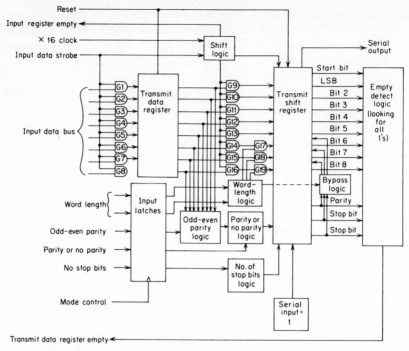

Fig. 10.2 Typical logical arrangement for the transmitter section of a UART.

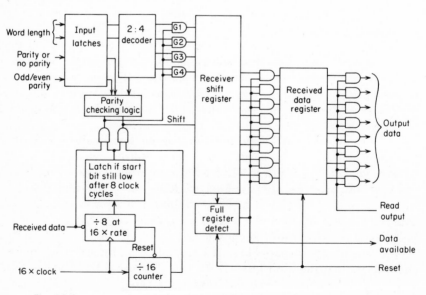

Fig. 10.3 A simplified block diagram of the receiver portion of a UART.

The word-length logic must be a factor since the parity calculations depend on the number of bits examined.

3. The selection of either 1 or 2 stop bits is done by injecting a HIGH state in appropriate sections of the transmit shift register.

When the input-data strobe falls, the shift logic begins to transmit bits out of the transmit shift register serially. Sixteen cycles of the ×16 clock are required for each bit shifted out the serial output terminal. At the bottom of the shift register HIGH bits are inserted as the word is shifted out. When the empty-detect logic determines that the shift register is completely filled with 1s, a transmit-data-register-empty signal is activated. This signal is used by the device controlling the UART to generate the next data word.

The input data bus is used for a third function in most UART devices. By addressing appropriately, the 8 data lines can read the status of many internal circuits to determine if the UART is in the proper mode. For example, the 8 status bits of the 6850 UART contain the following information:

Bit 0: Receiver data register full (see Fig. 10.3)

Bit 1: Transmit data register empty (see Fig. 10.2)

Bit 2: Data carrier detect (used in conjunction with a modem)

Bit 3: Clear to send (used with a modem)

Bit 4: Framing error (stop bit missing in a received data word)

Bit 5: Receiver overrun (one or more receiver characters lost)

Bit 6: Parity error

Bit 7: Interrupt request (can be used to determine that a received character is ready to be sent to the CPU)

Figure 10.3 shows an abbreviated block diagram of the receiver portion of a UART. This circuit is the inverse of the transmitter circuit. It receives serial data words and converts them into parallel. Its mode of operation is set up using the same five control inputs. Word length is controlled by inserting the serial input data into the shift register at selectable points using gates G1 to G4. The parity circuit checks the serial data as they pass by and activates status register bit 6 if an error is detected. No input is needed in the receiver for the 1-stop-bit or 2-stop-bit mode control used in the transmitter.

Received data first pass through circuitry which is looking for the negative edge of the start bit. This negative edge gates the ×16 clock in a ÷8 counter. If the start bit is still low after a count of 8, it has been determined that a valid start bit is present. Since the ×16 clock runs at 16 times the bit rate, 8 counts is the middle of the start bit. A latch is then set which resets a ÷16 counter. Each time this counter times out (16 cycles of the ×16 clock) we will be centered on a new data bit. At each of the bit center locations a shift pulse is generated which clocks the present value of the received data into the shift register.

This process therefore samples the received serial data at the middle of each bit period after it has stabilized.

When the start bit arrives at the full-register detector, the full-register flip-flop is set. This activates the data-available status line, which can be polled by the CPU. In some chips an interrupt is generated so that the CPU will know the UART needs attention.

10.2   UART APPLICATION: A CASSETTE MODEM

ALTERNATE NAMES   Cassette interface, FSK modem, frequency-shift-keyed modem, analog interface, FM cassette circuit.

PRINCIPLES OF OPERATION   The low-priced audio cassette recorder is one of the most popular storage mediums for small microprocessor systems. Before this storage medium could become widely accepted, however, a standard recording method had to be worked out. This was done in November 1975 at a symposium in Kansas City. As one might expect, the standard has been nicknamed the Kansas City Standard (KCS or KCS format).

The KCS can be implemented with any of the available single-chip UARTs. The UART is a digital device; i.e., both the parallel and serial interfaces handle digital data. The KCS is a standard which specifies how the UART serial input-output must be converted from or to an analog format, i.e., sine waves of several frequencies. Low-cost cassette recorders do a poor job of recording and reproducing digital data. The KCS works on a principle of frequency modulation where only two frequencies are used, 1200 and 2400 Hz. This modulation method is called *frequency shift keying* (FSK). It allows serial data to be transmitted at speeds up to 300 bauds. Tape-speed variations up to ±25 percent are possible with some of the better KCS implementations.

The basic KCS standard follows:

1. A mark (logic HIGH) bit consists of eight cycles at a frequency of 2400 Hz.

2. A space (logic LOW) bit consists of four cycles at a frequency of 1200 Hz.

3. Each recorded character consists of a start bit (logic LOW), 8 data bits, and 2 or more stop bits (logic HIGH).

4. The interval between characters is unspecified.

5. The 8 data bits are organized as least significant bit first, most significant bit last, and a parity bit (optional). The total number of data bits including parity cannot exceed 8.

6. If fewer than 8 data bits are used, the unused bit locations are left in the HIGH state (2400 Hz).

7. Data are organized in blocks of arbitrary length preceded by at least 5 s of marks (2400 Hz).

8. To minimize errors from splices and wrinkles at the beginning of the tape, the beginning of the first block of data will occur at least 30 s after the beginning of a clear leader.

9. The KCS is designed around the use of a ×16 clock for both transmit and receive. The ×16 clock is reconstructed from the received data using frequency-multiplication techniques (16 × 1200 Hz = 19,200 Hz is the standard ×16 clock used).

10. Transitions from mark to space and vice versa occur at the zero crossover point in the serial waveform. This eliminates excessive harmonic generation, which could be a source of noise.

There is a valid reason for recording eight cycles of 2400 Hz for a mark and four cycles of 1200 Hz for a space. This technique allows the receiver at playback to determine how fast the tape is going. If the tape is played back 10 percent slow, the 1200 Hz becomes 1080 Hz and the 2400 Hz becomes 2160 Hz. However, the receiver multiplies 1080 by 16 to obtain 17,280 Hz. This is used as the receiver's ×16 clock and is sent to the UART. The receiver system therefore slows down all processes by 10 percent and synchronizes to the slower tape speed.

The receiver circuit   Figure 10.4 is a typical receiver for a cassette interface modem. Figure 10.5 shows waveforms at various points in the circuit. The input to the circuit is taken from the "ear" output of the

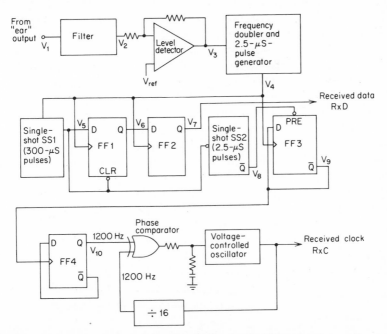

Fig. 10.4   The receiver portion of a modem designed to interface a low-cost audio cassette recorder to a UART and microprocessor system.

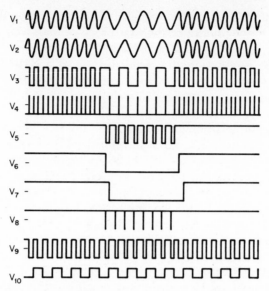

Fig. 10.5 Timing relationships of various waveforms in Fig. 10.4.

recorder. The signal is first filtered to remove noise outside the 1200-to 2400-Hz frequency range. A level detector, or slicer, is next utilized to change the sine-wave inputs into rectangular waveforms V_3 so that they can be digitally processed. This circuit uses positive feedback to produce steep edges on the rectangular waveform. The next block produces a short positive pulse V_4 for each transition (positive or negative) of the rectangular signal. These pulses will be spaced at 208 μs when the mark (2400-Hz) bit is received and at 416 μs when a space (1200-Hz) bit is received. We will have 16 pulses during the mark and 8 pulses during a space. These short pulses trigger single-shot SS1, which generates relatively long pulse widths V_5 of approximately 300 μs. SS1 is a retriggerable single-shot; i.e., if it is retriggered while an output pulse is present, the new pulse is tacked onto the first pulse with no gap in between. Therefore, if mark pulses occur at 208-μs intervals and the single-shot pulse width is 300 μs, the output will continuously remain HIGH. If a space occurs (416 μs), the single-shot times out, causing the waveform to go LOW for 116 μs. Whenever the output of SS1 goes LOW, it clears FF1, causing its output V_6 to also go LOW. If a mark is present, the output of FF1 remains in the HIGH state. We therefore have recovered digital data at the output of FF1. However, these data have short spaces and long marks. FF2 places symmetry back into the recovered data V_7 by using trigger pulses from the frequency doubler.

Single-shot SS2 generates a short negative pulse V_8 each time V_5 has a negative transition. Each of these pulses presets FF3 so that its $\overline{Q}$

output V_9 goes LOW. The next pulse from the frequency doubler V_4 toggles FF3. A study of V_4 and V_8 in Fig. 10.5 shows that V_9 is at 2400 Hz when V_4 is 4800 Hz. However, when V_4 is at 2400 Hz, the output V_9 is still at 2400 Hz since V_8 keeps presetting FF3 in between the slower rate pulses. The output V_9 is therefore always 2400 Hz, although the symmetry is poor in the space period. FF4 is a simple toggle flip-flop which places symmetry in the waveform and changes the frequency to 1200 Hz.

The last few blocks in Fig. 10.4 are a $\times 16$ frequency multiplier utilizing a phase-locked loop (PLL) chip and a $\div 16$ counter, both discussed in detail in Chap. 27. Essentially, this frequency multiplier has a voltage-controlled oscillator (VCO) tuned to approximately 19,200 Hz. This output goes to the clock input of the UART. The same output is also divided down by a four-stage counter so it becomes 1200 Hz. This frequency is compared with the 1200 Hz from FF4 in a phase comparator. If any difference in frequency or phase exists between these inputs, a correction voltage is sent over to the VCO. This voltage levels out at a voltage which provides a precise phase lock of the two inputs.

The transmitter circuit The transmitter portion of a cassette interface modem is very simple compared with the receiver. A typical circuit is shown in Fig. 10.6a. All this circuit must do is generate 2400 Hz when

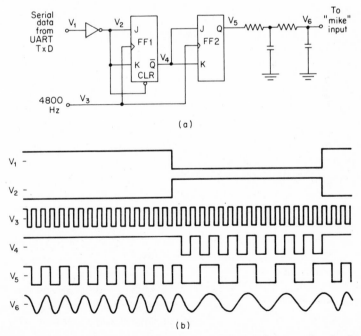

(a)

(b)

Fig. 10.6 A typical modem transmitter for a cassette interface: *(a)* the circuit and *(b)* its waveforms.

a HIGH (or mark) state is received from the UART and generate 1200 Hz when a LOW (or space) is received. The various waveforms for the circuit are shown in Fig. 10.6b.

A 4800-Hz clock is coupled to the clock input of the two flip-flops in the circuit. When the UART data output TD is LOW, V_2 is HIGH. A HIGH level on the JK inputs of FF1 causes it to divide by 2. Likewise, FF2 divides again by 2, and the output V_5 is at 1200 Hz. A low-pass filter clips off the square corners and makes a quasi sine wave at V_6.

When V_1 is HIGH, V_2 is LOW, causing FF1 to remain in the reset state. This keeps the JK inputs of FF2 HIGH, so that it divides the 4800-Hz clock by 2. The resulting 2400-Hz square wave is filtered and passed onto the cassette recorder mike input. We note that since changes from one output frequency to the other occur during zero crossover, high-order harmonics should not be prevalent.

10.3   UNIVERSAL SYNCHRONOUS RECEIVER-TRANSMITTER

ALTERNATE NAMES   USRT, USART, programmable communication interface, synchronous data communication device, synchronous serial-data adapter (SSDA).

PRINCIPLES OF OPERATION   Devices of this class provide a bidirectional serial interface for synchronous transfer of data. Most USRT chips can simultaneously transmit and receive data (full duplex). A block diagram of a Motorola 6852 synchronous serial data adapter (SSDA) is shown in Fig. 10.7. This is a second-generation USRT. The 6852 is designed to operate with an 8-bit microprocessor data bus. It can be programmed to handle numerous data formats by storing command information in three internal control registers. The sync code is also stored during the preliminary programming of the 6852. Up to three words of transmitted data can be stored in a first-in first-out (FIFO) set of registers. Received data can likewise be collected in three FIFO registers. The state of eight internal registers is always available by reading the status register. Thus we have eight write-only registers and four read-only registers in this synchronous serial interface chip.

Referring to Fig. 10.7, we observe that the $\overline{\text{RESET}}$ line interfaces only with control register 1. When this line goes LOW, both the transmitter and receiver sections are latched into a reset condition. Also, peripheral control bits PC1 and PC2 are cleared, the error-interrupt enable (EIE) is cleared, and the transmitter-data-register available (TDRA) is cleared. All these control bits can later be set by software control using the three control registers.

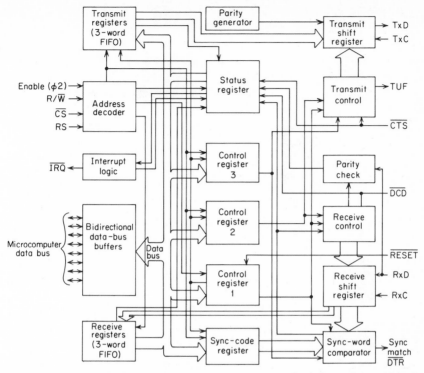

Fig. 10.7 Block diagram of the Motorola synchronous serial data adapter (SSDA). This is a second-generation USRT chip *(Motorola, Inc.)*.

Initialization The SSDA requires several programming steps before it can be used for transmitting or receiving. The first step is to load control word 1 (C1) into the SSDA. Table 10.1 shows that we require only RS = **0** and R/$\overline{W}$ = **0** to perform this command. The programming of the bits in C1 depends on the application. For example, if we are initially programming the SSDA (after a $\overline{\text{RESET}}$) for full duplex operation, C1 would probably be 00000000. This is done with a CLEAR command for the address of C1. Referring to Table 10.1, we observe that this command does the following:

Bit 0: Maintains receiver at inhibited state until later in initialization sequence.

Bit 1: Maintains transmitter at inhibited state.

Bit 2: Sync stripping not active (this will be done later).

Bit 3: Enables synchronization when bit 2 is later enabled.

Bit 4: Transmitter-interrupt-status bit and IRQ placed in standby state.

Bit 5: Receiver-interrupt-status bit and $\overline{\text{IRQ}}$ placed in standby state.

TABLE 10.1 SSDA Programming Model*

Register name and address				Register content							
RS	R/W̄	AC2	AC1	Bit 7	Bit 6	Bit 5	Bit 4	Bit 3	Bit 2	Bit 1	Bit 0
Control 1 (C1)				(AC2)	(AC1)	Receiver interrupt enable (RIE)	Transmit interrupt enable (TIE)	Clear receiver sync	Strip sync char.	Transmit reset (TxRs)	Receiver reset (RxRs)
0	0	X	X								
Control 2 (C2)				Error interrupt enable (EIE)	Transmit sync code if underflow (Tx sync)	Word-length select 3 (WS3)	Word-length select 2 (WS2)	Word-length select 1 (WS1)	1-byte–2-byte transfer (1B/2B)	Periph. cont. 2 (PC2)	Periph. cont. 1 (PC1)
1	0	0	0								
Control 3 (C3)								Clear transmit. underflow status (CTUF)	Clear C̄T̄S̄ status (CLR C̄T̄S̄)	1 or 2 sync char. control (1S/2S)	External-internal sync mode control (E/I sync)
1	0	0	1								
Status (S)				Interrupt request (IRQ)	Receiver parity error (PE)	Receiver overrun (Rx OV)	Transmit. underflow (TUF)	Clear to send (C̄T̄S̄)	Data-carrier detect (D̄C̄D̄)	Transmit-data-register available (TDRA)	Receiver data available (RDA)
0	1	X	X								
Sync code				D7	D6	D5	D4	D3	D2	D1	D0
1	0	1	0								
Receive data FIFO				D7	D6	D5	D4	D3	D2	D1	D0
1	1	X	X								
Transmit data FIFO				D7	D6	D5	D4	D3	D2	D1	D0
1	0	1	1								

*X is a don't-care logic level.

Bits 6 and 7: A code of **00** allows the addressing of control register 2 (C2) on the next SSDA command.

The second step in the initialization of the SSDA is a write operation into C2. To address this register we need RS $= 1$ and R/$\overline{\text{W}} = 0$ plus the AC1, AC2 bits previously left in C1. Again assuming full duplex operation, we write 10111101 into C2. This instruction sets up the SSDA as follows:

Bit 0: Selects the sync match mode.

Bit 1: Causes a 1-bit-wide pulse to appear at the SM/$\overline{\text{DTR}}$ terminal each time a sync match occurs.

Bit 2: A HIGH level on this bit chooses the single-word transmitter-receiver FIFO mode. If this bit was LOW, 2 words could be moved into or from the two FIFOs.

Bits 3 to 5: These 3 bits choose the transmit or receive word length, as shown in Table 10.2. Note that we have selected 8 bits plus odd parity in this example.

Bit 6: A LOW level causes the transmitter to send mark characters (HIGH logic level) *automatically* whenever data are not available for transmission from the transmit FIFO.

Bit 7: If this bit is HIGH, the error-interrupt circuitry is enabled. This allows the IRQ status bit to go HIGH and the $\overline{\text{IRQ}}$ output to go LOW when any of the following occur:

1. Receiver overrun (more than 3 words trying to load receiver FIFO)
2. $\overline{\text{DCD}}$ goes HIGH
3. Received-data parity error
4. $\overline{\text{CTS}}$ goes HIGH
5. Transmitter underflow(transmitter FIFO empty)

The third initialization step is to load control register 3 (C3). This is done by first writing in another C1 command of 01000000. Bits 7 and 6 (AC2 and AC1) of C1 now allow us to address C3 (see Table 10.1). If we also make RS $= 1$ and R/$\overline{\text{W}} = 0,$ the addressing is complete.

TABLE 10.2 Word-Length Selection in Control Word 2

Bit 5 WS3	Bit 4 WS2	Bit 3 WS1	Word length
0	0	0	6 bits + even parity
0	0	1	6 bits + odd parity
0	1	0	7 bits
0	1	1	8 bits
1	0	0	7 bits + even parity
1	0	1	7 bits + odd parity
1	1	0	8 bits + even parity
1	1	1	8 bits + odd parity

The data-bus content written into C3 for our example is 00001110. This word does the following:

Bit 0: A LOW selects the internal synchronization mode, whereby the sync character(s) are used to begin a new block of data.

Bit 1: A HIGH indicates that a single sync character is required to flag the start of a block of data.

Bit 2: Writing a HIGH into this location clears the $\overline{CTS}$ status bit and its associated interrupt. The USRT is then available for external transmitter control through the clear-to-send ($\overline{CTS}$) input, if required.

Bit 3: A HIGH written into bit 3 clears the transmit-underflow-status bit (CTUF) and its associated interrupt. This readies the device for transmission applications which use the transmit-underflow-status (TUF) bit.

Bits 4 to 7: Not used by the SSDA.

After all three control registers are programmed, the sync-code register can be loaded. This requires that we first write 10000000 into control register 1 so that AC2 and AC1 will address the sync-code register. We now select RS = 1 and R/$\overline{W}$ = 0 to complete the addressing. Simultaneous to RS = 1 and R/$\overline{W}$ = 0 the sync-code word is placed on the data bus to be written into the SSDA.

The last step in initialization is to release the transmitter- and receiver-reset control bits so that both sections of the SSDA will be ready for operation. We will now sequence through typical transmit and receive procedures.

Transmitter operation   The next control register 1 command should include AC2 = AC1 = 1 in order to activate the transmit data FIFO. The word on the data bus will then be parallel-loaded into the 3-byte transmit FIFO. Enable (ϕ2) pulses clock data into and through all three stages of the FIFO. The byte then transfers over to the transmit shift register, where it is serially transmitted at the rate of the transmit clock. Data move from the last register of the FIFO to the transmitter shift register during the last half of the last bit of the previous character. This transfer is initiated by the transmitter clock, whereas movement of bytes in the FIFO is clocked by the enable (ϕ2) pulses.

Data are transmitted least significant (LSB) bit first. The parity bit is optionally attached after the MSB. If the shift register becomes empty and a new byte is not available in the FIFO, a transmit underflow occurs. This sets bit 4 (TUF) in the status register and lowers the $\overline{IRQ}$ output. This interrupt is seen by the microprocessor, which in turn reads the status register to determine the needs of the SSDA. If the microprocessor notes that the TUF bit is set, another byte is sent over to the SSDA on the 8-bit data bus. However, if the microprocessor chooses not to send over another byte, the transmitter will begin shifting out marks (HIGH logic level) or sync characters, depending on the state of C2 bit

6. The underflow condition is also indicated by a positive pulse approximately one Tx Clk wide on the TUF terminal which coincides with the transfer of the last half of the last bit preceding each underflow character.

When transmitter underflow occurs and data are to be sent again, the transmitter-underflow-status bit should first be cleared. This is done with bit 3 of C3. Since we need AC2 = **0** and AC1 = **1** to program C3, C1 will probably need to be accessed first. The clear underflow command will also reset the $\overline{\text{IRQ}}$ output. AC2 must then be returned to the HIGH state so that the data bus will again have access to the transmit FIFO.

If the SSDA is to be automatically controlled externally by some device such as a modem, the clear-to-send $\overline{\text{CTS}}$ input can be used. The $\overline{\text{CTS}}$ output from the modem is connected directly to the SSDA $\overline{\text{CTS}}$ input. When this line is HIGH, the transmitter shift register is inhibited and reset. The TDRA status bit is also inhibited if $\overline{\text{CTS}}$ is HIGH and the internal sync mode is selected. If external sync is being used, TDRA is unaffected by $\overline{\text{CTS}}$ in order to provide transmit FIFO status for preloading and operating the transmitter under control of the $\overline{\text{CTS}}$ input.

The transmit FIFO can be cleared by setting the transmitter reset bit (TxRs). This also clears the TDRA status bit. After one $\phi 2$ pulse has occurred, the transmit FIFO becomes available for new data.

Receiver operation   Data and a synchronous clock are applied to the SSDA at the Rx Data and Rx Clk inputs. The data are a continuous stream of bits with no means of identifying character boundaries within the data stream. Synchronization must therefore be achieved at the beginning of a block of characters. After synchronization has been achieved, we assume it to be retained for all successive characters within the block. The sync character(s) received before synchronization is complete are not transferred to the receiver FIFO.

Once synchronization has been achieved, subsequent characters are automatically transferred from the receiver shift register into the receiver FIFO. The characters are clocked through the FIFO to the last empty location by $\phi 2$ pulses. The receiver-data-available-status bit (RDA) indicates when data are available to be read from the last FIFO location (location 3) when operation is in the 1-byte transfer mode. If the 2-byte transfer mode is utilized, the RDA status bit indicates that data are available when the last two FIFO register locations are full. This data-available status also causes an interrupt request ($\overline{\text{IRQ}}$ goes LOW) if the receiver-interrupt-enable (RIE) bit is set. The microcomputer then reads the SSDA status register, which will indicate that data are available to read in the last stage of the receiver FIFO. The $\overline{\text{IRQ}}$ and RDA status bits are reset by the FIFO read command. If two or more characters are resident in the receiver FIFO, a second $\phi 2$ pulse will cause the FIFO to update and the RDA and IRQ status bits will again be set. A third $\phi2$ pulse is required to clock the second word out of the FIFO.

Parity is automatically checked as data are received, and the parity-status condition is maintained with each character until the data are read from the receiver FIFO. Parity errors will cause an $\overline{IRQ}$ output if the error-interrupt enable (EIE) has been set. Since the parity bit is not transferred to the data bus, it must be checked in the status register. Thus, in the 2-byte transfer mode the parity error must be checked before reading the second word. This parity error is stored in the FIFO along with the second word, and it activates the status register as it reaches the third position.

If the receiver FIFO is completely full with 3 words and the shift register transfers in another word, the overrun status bit will be set. This also causes an interrupt output if the error-interrupt enable (EIE) has been set. The transfer of the overrunning character into the FIFO causes the previous character stored in position 1 to be lost. The overrun status bit is cleared by reading the status register followed by a read command of the receiver FIFO. Overrun cannot occur and be cleared without providing an opportunity to detect its occurrence in the status register.

A positive transition on the $\overline{DCD}$ input causes an interrupt output ($\overline{IRQ}$ goes LOW) if the EIE is set. This interrupt is cleared by reading the status register while $\overline{DCD}$ is HIGH followed by a receiver FIFO read command. The $\overline{DCD}$ status bit will subsequently follow the state of the $\overline{DCD}$ input when it goes LOW.

REFERENCES

1. Smith, R. L.: Understanding the UART, *Pop. Electron.*, June, 1975, p. 43.
2. Smith, L.: USART: A Universal μP Interface for Serial Data Communications, *EDN*, Sept. 5, 1976, p. 81.
3. Etcheverry, F. W.: Binary Serial Interfaces: Making the Digital Connection, *EDN*, Apr. 20, 1976, p. 40.
4. Peschke, M., and V. Peschke: BYTE's Audio Cassette Standards Symposium, *Byte*, February 1976, p. 72.

Chapter 11

Parallel Communications Microcircuits

INTRODUCTION

Chapter 10 discussed digital communication devices which transmit data over a single wire and receive data over another single wire. The present chapter covers a similar class of microcircuits, i.e., those which interface a microcomputer to the outside world using a bidirectional data bus of four or more parallel lines. External equipment cannot be tied directly to the microprocessor data bus because data are continually changing on these lines. A device is required which captures all data-bus signals at a specific time when the data bus is stable and contains the correct information. This process is implemented in a number of ways by manufacturers, but we will discuss only a few of the simpler techniques in this chapter.

11.1 BUFFER GATE

ALTERNATE NAMES Quad buffer, hex buffer, octal buffer, bus driver, tristate buffer, inverting buffer, noninverting buffer, logic translating buffer.

PRINCIPLES OF OPERATION At least six types of buffer gates are used in parallel-bus interfaces. Figure 11.1 shows the logic symbols for a few of the common buffer gates available with four, six, or eight gates to a package. Data are transferred through the gate only when the enable line (EN, the line going into the side of the gate symbol) is

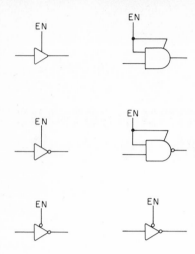

Fig. 11.1 A few of the common inverting and noninverting buffer gates used in parallel microprocessor interfaces. The line on top of each gate is the tristate enable line.

enabled. If a LOW enable ($\overline{\text{EN}}$) is required, the small circle is used in the symbol.

The outputs of these gates are in the high-impedance state when the enable line is OFF. In this state the outputs are then free to assume the logic level of other gates tied to this line. Buffer gates of the types shown in Fig. 11.1 are typically used as inputs to the microprocessor data bus. All devices driving this data bus must have tristate capability so that only the addressed device is able to place data on the bus.

Figure 11.2 shows a representative application for buffer gates in a microprocessor system. A buffer-gate package typically has one or two enable inputs. A single line, of course, would enable all gates in the package simultaneously. If two enable lines are brought out, at least three configurations are available: (1) the two enable lines are ORed, (2) the two enable lines are ANDed, or (3) each enable line controls half the gates. If an OR (or NOR) or AND (or NAND) is used internally, it can be utilized to perform some of the address decoding. For example, in Fig. 11.2 the AND in this circuit decodes address 8005 (hex) by taking the AND of address bit 15 and the sixth output of a 1-of-8 decoder which is attached to the three lowest-order address bits. Since buffer gates are typically used as inputs to the microprocessor system, the line R/$\overline{\text{W}}$ must be high if the gate array is to be enabled. This can be done in a number of ways. If the 1-of-8 decoder has an enable or strobe input, the R/$\overline{\text{W}}$ can be applied to that input.

Buffer gates can also be used as microprocessor data-bus outputs. However, the outputs will contain the correct information for only several microseconds (or less). In most applications the output information must be stored (or latched) so that the external circuitry has time to perform its function. This topic is discussed in Sec. 11.2.

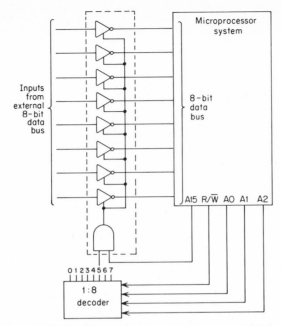

Fig. 11.2   An array of eight buffer gates used to transfer an external byte of information onto the microprocessor data bus. This transfer takes place when the address bus contains 8005 (hex).

Buffer gates are also useful for conversion from one type of logic level to another. This is especially required when going from CMOS to TTL since most CMOS output circuits will handle only one or two low-power TTL inputs. These same buffers are also useful for interfacing two systems which operate at different V_{CC} levels. Typical devices in this class are the 74C901 inverting TTL buffer and the 74C902 noninverting buffer.

11.2   LATCH–*D* FLIP-FLOP

ALTERNATE NAMES   Input-output port, parallel-output port, parallel peripheral driver, quad latch, quad *D* flip-flop, hex latch, hex *D* flip-flop, octal latch, octal *D* flip-flop.

PRINCIPLES OF OPERATION   Latches and *D* flip-flops are widely used as parallel-output ports for microcomputer systems. These devices are readily available with four, six, or eight to a package. Latches differ from *D* flip-flops in one minor respect. A latch has a LATCH ENABLE input. When this terminal is HIGH, the latch outputs will follow the *D* inputs. When LATCH ENABLE goes LOW, the data on the *D* inputs, which

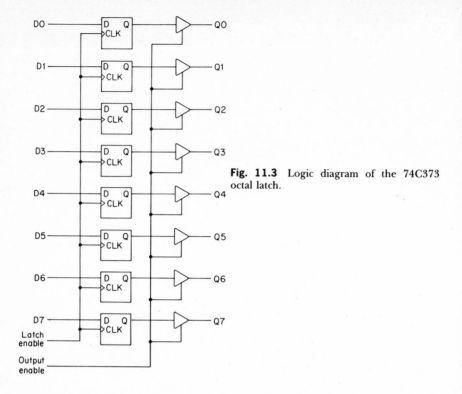

Fig. 11.3 Logic diagram of the 74C373 octal latch.

satisfy setup and hold-time requirements, are retained on the outputs until LATCH ENABLE returns to the HIGH state. The D flip-flop, however, is an edge-triggered device. A clock input is utilized to move data from the D inputs to the corresponding output. D flip-flops sensitive to either positive or negative clock edges are available.

Figure 11.3 shows the logic diagram for the 74C373 octal latch. The logic diagram for the 74C374 octal D flip-flop is identical except that the LATCH ENABLE in this case is called the clock input. The LATCH ENABLE is normally maintained in the LOW state. Whenever data stored in the eight latches are to be updated, the LATCH ENABLE line momentarily receives a positive pulse and the data inputs are latched over to the outputs. The LATCH ENABLE signal is usually derived from an address decoder which uses some of the address lines, the R/$\overline{\text{W}}$ line, and a clock signal. The address decoder sometimes requires three to five equivalent gates in series to handle all the needed inputs. The time delays of all these gates must be carefully evaluated to make sure the LATCH ENABLE signal arrives slightly after the D inputs have stabilized but not after the D inputs are changing to something else.

11.3 PARALLEL PERIPHERAL INTERFACE

ALTERNATE NAMES Peripheral interface adapter (PIA), programmable peripheral interface (PPI), universal peripheral interface.

PRINCIPLES OF OPERATION Every microcomputer chip set has programmable LSI devices available for parallel interfacing to the outside world. Although each manufacturer has a slightly different way of imple-

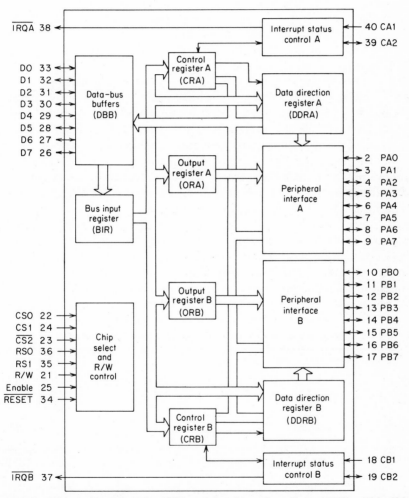

Fig. 11.4 A block diagram of the 6820 peripheral interface adapter (PIA). This Chip is made by several companies which manufacture the 6800 microcomputer chip set (Motorola, Inc., and American Microsystems, Inc.).

menting a parallel interface, one basic characteristic is common to all: data are transported in or out of the microcomputer system on a parallel bus whenever the correct address appears on the chip. An R/$\overline{W}$ line is also required to inform the device which way data are flowing. Most of these devices can perform some of the address decoding so that external gates are kept to a minimum.

The 6820 peripheral interface adapter (PIA) is a representative parallel input-output device made by several companies. As shown in Fig. 11.4, this chip has an 8-bit data-bus interface on the microcomputer side and two bidirectional 8-bit ports on the peripheral side. Four control-interrupt lines are also provided on the peripheral side. Two of these lines are bidirectional, and two are inputs to the PIA.

We note from Fig. 11.4 that each of the two peripheral data buses has a group of supporting logic. Each side has a data-direction register which controls the direction of data flow (to or from the microprocessor). Each side has a control register which sets up the mode of operation for that bus. Each side has an interrupt input and a bidirectional control line. Finally, each side has an output register (a set of latches) which temporarily stores the bus content until changed by a new command.

Before using the PIA the control registers CRA and CRB must be programmed so that the correct mode of operation is set up. The data-direction registers are next programmed to fix the direction of data flow. The PIA is then ready to be used.

The two input lines CA1 and CB1 are used primarily as interrupt inputs. They affect the status of PIA interrupt flags in the two control registers and also the PIA interrupts ($\overline{\text{IRQA}}$ and $\overline{\text{IRQB}}$). These flags and interrupts can each be programmed to respond to CA1 and CB1 inputs in four different ways, as shown in Table 11.1. Each input can be made positive- or negative-edge-sensitive according to the status of

TABLE 11.1 Four Modes of Operation for Control Inputs CA1 and CB1

Control-register inputs		Input CA1 (CB1)	Interrupt flag CRA7 (CRB7)	Interrupt to microprocessor $\overline{\text{IRQA}}$, ($\overline{\text{IRQB}}$)
CRA1 (CRB1)	CRA0 (CRB0)			
0	0	↓	Set HIGH on ↓ of CA1 (CB1)	Always HIGH
0	1	↓	Set HIGH on ↓ of CA1 (CB1)	Goes LOW when CRA7 (CRB7) goes HIGH
1	0	↑	Set HIGH on ↑ of CA1 (CB1)	Always HIGH
1	1	↑	Set HIGH on ↑ of CA1 (CB1)	Goes LOW when CRA7 (CRB7) goes HIGH

TABLE 11.2   Four Modes of Operation for CA2 and CB2 when Used as Inputs*

Control-register inputs		Input	Interrupt	Interrupt to
CRA4 (CRB4)	CRA3 (CRB3)	CA2 (CB2)	flag CRA6 (CRB6)	microprocessor $\overline{IRQA}$, $(\overline{IRQB})$
0	0	↓	Set HIGH on ↓ of CA2 (CB2)	Always HIGH
0	1	↓	Set HIGH on ↓ of CA2 (CB2)	Goes LOW when CRA6 (CRB6) goes HIGH
1	0	↑	Set HIGH on ↑ of CA2 (CB2)	Always HIGH
1	1	↑	Set HIGH on ↑ of CA2 (CB2)	Goes LOW when CRA6 (CRB6) goes HIGH

*CRA5 (or CRB5) must be LOW.

two control bits.   In two of the four modes the $\overline{IRQ}$ output lines are not affected.   In the other two modes both interrupt flags and $\overline{IRQ}$ output lines are active.

The other two control lines CA2 and CB2 can be made to perform exactly like CA1 and CB1 if CRA5 (or CRB5) is in the LOW state. Table 11.2 shows the modes of operation for these two lines if they are used as inputs.

Control lines CA2 and CB2 can also be used as outputs if CRA5 (or CRB5) is HIGH.   In this case, however, CA2 and CB2 operate quite differently, as shown in Tables 11.3 and 11.4.

Section A peripheral bus   Each of the peripheral data lines can be independently programmed to act as an input or an output.   This programming is performed by writing a 1 into the corresponding data-direction-

TABLE 11.3   Four Modes of Operation for CA2 when Used as an Output*

Control-register inputs		Conditions needed to clear CA2	Conditions needed to set CA2
CRA4	CRA3		
0	0	LOW on ↓ of enable after a microprocessor read of PIA side A data	HIGH on ↑ or ↓ transition of CA1
0	1	LOW after a microprocessor read of PIA side A	HIGH on ↓ transition of enable
1	0	LOW on ↓ of CRA3 as a result of a write into control register A	LOW if CRA3 is LOW
1	1	HIGH if CRA3 is HIGH	HIGH on ↑ of CRA3 as a result of a write into control register A

*CRA5 must be HIGH.

TABLE 11.4   Four Modes of Operation for CB2 when Used as an Output*

Control-register inputs		Conditions needed to clear CB2	Conditions needed to set CB2
CRB4	CRB3		
0	0	LOW on ↑ of enable after a microprocessor write into PIA B data register	HIGH when interrupt flag CRB7 is set by active transition of CB1 input
0	1	LOW on ↑ of enable after a microprocessor write into PIA B data register	HIGH on ↑ of enable
1	0	Follows CRB3	Follows CRB3
1	1	Follows CRB3	Follows CRB3

* CRB5 must be HIGH.

register bit for those lines which are to be outputs. A LOW state in a data-direction-register bit makes the corresponding A line an input. If the data-direction register contains all 0s, then during a microprocessor-read-peripheral-data instruction the data on each A peripheral line appear directly on the corresponding microprocessor-data line. Likewise, in the write mode a microprocessor write into the PIA causes an immediate transfer of the data bus over to the A peripheral bus.

Section B peripheral bus   The B peripheral data lines perform similarly to the A side except for several minor points. As inputs, the B peripheral lines do not source current as a TTL device, but they have a high input impedance. As outputs, the B lines can source up to 1 mA for directly driving the base of a power transistor. A standard TTL output will typically source only 0.1 mA.

Both the A and B peripheral buses can be read by the microprocessor even when they are programmed as outputs. If the outputs are so heavily loaded that incorrect data are sent to the peripheral, the microprocessor can determine these incorrect data by reading the output bus.

Microprocessor addressing of the PIA   The 6820 has eight control inputs from the microprocessor system. Three of these lines, CS0, CS1, and $\overline{CS2}$, perform part of the address decoding for the PIA. They replace two external-address decoding gates. The PIA cannot be accessed (read or write) unless CS0 = CS1 = 1 and $\overline{CS2}$ = 0 simultaneously.

The R/$\overline{W}$ line must be HIGH whenever the PIA is being read and LOW during a write operation. The $\overline{RESET}$ line must be maintained in the HIGH state for normal operation. If it is momentarily brought LOW, all internal PIA registers are cleared. The enable line is normally one of the microprocessor system clock lines. The enable line must go LOW after the address and R/$\overline{W}$ lines have stabilized. This guarantees that a correct address will be clocked into the PIA.

TABLE 11.5 Addressing the Six PIA Registers

RS1	RS0	Control-register bit*		Internal register selected
		CRA2	CRB2	
0	0	1	X	Peripheral register A
0	0	0	X	Data direction register A
0	1	X	X	Control register A
1	0	X	1	Peripheral register B
1	0	X	0	Data direction register B
1	1	X	X	Control register B

* X = don't-care state.

The control lines RS0 and RS1 are utilized to access the various registers in the PIA. These lines, in conjunction with bits previously stored in the control registers, allow the microprocessor to select one of six PIA registers. Table 11.5 indicates the programming requirements for these registers.

Whenever RS1 = **0** and RS0 = **1,** we select control register A. The control-register bits have no effect in this selection process since it would be illogical for a bit in a register to address itself. Likewise, unconditional access to control register B is obtained by letting RS1 = RS0 = **1.**

The complete format for the two control registers is shown in Table 11.6. The details of each group of bits have previously been shown in Tables 11.1 to 11.5.

PIA programming example Let us assume that we are using the A peripheral bus to send or receive data for a machine. The machine also requires the CA2 control signal to start some operation. The input control signal CA1 is used to alert the microprocessor that the operation is complete. Figure 11.5 is a schematic of this 10-wire PIA-machine interface.

After a system reset has taken place, several setup instructions are required before the PIA can begin the normal data flow. The following sequence is a typical setup procedure for this type of PIA.

1. System RESET gets LOW momentarily, setting all PIA registers to zero.

TABLE 11.6 Control-Word Format

	Bit number							
	7	6	5	4	3	2	1	0
CRA	IRQA1	IRQA2	CA2 control			DDRA access	CA1 control	
CRB	IRQB1	IRQB2	CB2 control			DDRA access	CB1 control	

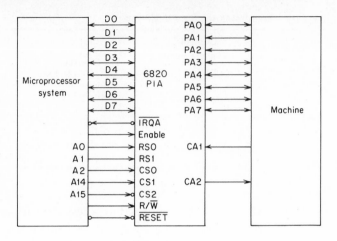

PIA Register	Address (HEX)
Data direction A and peripheral A	4004
Control A	4005
Data direction B and peripheral B	4006
Control B	4007

Fig. 11.5   Using a 6820 PIA to monitor and control a machine.

2. The microprocessor writes a 00111000 into the *A* control register (hex address 4005).   This command makes the CA2 output HIGH (its inactive state), disables the CA1 input, and sets CRA2 equal to zero so that data-direction register A can next be addressed.

3. The microprocessor writes a 11111111 into the A data-direction register (hex address 4004 with CRA2 cleared). These 1s in the data-direction register cause the contents of the peripheral register to be outputs.

4. Sending data from the microprocessor through the PIA to the machine requires that we again access the control register and change CRA2 to the HIGH state.   The next data written into hex 4004 will then be directly transferred over to the peripheral data bus.

5. Suppose we want to activate CA2 at this point to make the machine process the 8-bit word just sent to it.   Since CA2 is HIGH in the standby state, we need a command to make it go LOW momentarily. This is performed by writing a **0** into CRA3.

6. At the same time as the above operation we also write a **1** into CRA0.   This sets up control input CA1 so that it will look for a negative transition.

7. When the machine sends over a negative transition on CA1,

the flag bit in CRA7 is set to the HIGH state, and the PIA $\overline{\text{IRQA}}$ (interrupt) output goes LOW.

8. The $\overline{\text{IRQA}}$ line from the PIA is usually wire-ORed with other $\overline{\text{IRQ}}$ lines. Therefore, the microprocessor must check the status of all devices having an $\overline{\text{IRQ}}$ output. This PIA is checked by performing a read operation on control register A. The microprocessor will observe that CRA7 is HIGH. (This read operation resets CRA7 to the LOW state and sets $\overline{\text{IRQA}}$ back to the HIGH state.)

9. Now that the microprocessor has found out that CA1 is LOW, it must read the peripheral data bus from the machine. This first requires that CRA2 be changed to a **0** so that the data-direction register can be accessed. We next write 00000000 into the data-direction register, causing the peripheral data bus to change into inputs.

10. Next we change CRA2 to the HIGH state so that the peripheral data register can be accessed. A read operation on that register will then transfer the 8-bit word from the machine over to the microprocessor.

Chapter 12

BCD Circuits

WHY BCD CIRCUITS ARE NECESSARY

Binary-coded-decimal (BCD) circuits exist merely to interface people with binary circuits. If our number system used a natural base such as 8 or 16 instead of base 10, BCD circuits would not be necessary. Many BCD-code-conversion microcircuits have been developed to simplify this person-to-binary interface. However, some system designers prefer to operate their entire system in a BCD code, thereby eliminating code conversions at the input and output. These types of systems can use many of the BCD data-processing microcircuits such as BCD adders, BCD subtractors, BCD A/D converters, etc. The number of BCD circuit types is continually growing, and many designers find the all-BCD approach quite reasonable as long as a slight increase in the number of wires between chips can be tolerated.

12.1   BCD CODES AND CONVERSIONS

Dozens of BCD codes are possible, but only the natural 8-4-2-1 code as shown in Table 12.1 has widespread use. Two digits, or 8 bits, of BCD information are tabulated; 14 of the 100 possible bit combinations are shown. Each BCD digit uses 4 bits to represent decimal numbers from 0 to 9. The second BCD digit has 4 bits to represent decimal numbers from 00 to 90 in tens. A decimal number with N digits requires $4N$ bits to write the number in BCD. If the decimal number is converted to binary, approximately 13 percent fewer bits are required to represent the number. The IC package count in a totally binary system would also be approximately 13 percent less than in a BCD system. However,

TABLE 12.1   The Natural BCD Code and Its Decimal Equivalent

| | Binary coded number | | | | | | |
| | Most significant digit | | | | Least significant digit | | |
Decimal number	MSB	Bit 3	Bit 2	LSB	MSB	Bit 3	Bit 2	LSB
0	0	0	0	0	0	0	0	0
1	0	0	0	0	0	0	0	1
2	0	0	0	0	0	0	1	0
3	0	0	0	0	0	0	1	1
4	0	0	0	0	0	1	0	0
5	0	0	0	0	0	1	0	1
6	0	0	0	0	0	1	1	0
7	0	0	0	0	0	1	1	1
8	0	0	0	0	1	0	0	0
9	0	0	0	0	1	0	0	1
10	0	0	0	1	0	0	0	0
11	0	0	0	1	0	0	0	1
36	0	0	1	1	0	1	1	0
99	1	0	0	1	1	0	0	1

if the BCD-to-binary input devices and the binary-to-BCD output devices are considered, the difference in package count may be insignificant.

12.2   BCD SYSTEM FORMAT

A BCD system can be designed around four different formats. These formats are pictorially represented in Fig. 12.1 and summarized as follows:

1. *Bit-serial, digit-serial* (Fig. 12.1a). Only one line is needed to transfer data, but the time required to transfer N digits is $4N$-bit periods. This is the lowest cost approach of the four formats. It is extensively used in calculators and low-cost instrumentation.

2. *Bit-serial, digit-parallel* (Fig. 12.1b). We have included this format only to show all four format possibilities. This method is not used because it would require N serial BCD processing circuits at each point in the circuit.

3. *Bit-parallel, digit-serial* (Fig. 12.1c). Most computers and microprocessors use this format when operating on BCD data. Computer systems use the same format for all binary calculations, where a binary word replaces the digit. The word could be from 4 bits wide to over 100 bits wide.

4. *Bit-parallel, digit-parallel* (Fig. 12.1d). This format is very expensive and is used only within systems between various BCD chips. However, it is the fastest format, requiring only a 1-bit period for all data to be transferred.

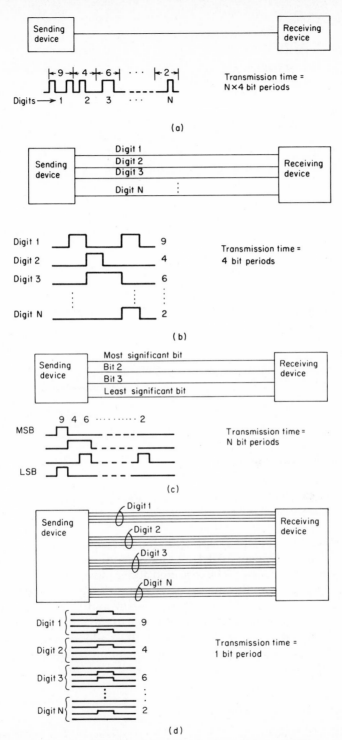

Fig. 12.1 Four possible formats for transfer of information in a BCD system: *(a)* bit serial, digit serial; *(b)* bit serial, digit parallel; *(c)* bit parallel, digit serial; and *(d)* bit parallel, digit parallel.

12.3   BCD ADDER

ALTERNATE NAMES   Adder, NBCD adder, full adder, 4-bit adder, parallel-add–serial-carry circuit.

PRINCIPLES OF OPERATION   Referring back to Sec. 3.11 and Fig. 3.26, we recall that a 1-bit full adder has three inputs and two outputs. Logic inputs A and B are the quantities to be added. Logic input C_i is the carry input from a previous adder. The output S is the sum of A, B, and C_i:

$$\text{Sum} = S = C_i\ (AB + \overline{AB}) + \overline{C_i}\ (A\overline{B} + \overline{A}B) = C_i\ (A \odot B) + \overline{C_i}\ (A \oplus B)$$

The output C_o is the carry resulting from the single bit addition of A, B, and C_i:

$$\text{Carry} = C_o = AB + C_i\ (A + B)$$

Figure 3.26b showed how a series of 1-bit adders could be cascaded together to form an N-bit adder. This can be carried on endlessly for

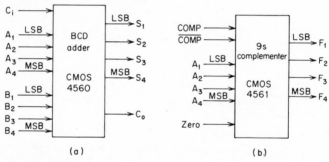

(a)   (b)

Fig. 12.2   Simplified logic representations used for (a) a 4-bit BCD adder and (b) a 4-bit 9s complementer used to make a BCD substracter (see Fig. 12.3).

binary operation. For BCD operation the cascading procedure immediately runs into problems. All 4-bit binary numbers 1010 through 1111 are not valid BCD codes. The cascaded adder must therefore be formed into 4-bit groups. As the sum in each group rises to 10, the sum must have 6 added to it so that it will reset to 0 and send a carry to the next 4-bit group. A 4-bit BCD adder with carry in and carry out utilizes approximately 60 gates. Data sheets for BCD adders can be consulted if one is interested in the complete logic diagram. Figure 12.2a shows the simplified logic representation used for a 4-bit BCD adder.

12.4   BCD SUBTRACTER

ALTERNATE NAMES   Subtracter, NBCD subtracter, full subtracter, 4-bit subtracter, adder with complementer.

PRINCIPLES OF OPERATION Subtraction of BCD numbers is performed by making one BCD number negative and adding it to another BCD number. The number which was negated is then effectively subtracted from the other number. The 9s-complementer device is used to negate a BCD number. Figure 12.2*b* shows the simplified logic diagram of such a device (Motorola MC14561B). Table 12.2 indicates the task required of the 9s complementer.

The 9s complement of a single-digit BCD number is found by subtracting the number from 9. This is easily verified in Table 12.2. We next use the table to see how subtraction is implemented. Suppose we wish to subtract 2 from 7:

$$\begin{array}{r} 7 \\ -2 \\ \hline 5 \end{array}$$

Using the 9s complement for −2, we would get

$$\begin{array}{r} 7 \\ +\ 7 \\ \hline 14 \end{array}\qquad \text{9s complement of 2}$$

$$\begin{array}{r} +\ 1 \\ \hline 5 \end{array}\qquad \text{end-around carry of 1 in 14}$$

The 1 in the 14 must be removed and added to the 4 to obtain the correct answer. This is called *end-around carry*. Examination of a few more single-digit subtractions using 9s-complement arithmetic reveals that the result is always between 10 and 19 (as long as the result is positive). Thus, the 1 is always available for the end-around carry.

Figure 12.3 shows the logic diagram for a two-digit adder-subtracter using two BCD adders and two 9s complementers. The circuit has an input called $\overline{\text{ADD}}/\text{SUBTRACT}$. When this line is LOW the circuit adds

TABLE 12.2 9s Complementer Input-Output Codes

Decimal number	4-bit BCD code input				9s-complementer output			
	MSB			LSB	MSB			LSB
0	0	0	0	0	1	0	0	1
1	0	0	0	1	1	0	0	0
2	0	0	1	0	0	1	1	1
3	0	0	1	1	0	1	1	0
4	0	1	0	0	0	1	0	1
5	0	1	0	1	0	1	0	0
6	0	1	1	0	0	0	1	1
7	0	1	1	1	0	0	1	0
8	1	0	0	0	0	0	0	1
9	1	0	0	1	0	0	0	0

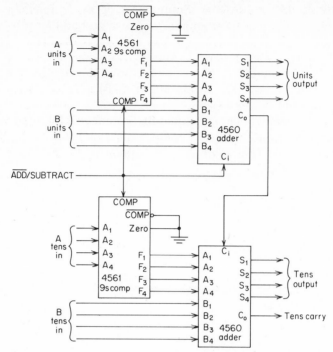

Fig. 12.3 An adder-subtracter circuit utilizing two 4-bit BCD adders and two 9s complementers.

the two-digit number A to the two-digit number B. When ADD/SUBTRACT is HIGH, the circuit subtracts the two-digit number A from the two-digit number B. A HIGH on this line also causes a 1 to be added to every sum through the C_i input of the units adder. This satisfies the end-around carry whenever the circuit is in the subtract mode.

In the add mode the output of Fig. 12.3 has a range of 0 to 198, and each input has a range from 0 to 99. The carry output from the tens adder provides the 1 output for numbers from 100 to 198. In the subtract mode the carry output is not required because the output range is 0 to 99 and the range of each input is 0 to 99. The circuit shown will not handle negative inputs or outputs. This requires some additional circuitry which compares input signs and magnitudes and then attaches the appropriate sign on the output. An additional bit (a sign bit) is required on all input or output numbers to handle negative numbers.

Figure 12.3 is an example of a bit-parallel, digit-parallel BCD circuit. The total output occurs during the same bit period as the input. No sequential or memory-storage operations are required. This method is very fast, but if one could trade off some speed for economy, a bit-

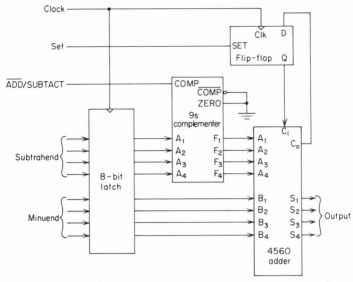

Fig. 12.4 A bit-parallel, digit-serial adder-subtracter circuit. Three clock cycles are required for each total result.

parallel, digit-serial approach, as shown in Fig. 12.4, could be used. This second type of BCD circuit is easily interfaced with bus-oriented digital systems such as microcomputers. An 8-bit microcomputer data bus could be tied directly to the eight inputs of the latch. The data bus would first present the two unit 4-bit words at the latch input. A fraction of a microsecond later the clock signal (under computer control) will cause the two 4-bit BCD words to be latched. A few nanoseconds later the units sum (or difference depending on the state of the ADD/SUBTRACT line) will appear at the output of the adder, along with a possible carry, which is stored in a D-type flip-flop. The microcomputer (or some other device or display) reads the output of the adder. Soon thereafter the microcomputer places the two BCD tens words at the latch inputs and the clock is again activated. The carry bit from the units sum was saved to use in the tens sum.

12.5 BINARY-TO-BCD CONVERSION

ALTERNATE NAMES Code converter, monolithic converter, binary-to-BCD ROM.

PRINCIPLES OF OPERATION A number of circuit techniques have been developed for converting a binary word to one or more BCD words. Most of these circuits use adders, counters, shift registers, gates, etc. If storage devices are used, the method typically requires many clock cycles to perform a conversion. However, if a PLA or ROM is utilized,

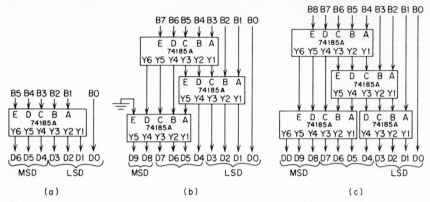

Fig. 12.5 Logical connections for *(a)* 6-bit, *(b)* 8-bit, and *(c)* 9-bit binary-to-BCD conversion using the 74185A ROM.

the code conversion is complete after only several gate delays. The 74185A, as shown in Fig. 12.5*a*, is probably the most popular ROM type of binary-to-BCD converter. This device, by itself, can convert a 6-bit binary word (maximum count = 63) into two BCD words. Since the output count cannot be greater than 63, the most significant output word has only 3 bits. Table 12.3 shows the input-output codes of a 6-bit binary-to-BCD converter. Since this table would be 63 entries long, only a few of the conversions are shown. We note that the least significant bit does not need to go through the ROM since this bit is unchanged in a binary-to-BCD conversion. The ROM therefore requires only 32 6-bit words to convert all binary numbers from 0 through 63.

The 74185A is cascadable to any size input or output word length. Figure 12.5*b* and *c* indicates how 8- and 9-bit binary numbers are converted into 3½- and 3¾-digit BCD numbers. Each 74185A contributes two gate delays (approximately 20 ns). The conversion time for the circuits in Fig. 12.5*b* and *c* will therefore be approximately 60 ns.

Microprocessor binary-to-BCD converter   Systems which contain a microprocessor can convert binary data into BCD data using software. Many algorithms have been developed for this task. Figure 12.6 portrays the flowchart for a straightforward method of 8-bit binary-to-BCD conversion. This program allows binary numbers up to 255 to be converted into 2½ BCD digits. The result ends up in three microprocessor registers H, T, and U, representing the number of hundreds, tens, or units, respectively.

The program starts by clearing H, T, and U, after which the number N to be converted is loaded into the accumulator. A decimal 100 is subtracted from the accumulator; then a test is made to see whether the accumulator is **0**. If so, N must have been equal to a decimal 100.

TABLE 12.3   Input-Output Codes for a 6-Bit Binary-to-BCD Converter

Decimal equiv-alent	Binary input						BCD output						
							MSD			LSD			
	B5	B4	B3	B2	B1	B0	D6	D5	D4	D3	D2	D1	D0
0	0	0	0	0	0	0	0	0	0	0	0	0	0
1	0	0	0	0	0	1	0	0	0	0	0	0	1
2	0	0	0	0	1	0	0	0	0	0	0	1	0
3	0	0	0	0	1	1	0	0	0	0	0	1	1
4	0	0	0	1	0	0	0	0	0	0	1	0	0
5	0	0	0	1	0	1	0	0	0	0	1	0	1
6	0	0	0	1	1	0	0	0	0	0	1	1	0
7	0	0	0	1	1	1	0	0	0	0	1	1	1
8	0	0	1	0	0	0	0	0	0	1	0	0	0
9	0	0	1	0	0	1	0	0	0	1	0	0	1
10	0	0	1	0	1	0	0	0	1	0	0	0	0
11	0	0	1	0	1	1	0	0	1	0	0	0	1
12	0	0	1	1	0	0	0	0	1	0	0	1	0
....													
19	0	1	0	0	1	1	0	0	1	1	0	0	1
20	0	1	0	1	0	0	0	1	0	0	0	0	0
21	0	1	0	1	0	1	0	1	0	0	0	0	1
....													
31	0	1	1	1	1	1	0	1	1	0	0	0	1
....													
41	1	0	1	0	0	1	1	0	0	0	0	0	1
....													
51	1	1	0	0	1	1	1	0	1	0	0	0	1
....													
62	1	1	1	1	1	0	1	1	0	0	0	1	0
63	1	1	1	1	1	1	1	1	0	0	0	1	1

The H counter is then incremented to **1** and the program stops with H = **1**, T = **0**, and U = **0**. If the accumulator is greater than **0** after the subtraction, N must be greater than 100. The H counter is incremented, and 100 is again subtracted from the accumulator. Whenever the accumulator goes below **0**, the number of hundreds has been counted and the program proceeds to the tens counter. However, since the accumulator is less than **0** (from the last subtraction of 100), 100 is first added back.

The tens counter operates identically to the hundreds counter. When this part of the routine is complete the T counter will contain the number of tens. As before, the accumulator contains a negative number when this routine is complete and so 10 must be added back to the accumulator. The final number remaining in the accumulator is the number of units. A transfer from the accumulator to U therefore completes the binary-to-BCD conversion.

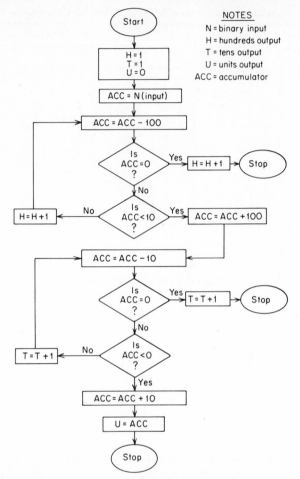

Fig. 12.6   A flowchart for 8-bit binary-to-BCD conversion.

12.6   BCD-TO-BINARY CONVERSION

ALTERNATE NAMES   Code converter, monolithic converter, BCD-to-binary ROM.

PRINCIPLES OF OPERATION   As with binary-to-BCD converters, the literature contains several circuits which perform BCD-to-binary conversion. Most of these circuits are slow because clocking of storage elements is required. The fastest methods use code-conversion ROMs. The 74184 is such a device. It is nearly identical to the 74185A mentioned in the last section except that the input-output codes of these devices are reversed. Therefore, Table 12.3 can be used for both ROMS if the input-output headings are reversed.

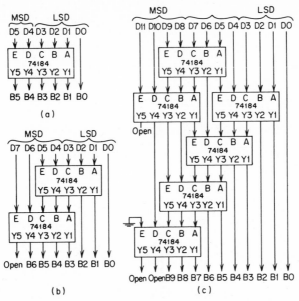

Fig. 12.7 Circuits for BCD-to-binary conversion using the 74184 ROM: *(a)* 6-bit BCD input, *(b)* 8-bit BCD input, and *(c)* 12-bit BCD input.

Figure 12.7*a, b,* and *c* shows how the 74184 can be utilized as a 6-, 8-, or 12-bit BCD-to-binary converter. This device has five inputs and six outputs, whereas a BCD code has more bits than its equivalent binary code. These converters will therefore have a slightly larger package count (for a given input word size) compared with binary-to-BCD converters.

Microprocessor BCD-to-binary converter A flowchart summarizing a microprocessor BCD-to-binary algorithm is shown in Fig. 12.8. The flowchart is presented in a simple format with no subroutines to complicate the discussion.

The routine starts by clearing two registers called A and B, which will contain the 16-bit result at the conclusion of the routine. Register A holds the most significant half of this 16-bit result. BCD input data are loaded into the four lower bits of each of the following registers:

$$TT = \text{number of ten thousands}$$
$$TH = \text{number of thousands}$$
$$H = \text{number of hundreds}$$
$$T = \text{number of tens}$$
$$U = \text{number of units}$$

Registers C and D are loaded with 2710 (hex), which is equivalent to 10,000 decimal. A loop is entered where TT is continually examined

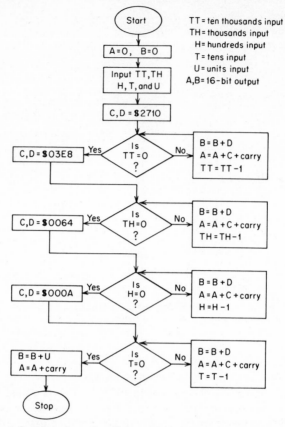

Fig. 12.8 A flowchart for a five-digit-to-16-bit BCD-to-binary computer program.

to see if it is zero. If not, 2710 (hex) is added to A and B and TT is decremented. When we add C to A, the carry from B = B + D must be used. When TT reaches zero, the C and D registers are loaded with 03E8 (hex), which is equivalent to 1000 decimal. The same procedure used in the ten thousands is repeated for thousands, hundreds, and tens. When units are reached, the units merely need to be added to A and B because all numbers 9 and lower have the same hexadecimal and decimal notation.

REFERENCES

1. Geiger, D. F.: Binary to BCD Conversion with μP's, *EDN*, Oct. 5, 1976, p. 110.
2. Tabb, J. A., and M. L. Roginsky: Microprocesor Algorithms Make BCD-Binary Conversions Super-fast, *EDN*, Jan. 5, 1977, p. 46.

Chapter 13

The Microprocessor and Microcomputer

MICROPROCESSORS VS. MICROCOMPUTERS

A common question asked about microprocessor devices is: How do they differ from microcomputer devices? The difference between microprocessors and microcomputers is not clearly defined, and the definition of these devices has changed over the years. In general, a microprocessor processes information while a microcomputer processes information, stores volatile and nonvolatile information, and provides serial and parallel user interfaces for this information. Although a microcomputer chip is sometimes referred to as a "computer on a chip," this is a misnomer because no matter how much logic is put on the chip, large devices such as a keyboard, display, and power supply are still required. In fact, most microcomputer systems the author has seen required several dozen other chips to interface the system with some outside process. The limited number of pins available on a microcomputer package is a major problem. Although chips containing millions of gates are predicted for the 1980s, the real problem in using these devices will be interfacing them to the outside world.

Figure 13.1 defines a microprocessor as the central device in a microcomputer. Some microcomputer chips contain fewer blocks than shown, and some contain more. Many special-purpose microcomputers contain on-chip input-output interfaces valid for only a small class of applications. These interfaces may be analog-to-digital or digital-to-analog converters, keyboard scanners, video monitor circuitry, alphanumeric driver circuits,

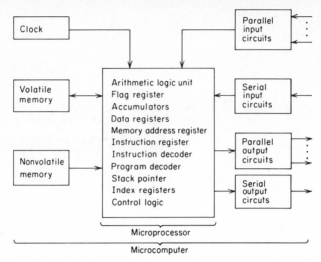

Fig. 13.1   Simplified block diagram showing the microprocessor as the major component in a microcomputer.

relay or lamp drivers, etc.   The potential list of specialized microcomputers is endless.

The remainder of this chapter will emphasize the list of items shown in the microprocessor block in Fig. 13.1.   The other parts of the figure, i.e., the parts that make up a complete microcomputer, are discussed in separate chapters.

13.1   ELEMENTS OF A MICROPROCESSOR

The arithmetic logic unit   This discussion will proceed like that of the design of a microprocessor.   That is, we begin with the basic block of the microprocessor, the arithmetic logic unit (ALU), and build upon this device until the microprocessor is complete.   The ALU was around long before the microprocessor.   For example, the 74181 is a 4-bit ALU with 75-gate complexity.   It performs 16 separate operations on two 4-bit input words with results appearing on a 4-bit output word plus four miscellaneous outputs.   The function select is made on four other input lines.   Figure 13.2 shows the logic representation of this device.   This device is cascadable up to any input-output word size. The 16 functions performed by the 74181 are listed in Table 13.1. If control input $M = 0$, arithmetic and logic operations are implemented. With $M = 1$ the output word F represents only logical combinations of words A and B.   When utilizing a device which performs both "plus" and OR, we must be careful not to use the $+$ symbol for both.   In the following discussion we use $+$ for OR and "plus" for addition.

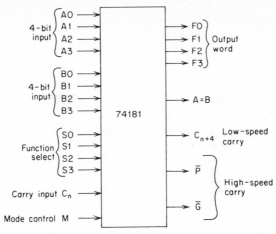

Fig. 13.2 Logic representation of the 74181 ALU.

The basic operations of a general-purpose ALU can be summarized as shown in Table 13.2.

Registers The ALU in a microprocessor is capable of little more than the operations listed in Table 13.2. A microprocessor gains its additional power by attaching several storage registers to the ALU, as shown in Fig. 13.3. The ALU can operate on data from only one or two of the registers at a time. Some of these registers, more versatile than the rest, are labeled *accumulators*.

TABLE 13.1 Arithmetic Operations of the 74181 ALU*

Function select				Arithmetic output function $M = C_n = 0$	Logic output function $M = 1$
S3	S2	S1	S0		
0	0	0	0	$F = A$	$F = \overline{A}$
0	0	0	1	$F = A + B$	$F = \overline{A + B}$
0	0	1	0	$F = A + \overline{B}$	$F = \overline{A} \cdot B$
0	0	1	1	$F = -1$ (2s complement)	$F = 0000$
0	1	0	0	$F = A$ plus $A \cdot \overline{B}$	$F = \overline{A \cdot B}$
0	1	0	1	$F = (A + B)$ plus $A \cdot \overline{B}$	$F = \overline{B}$
0	1	1	0	$F = A - B - 1$	$F = A \oplus B$
0	1	1	1	$F = A \cdot \overline{B} - 1$	$F = A \cdot \overline{B}$
1	0	0	0	$F = A$ plus $A \cdot B$	$F = \overline{A} + B$
1	0	0	1	$F = A$ plus B	$F = \overline{A \oplus B}$
1	0	1	0	$F = (A + \overline{B})$ plus $A \cdot B$	$F = B$
1	0	1	1	$F = A \cdot B - 1$	$F = A \cdot B$
1	1	0	0	$F = A$ plus $A = 2A$	$F = 1111$
1	1	0	1	$F = (A + B)$ plus A	$F = A + \overline{B}$
1	1	1	0	$F = (A + \overline{B})$ plus A	$F = A + B$
1	1	1	1	$F = A - 1$	$F = A$

*+ = logical OR, · = logical AND, $\oplus$ = logical XOR.

TABLE 13.2 Basic Operations of a General-Purpose ALU*

Function	Description (in any operation A and B can be interchanged)
$F = A$ plus 1	Increment A (or B)
$F = A - 1$	Decrement A (or B)
$F = A$ plus B	Add A and B
$F = A - B$	Subtract B from A
$F = A \cdot B$	Compute logical AND of inputs
$F = A + B$	Compute logical OR of inputs
$F = A \oplus B$	Compute logical XOR of inputs
$F = 2A$	Shift A left
$F = A/2$	Shift A right
$F = $	Rotate A left
$F = $	Rotate A right
$F = \overline{A}$	Complement A
$F = 0 \cdot A$	Clear A

*Assume two input buses A and B and an output bus F.

Devices called *multiplexers* transfer selected register contents into either ALU input port. Another multiplexer moves the ALU output to a destination register, where it is held until needed by some outside circuit. All these operations are controlled by a sequential circuit which is attached to all registers and multiplexers. If we are talking about an 8-bit microprocessor, the registers, accumulators, and multiplexers are 8-bit-wide parallel-in parallel-out devices.

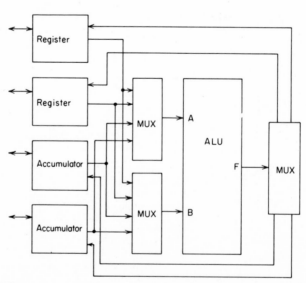

Fig. 13.3 When registers, accumulators, and multiplexers are added to the ALU, a simple microprocessor begins to form. Each line shown represents a data bus of 8 or 16 lines.

The next step in our development of a full-scale microprocessor is to add a program counter, condition code register, instruction register, instruction decoder, and memory. Since these devices are all interrelated, they must be added as a group. Figure 13.4 briefly shows the interrelationship of these circuits.

Program counter A typical 8-bit microprocessor has a 16-bit program counter. This counter keeps track of the sequence of operations for all microprocessor circuits. The number in the program counter is the line number (or memory address) being executed in the program running the microprocessor. A 16-bit program counter allows programs to be written up to a maximum of 65,535 steps long. Since most programs are only a few hundred or a few thousand steps long, the 65,535 limit is seldom reached.

Memory and instruction register As indicated by Fig. 13.4, the program counter does not directly sequence the ALU through various operations. As the program counter sequences through the numbers associated with a program stored in memory, the data word at each corresponding memory location is sequentially presented to the instruction register. Here it is temporarily stored while the instruction decoder determines what operation the ALU and/or the program counter must perform. If the instruction decoder discerns a branch or jump instruction, the program counter is changed to a new value and the memory is read at that location. The ALU is not accessed in this case. However, if the instruction decoder finds an instruction like those in Table 13.1 or 13.2, the ALU and its associated registers are placed in operation. Data move in and out of the ALU on a data bus having 4, 8, 16 (or more) parallel lines. This is the same data bus used by memory. Information can flow in

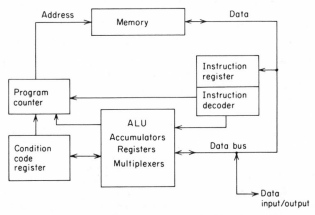

Fig. 13.4 Development of the microprocessor, showing how the program counter, condition code register, instruction code register, instruction decoder, and memory support the ALU and its registers.

either direction on this bus as controlled by sequencing logic. In most systems this bus is tristate. All receivers and transmitters connected to the bus are normally deactivated. When a particular instruction is being executed, the instruction decoder decides which transmitter and which receiver are activated. During that instruction execution, the bus carries information only between those two devices.

Condition code register While the ALU is performing its task, a special-purpose register called the *condition code register* (CCR) becomes activated. Each bit in this register represents a terse summary of various types of results possible during ALU operations. For example, if an ALU register goes to an all-zero state, a CCR bit, called the *zero bit* (Z bit), is set to the HIGH state. If the result was not **0**, $NZ = 1$. If a carry was generated, $C = 1$. If no carry was generated, we obtain $NC = 1$. This register is also called the status or flag register by various manufacturers. A number of other types of flags other than those above are possible. Table 13.3 summarizes some of the more commonly used flags. Since most 8-bit microprocessors use an 8-bit flag register, the list in Table 13.3 is usually scaled down somewhat. Some microprocessors use only six flags.

Most of the branch instructions depend on the results stored in the condition code resister. For example if a branch-if-not-equal (BNE) instruction is being processed, the instruction decoder must allow the ALU to complete its operation before instructions are sent over to the program counter. If the ALU results are not equal to **0**, the instruction decoder provides a new starting address (number) for the program counter. If the results of the ALU are equal to **0** the program counter is told to advance its count by **1**. The section of logic which decodes the condition code register and sends control signals to appropriate

TABLE 13.3 Condition-Code-Register Flags

Flag-bit name	Description
Z	Results $= 0$
NZ	Results $\neq 0$
C	Results produce a carry
HC	Results produce a half carry (BCD operations)
NC	Results produce no carry
B	Results produce a borrow
NB	Results produce no borrow
E	Two input words equal
GT	One input greater than another
P	Results positive
N	Results negative
PA	Parity is odd
OV	Operation overflows register
I	Interrupt mask bit

sections of the microprocessor, called *conditional branch logic,* is one of the most complex parts of the device.

The control section Microprocessors contain some supervisory logic called the *control section.* This section usually consists of the program counter, instruction register, and instruction decoder. The control section links together all diverse registers in the microprocessor to perform the requirements of a particular instruction. Each instruction of the user's program might require up to 10 internal steps to implement. Each step typically uses one cycle of the system clock. These internal steps are called *microcycles.* A microinstruction is required to direct each microcycle. Some microcycles hold intermediate results in registers not shown in Fig. 13.4. For example, we sometimes find a block between the program counter and memory called the *memory address register* (MAR). Between the memory and instruction register another register, called the *memory data register* (MDR), is sometimes used. A set of instructions is stored in the microprocessor which remembers all the particular microinstructions required to complete each external instruction. These microcycle instructions are stored in a read-only memory called the *control ROM* or *micro ROM.* Most microprocessors have this ROM on the chip, but some multiple-package microprocessors have a separate ROM chip, allowing designers to change some of the microprocessor instructions to fill their particular requirements. Microprocessors of this type are referred to as *microprogrammable.* Since sequential addressing of the micro ROM is required, a small program counter, called the *micro PC,* is utilized. This counter need be only 3 or 4 bits wide.

Index register As an aid to efficient programming, one or more index registers have been included in most newer microprocessors. In 8-bit microprocessors the index register is typically 16 bits wide. The usefulness of the index register is fully realized only through good programming. It allows a programmer to scan through one section of memory automatically, e.g., picking up entries from a table of data, while the program counter is executing instructions in the main program. In effect, two program counters are operating simultaneously while the index register is being utilized. The 16-bit index register can address the full 65,535-word memory. We discuss this register further in Section 13.2.

Stack pointer This is another register with potential fully realized only by good programmers. The stack pointer is also 16 bits wide in most 8-bit microprocessors and can therefore address (point to) any section of memory space. The "stack" pointed to by the stack pointer is any prescribed group of memory locations in RAM external to the microprocessor. In a microcomputer chip all or part of the stack may be on the chip. The stack provides an easily programmable means for storage and retrieval of successive data words. Its primary use is for storing blocks of data (several data words) when the microprocessor receives an interrupt command from an external device. The interrupt line on the microprocessor is usually hard-wired to some peripheral device. When this device needs the services of the microprocessor, it pulses the interrupt line, causing the microprocessor to stop. The microprocessor then stores the program counter on the stack so that it can remember where to restart the program after the peripheral device has been serviced. In response to an interrupt the microprocessor also stores the contents of the flag register, all accumulators, all index registers, etc. When the peripheral routine is complete, all these registers are reinstated and the program resumes.

The stack is also used to remember the old program counter address when the program jumps to a subroutine. At the end of the subroutine a return-from-subroutine instruction directs the microprocessor to restart the program counter at one position past the number stored in the stack.

Input-output Microprocessors have an internal data bus for transfer of information between various registers. Bidirectional buffers are required to transfer this low-level bus to or from the higher-power-level outside circuits. Most of the internal registers also contain a set of bidirectional buffers for moving data to or from the device over the internal data bus. A typical bidirectional buffer is shown in Fig. 13.5. This type of buffer is also used outside the microprocessor to move data between the microprocessor and other devices. A transfer of data from a register inside the microprocessor to a device outside the microprocessor might therefore require the use of three bidirectional buffer arrays, as shown in Fig. 13.6. The direction of information flow on the internal and external data buses is controlled automatically via soft-

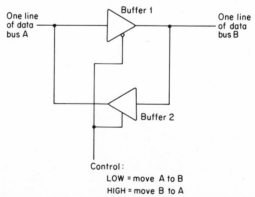

Fig. 13.5 An individual bidirectional gate of the type used inside a microprocessor for interfacing the internal data bus to the outside and for transferring data to various devices outside the microprocessor.

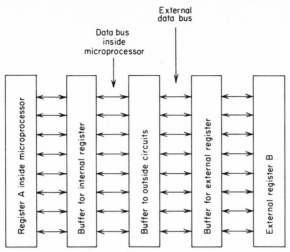

Fig. 13.6 A microprocessor and its external system require many bidirectional buffers to move data back and forth between various types of registers. Some applications require data to flow in only one direction.

ware. For example, in Fig. 13.6 suppose the instruction "store A in B" is encountered. All three bidirectional gates would simultaneously transfer information to the right, and within several microseconds we would have $A = B$.

External memory A microprocessor stores nearly all its permanent and temporary data or instructions in devices outside its package. A microcomputer chip might store all or part of the temporary data and instructions inside the chip. Likewise, many microcomputer devices also contain read-only memories (either permanently programmed or user-alterable) for holding nonvolatile data and instructions. In either case, programming of the microprocessor or microcomputer proceeds along identical lines. Design of the external-memory hardware, however, is much more difficult than having it already available on the microcomputer chip.

Many factors must be considered when interfacing external memory to a microprocessor. Designers must first decide which area in memory space is to be used by the block of external memory. They must then design addressing circuitry that will properly time the address bus with the data bus, read-write line, and several other signals. This was explored in greater depth in Chaps. 5 and 6.

13.2 MICROPROCESSOR-MICROCOMPUTER SOFTWARE

Every type of microprocessor and microcomputer understands a slightly different language. All these languages are merely extensions of the

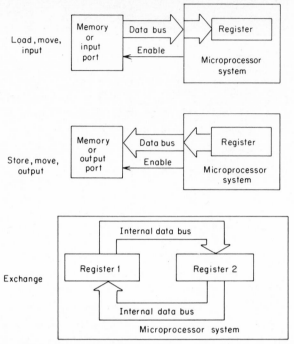

Fig. 13.7   Simple illustrations summarizing a few important types of data movement.

basic computer instructions listed in Table 13.2.   From a programmer's point of view, microprocessor instructions can be conveniently sorted into three classes:

1. Data movement: input-output, load, store, exchange, move
2. Data manipulation: add, subtract, multiply, divide, complement, clear, shift, rotate, increment, decrement, AND, OR, XOR
3. Program manipulation: jump, skip, branch, call, return, halt

We will expand on each of these classes.

Data movement   Data can be moved about in a variety of ways, inside or outside the microprocessor.   Figure 13.7 briefly illustrates a few of the important types of data movement.   Some types of data transfer require more than one instruction, but in most microprocessors the majority of data transfers can be performed with one instruction. Data to be moved reside in internal registers, internal memory, or external memory.   Specialized external devices such as input-output ports, timers, number crunchers, etc., all appear as external memory to the microprocessor.   Data movement can therefore be classified as one of four types:

1. Register to register

2. Register to memory (including output ports)

3. Memory to register (including input ports)

4. Memory to memory

All microprocessors will handle the first three data transfers listed above with one instruction. Two instructions, such as number 3 followed by number 2, were required to achieve instruction 4 in the first- and second-generation microprocessors.

Many variations of the four basic data-movement instructions are available from various microprocessors. The Z80 microprocessor, a third-generation device, has 7 of the type 1 data-movement instructions, 9 of type 2, 17 of type 3, and 5 of type 4. It also has 10 exchange instructions, each of which is a type 1 instruction done in both directions simultaneously. But this is not all. Some of the instructions such as LDr,r' have 64 possible combinations of any r with any r'. This instruction loads any of the eight r registers with the content of any of the eight r' registers. When all the possible combinations within each of the instructions are considered, a grand total of 188 data-movement instructions are possible with the Z80. The fourth-generation microprocessors (most of which are now called microcomputers) offer an even more flexible instruction set.

Addressing modes Before discussing the other two classes of instructions we need to explore the concept of addressing modes. Most 8-bit microprocessors and microcomputers use a 16-bit address word capable of addressing $2^{16} = 65,536$ memory locations. If an instruction refers to a memory address using the *extended addressing mode,* all 16 bits of the memory address are attached to the 8-bit instruction. This is called a 3-byte instruction and is the least efficient method of finding an address. When the address immediately follows the instruction, we refer to this mode as *immediate addressing.* If the required address is in the first 256 words of memory, the *direct addressing mode* can be used. This mode requires one 8-bit word following the instruction. An 8-bit word can select 256 memory locations in the first page of memory, i.e., the first 256 words of memory space.

The technique of *indexed addressing* also allows a 16-bit address to be generated without providing all 16 bits following the instruction. The 16-bit index register keeps track of the memory address. This register can be loaded, incremented, or decremented using other instructions. For example, suppose we want the A accumulator to be successively loaded with numbers from a section of memory starting at B100 and ending at B3FF. After A is loaded, a subroutine is entered where the data are processed. If the 6800 microprocessor is used, the program might be set up as follows, where # means immediate data and $ means hexadecimal data:

```
        LDX       #$B100      Load index register with B100
LOOP    LDAA      0,X         Load A with data at address
                                equal to value of X
        JSR       NAME        Jump to subroutine where A is
                                used
        INX                   Increment index register
        CPX       #$B3FF      Compare X with B3FF
        BNE       LOOP        Go to LOOP if X ≠ B3FF
```

Although the 8080 microprocessor does not have an index register, its *register indirect-addressing mode* allows a programmer to perform indexed addressing. For example, the above program could be executed as follows using the 8080:

```
        LXIB      #$B100      Load register pair BC with B100
LOOP    LDAX      B           Load A indirect using address in
                                BC
        CALL      NAME        Jump to subroutine where A is
                                used
        INXB                  Increment register pair BC
        LXIH      #$ − B3FF   Load register pair HL with
                                −B3FF
        DADB                  HL = HL + BC
        JNC       LOOP        Go to loop if BC ≠ −HL
```

In the above example we have used the register pair BC as a 16-bit index register. The 8080 does not have a 16-bit compare or subtract instruction, but it does have a 16-bit add called DAD rp. This instruction adds register pair rp (here we use BC) to the contents of the HL register pair. In this case we start with HL = −B3FF. If BC = B3FF, a carry is generated and the jump instruction is disregarded.

Data manipulation   The real power of a microprocessor instruction set is established if it has a large and versatile set of data-manipulation instructions. Many software and hardware designers choose a microprocessor by its ability to perform mathematical and logical calculations with a minimum number of instructions. Referring back to Table 13.2, we note that all basic ALU instructions fit into the category of data-manipulation instructions. Each new generation of microprocessors (or microcomputers) has new types of data-manipulation instructions. Early devices had only the basic ALU instructions listed in Table 13.2. An expanded listing of these instructions available in second- and third-generation microprocessors is shown in Tables 13.4 and 13.5. Some of the instructions, such as multiply and divide, require a large number of microinstructions within the basic instruction. Although not explicitly shown in the tables, third-generation microprocessors also have many instructions which repeat an indexed process until a given counter goes to 0. In the Z80 microprocessor, for example, the block-transfer and block-search group of instructions belong to this class. These types of instructions reduce the program size but still require many steps to complete. The main improvement in each generation of microproces-

**TABLE 13.4 Arithmetic Data-Manipulation Instructions
Available in Second- and Third-Generation Microprocessors**

Name	Description of instruction options
Add	Carry bit not added
	Add carry bit
	Add register to accumulator
	Add one accumulator to another
	Add immediate data to accumulator
	Add memory data to accumulator
	Direct addressing
	Extended addressing
	Relative addressing
	Indexed addressing
	Indirect addressing
	Multiple-precision capability (carry bit)
	BCD-add capability (decimal adjust)
Subtract	Carry (borrow) bit not included
	Subtract using carry (borrow) bit
	Subtract register from accumulator
	Subtract one accumulator from another
	Subtract immediate data from accumulator
	Subtract memory data from accumulator
	Direct addressing
	Extended addressing
	Relative addressing
	Indexed addressing
	Indirect addressing
	Multiple-precision capability (carry bit)
Clear	Clear accumulator
	Clear register
	Clear memory location
	Direct addressing
	Extended addressing
	Indexed addressing
Multiply	Multiply register by another register
Divide	Divide register pair by another register
Increment	Increment accumulator
(decrement)	Increment register
	Increment memory location
	Extended addressing
	Indexed addressing
	Increment stack pointer
	Increment register pair
	Increment index register
	Increment and skip if 0
Negate	Complement accumulator and add 1

sors is in the option list for each data-manipulation instruction. Addressing modes are being continually added or improved so that routines can be written more compactly. Two or more index registers in some microprocessors allow several indexed operations to be carried on simultaneously.

Program manipulation Data-manipulation and data-movement instructions are concerned primarily with the data contained in registers, accu-

TABLE 13.5 Logic Data-Manipulation Instructions Available in Second- and Third-Generation Microprocessors

Logic instruction name	Description of instruction options
Complement	Exchange 1s and 0s in accumulator
Rotate	Rotate right, including carry
	Rotate left, including carry
	Rotate accumulator
	Rotate register
	Rotate memory location
	Extended addressing
	Indexed addressing
	4-bit rotate
	Rotate register pair
Shift	Arithmetic shift left, 0s into LSB
	Arithmetic shift right, keep MSB unchanged
	Arithmetic shift, any accumulator
	Any register
	Any memory location
	Extended addressing
	Indexed addressing
	Logical shift right, 0s into MSB
AND	Accumulator AND register
	Accumulator AND immediate data
	Accumulator AND memory location
	Direct addressing
	Extended addressing
	Indexed addressing
OR	Accumulator OR register
	Accumulator OR immediate data
	Accumulator OR memory location
	Direct addressing
	Extended addressing
	Indexed addressing
XOR	Accumulator XOR register
	Accumulator XOR immediate data
	Accumulator XOR memory location
	Direct addressing
	Extended addressing
	Indexed addressing
Compare	Compare accumulator with register
	Compare accumulator with immediate data
	Compare accumulator with memory data
	Direct addressing
	Extended addressing
	Indexed addressing
	Compare accumulator with register, decrement another register

mulators, and memory locations. These two classes of instructions affect the status of the program counter only indirectly. Program-manipulation instructions, however, directly control the actions of the program counter. These instructions change the order in which the program instructions are carried out. Program-manipulation instructions are either unconditional or conditional. *Unconditional program-manipulation instructions* cause the program counter to assume a value specified in the second word (or second and third words) of the instruction whenever the instruction is encountered in the program. *Conditional program-manipulation instructions* cause the program counter to assume a value specified in the second word (or second and third words) of the instruction only if given conditions are met. These conditions are specified by particular bits in the condition code register or by the content of a given general-purpose register.

The jump instruction changes the program-counter content to the value specified in the second and third words of the instruction. From our earlier discussion of addressing modes we recall that this is the extended addressing mode. If this instruction is conditional, it is ignored, i.e., skipped, until the conditions are met. An indexed-jump instruction loads the program counter with a number found in one of the index registers. This addressing mode allows the jump destination to be dynamically modified as the program proceeds.

The jump relative or branch instructions are two different names for an instruction which requires only one additional word to specify the new value for the program counter. In 8-bit microprocessors this additional word specifies an address within ±127 words of the present program-counter number. This second word is often referred to as the *relative offset*. It uses signed-2s-complement notation; i.e., the most significant bit is 1 for negative offsets.

The call and JSR instructions are extended address commands which cause the program counter to jump to the beginning of a subroutine. The RET or RTS instructions at the end of the subroutine are also extended address commands which return the program counter back to one instruction past the call or JSR instruction leading into the subroutine. Some microprocessors also offer conditional call and RET instructions.

REFERENCES

1. Nemec, J., and S. Y. Lau: Bipolar μP's: An Introduction to Architecture and Applications, *EDN,* Sept. 20, 1977, p. 63.
2. Weiss, C. D.: Software for MOS/LSI Microprocessors, *Electron. Des.,* Apr. 1, 1974, p. 50.
3. Ungermann, R., and B. Peuto: Get Powerful Microprocessor Performance by Using the Z80, *Electron. Des.,* July 5, 1977, p. 54.

Chapter 14

Microprocessor-controlled A/D Converters

OBJECTIVES OF APPLICATION

Software control of analog-to-digital converters provides a wide range of application possibilities to systems designers. This chapter discusses two applications utilizing the Teledyne 8700 incremental charge-balancing A/D converter. In the first application converted samples are averaged and compared with other data stored in software. Program control depends on the outcome of each averaged sample compared with previous averaged samples. This particular program was developed for a system requiring precise determination of the amplitude of a noisy signal. The peak measurement in a given time interval is stored and its magnitude used to control the program direction. This A/D converter system utilizes the 6800 microprocessor. The second application shows how this same A/D converter could interface with an 8080 microprocessor. The interrupt mode of operation is used in the second application, whereas the free-running mode is used in the first.

14.1 FREE-RUNNING AVERAGING A/D CONVERTER

ALTERNATE NAMES Averaging dc voltmeter, integrating voltmeter, sampling ADC.

PRINCIPLES OF OPERATION As indicated in Fig. 14.1, the A/D chip interfaces with the microprocessor data bus through two 80C95 hex CMOS tristate buffers. Each time the address decoder generates a nega-

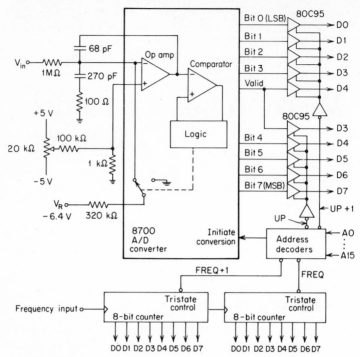

Fig. 14.1   An A/D converter controlled by a microprocessor.

tive pulse on UP or UP + 1 each buffer transfers the A/D converter outputs to the data bus.

The A/D output data are not valid unless their data valid line is HIGH. Figure 14.2 shows that this condition does not exist for about 1.25 to 1.8 ms after the initiate conversion (IC) line is pulsed by the microprocessor. This is called the *clocked mode* since the A/D converter remains idle until the microprocessor requests a conversion. The software must be written so that the output data are not read until the valid line has gone LOW and returned to the HIGH state. If the output data are read within 1 ms after the rising edge of IC, the data will be old.

If the A/D converter is used in the free-run mode, as in this example, the sampling interval of the microprocessor must be greater than 1.8 ms to assure independent samples. Assume that this circuit is part of a large system controlled by a single microprocessor. The A/D converter is only one device of many being sampled on a periodic basis. Assume that the sampling interval of any particular device is 10 ms minimum. This assures independent samples of the A/D converter since only one of every six converter samples will be utilized. As shown in Fig. 14.3, the A/D converter must be sampled only when valid is HIGH. One approach would be to read the valid line just before reading

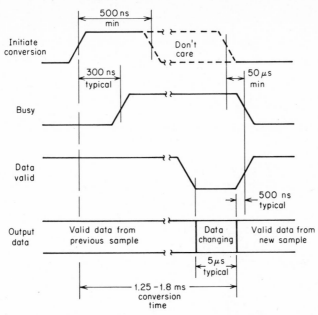

Fig. 14.2   Timing waveforms for the 8700 A/D converter operating in the clocked mode.

the converter.   If valid = **1,** the measurement is made.   Otherwise valid is reread until it indicates that a complete conversion is available for transfer to the microprocessor.   However, this approach has a basic problem.   After valid is read, it takes 5 to 10 μs for the microprocessor to decide if valid = **1** and make the decision to read the A/D converter data.   During this 10-μs interval the data bus can start changing. A better approach would be to read the data bus and the valid line simultaneously using a load index register (LDX) instruction.

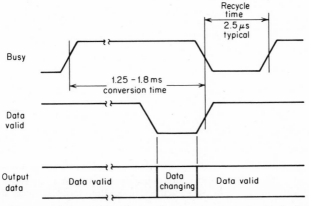

Fig. 14.3   Timing waveforms of the 8700 A/D converter operating in the free-run mode.

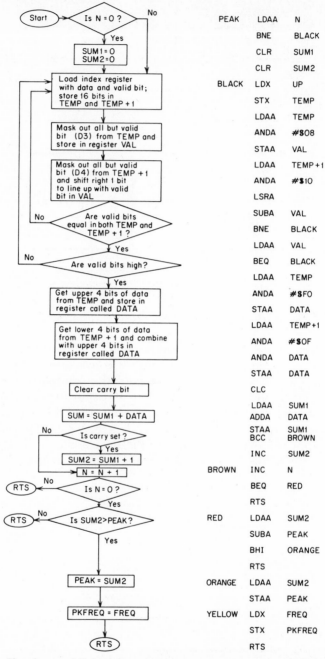

Fig. 14.4 Flowchart and listing of a subroutine to average an input voltage over 256 samples, compare with previous peaks, and store another variable which identifies the location of the peak.

The LDX instruction is 5 μs long assuming a 1-MHz clock. The 16-bit index register is loaded in two 8-bit bytes during the fourth and fifth microseconds (see Motorola 6800 data sheet). Suppose we load 4 bits of the A/D converter data along with the valid bit during the fourth microsecond. On the fifth microsecond we load the other 4 data bits along with the valid bit again. If both readings of the valid bit are HIGH, the data word has a 100 percent probability of being correct. The software then combines the two halves of the data word and processes the result.

The flowchart and program listing of the microprocessor-controlled A/D converter subroutine is shown in Fig. 14.4. Both the listing and the flowchart should be followed as we describe the subroutine. The subroutine computes the average of 256 samples from the A/D converter. An 8-bit counter labeled N keeps track of the sample number. As the subroutine is entered, N is checked to see if it is 0. If so, a previous 256 average has been completed and a new average is to be started. This is done by clearing out SUM1 and SUM2, i.e., the 16-bit register holding the sum of the samples. Each new sample is added to SUM1. Whenever SUM1 overflows, SUM2 is incremented. After 256 additions into SUM1 the 16-bit register represented by SUM2 and SUM1 contains the total number. The average sample is merely the total divided by 256. This could be obtained by shifting the 16-bit total to the right 8 places, but this division is not required if we use the upper 8 bits directly; i.e., SUM2 is the average value of the 256 samples.

The 16-bit index register is loaded with 8 bits of data and the valid bit (twice), as shown in Fig. 14.5. After the index register is loaded, it is immediately stored in TEMP and TEMP + 1. Register TEMP contains the valid bit (D3) and the upper 4 bits of the data word (D4 to D7). TEMP is loaded into the A accumulator, where the data word is removed by ANDing with $08. If valid is HIGH, D3 will be HIGH. The result is stored in a register called VAL. Register TEMP + 1 is next

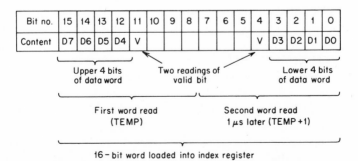

Fig. 14.5 Arrangement of data and valid information loaded into the 16-bit index register.

loaded into the A accumulator. This register contains the valid bit again (D4) and the lower 4 bits of the data word (D0 to D3). The data word is masked out by ANDing with $10. The result is shifted right one place so that the valid bit is in the same bit position as the valid bit in TEMP. We now subtract the register called VAL from accumulator A. If both valid bits are HIGH, the result is 0 and the program continues. If one of the valid bits is LOW, accumulator A will contain a nonzero number. In this case the program returns to black, where the A/D sample is read again. The program also returns to black if both valid bits are LOW.

If both valid bits are HIGH, the program reconstructs the 8-bit data word by using a $F0 mask on TEMP and a $0F mask on TEMP + 1. The results of each masking are combined and stored in a register called DATA. This register is next added to the SUM1 register. If the SUM1 register overflows, register SUM2 is incremented one count. Counter N is also incremented at this time and tested for zero content. If $N \neq 0$, the subroutine is finished for this loop of the main program (256 passes through the subroutine are required for each averaged sample). If $N = 0$, the averaging is complete and SUM2 contains the average of all 256 A/D conversion samples. This average (SUM2) is compared with PEAK (the largest average found at some earlier time when this subroutine was utilized). If SUM2 $\leq$ PEAK, the subroutine is terminated. If SUM2 > PEAK, SUM2 replaces PEAK. Another variable is recorded at this time to identify the time (or some other identifier such as operating frequency) where this new PEAK occurred. In the author's system the A/D conversion was performed on the AGC of a sophisticated receiver and PKFREQ was the tuning frequency where the peak AGC occurred. This subroutine facilitated automatic tuning of the receiver.

14.2 SOFTWARE-INITIATED A/D CONVERTER

This section describes a method of interfacing an 8-bit A/D converter to the 8080 microprocessor using the interrupt mode of operation. We assume the Teledyne 8700 A/D converter is used.

PRINCIPLES OF OPERATION The system described below provides one A/D converter sample for one initiate-conversion (IC) signal from the microprocessor. Since the conversion requires over 1 ms to perform, the microprocessor proceeds with other calculations while the A/D converter responds to the IC command. When conversion is complete, an interrupt (data valid) is transferred to the microprocessor and the 8-bit converter output is read.

The data-valid output is normally HIGH, indicating that the converter output lines are valid, except for approximately 5 μs before the end of the conversion time when these outputs are updated. The output latches maintain the data from the previous conversion while the next conversion is being performed. The initiate-conversion input line allows the A/D converter to be operated under system control. A positive-going pulse of at least 500 ns duration causes the conversion to begin.

Communications between an 8080 microcomputer system and the outside world are via input-output (I/O) ports addressed by the address bus. I/O instructions utilize 8-bit addresses which simultaneously appear on both the upper 8 bits and the lower 8 bits of the 16-bit address bus. In addition to the 24 address and data lines, the 8080 system communicates with the A/D converter via a set of control signals called $\overline{\text{IN}}$ and $\overline{\text{OUT}}$. As shown in Fig. 14.6, data-bus buffering and the generation of $\overline{\text{IN}}$ and $\overline{\text{OUT}}$ are performed with the 8228 chip. A logic LOW on the $\overline{\text{IN}}$ line enables the input port corresponding to the address on the address bus at that time. The $\overline{\text{OUT}}$ line performs a similar function for data coming out of the microprocessor system.

The conversion is started on command of the 8080 using the IC input of the 8700. When the conversion is complete, the 8700 requests an interrupt. The interrupt-service routine transfers the current data from all working registers to the stack memory, and the A/D input port is read. Another IC command is issued and the cycle repeats.

In most microprocessor systems the data bus is shared by many devices such as memory, I/O ports, etc. If the 8228 bidirectional bus driver-receiver is used, the data-bus information is inverted throughout the system. Therefore, 80L98 inverting buffers are provided at the A/D converter outputs to invert the data-bus lines, to boost their power level, and provide a three-state capability. Thus, the 8700 is electrically removed from the 8080 system data bus when its input port is not selected.

Each port of the microprocessor system is assigned an address. Either the upper 8 bits or the lower 8 bits of the address bus must be decoded for each port. In Fig. 14.6 the output of the 7430 gate is LOW only when address lines $\overline{\text{A0}}$ to $\overline{\text{A7}}$ are all LOW. This corresponds to an I/O port having an address of FF (hex). This same port address is used for input or output from and to the 8700 A/D converter. The initiate-conversion pulse is generated using the $\overline{\text{OUT}}$ instruction to output port FF twice. On the first $\overline{\text{OUT}}$ instruction 80 (hex) is loaded onto the data bus. The D flip-flop FF2 sees the 80 inverted to a LOW state on data bus line $\overline{\text{D7}}$, and the flip-flop $\overline{Q}$ output latches into the HIGH state. A second output instruction, sending zero to the same port, will clear the flip-flop $\overline{Q}$ output and remove the IC signal. Since each output instruction requires approximately 5 μs, this will be the approximate width of the IC pulse. After this pulse is issued, the microprocessor is free to perform other tasks while the A/D converter performs its conversion.

When the 8700 completes the A/D conversion, the data-valid line

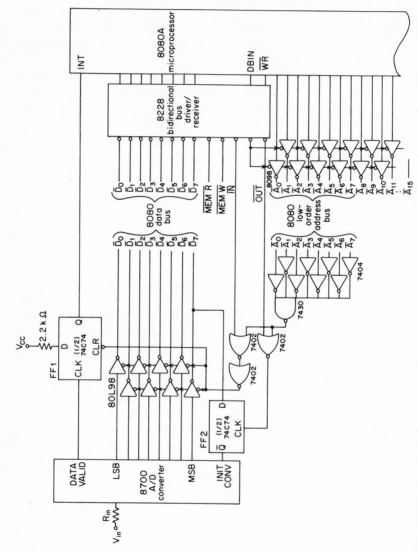

Fig. 14.6 Digital interface of an 8700 A/D converter and an 8080-base microprocessor system. See Fig. 14.1 for the analog connections to the 8700.

TABLE 14.1   Program Listing of 8700 A/D Converter Interrupt-Servicing Subroutine

START	MVI	A, 80H	Initiate conversion:
	OUT	FFH	output HIGH level
	MVI	A, 0	then
	OUT	FFH	output LOW level
	RET		
INTERRUPT	PUSH	B	
	PUSH	D	
	PUSH	H	
	PUSH	PSW	
	IN	FFH	
	MOV	B, A	
	MVI	A, 80H	
	OUT	FFH	
	MVI	A, 0	
	OUT	FFH	
	POP	PSW	
	POP	H	
	POP	D	
	POP	B	
	RET		

goes HIGH.   This signal triggers FF1, clocking a logic HIGH from the D input onto the microprocessor interrupt-request (IRQ) line.   The interrupt-service routine is initiated when IRQ goes HIGH.   The microprocessor is directed to read the 8700 data at input address FF (hex). This action causes the $\overline{\text{IN}}$ line to go LOW, which in turn activates the 80L98 buffer and clears FF1.   After the A/D converter data have been read and stored in memory, the microprocessor system is free to issue another initiate-conversion command whenever needed.

The software program to operate the 8700 A/D converter is shown in Table 14.1.

REFERENCE

1. Guzeman, D.: Interfacing the 8700 A/D Converter with the 8080 µP System, *Teledyne Semiconductor Appl. Note* 8, August 1976.

Chapter 15

Microcomputer-based Traffic Systems

RANGE OF APPLICATIONS

The most obvious need for microprocessors in traffic systems is for intelligent control of traffic at street intersections under varying traffic conditions. This chapter will present a microprocessor-based system for control of traffic at a main-street–side street intersection. Other potential applications for microprocessors or microcomputers in traffic control systems include

  Traffic control at the intersection of two main streets
  Traffic control at five- or six-entrance intersections
  Freeway on-ramp traffic regulation
  Freeway speed control under varying traffic, weather, and lighting conditions
  Synchronization of traffic lights on expressways and major roads
  Counting and monitoring traffic for highway planning
  Control of traffic signals under emergency situations

15.1 MAIN-STREET–SIDE-STREET TRAFFIC CONTROLLER

ALTERNATE NAMES Traffic light timer, intersection traffic regulator, microprocessor-controlled traffic signal, traffic sequencer.

PRINCIPLES OF OPERATION A solution to the problem of equitably regulating the flow of traffic at an intersection where one street has much more traffic than the other is presented in the following pages.

The busy street is called *main* and the less busy street is called *side*. We present this application in three sections: (1) system definition, a statement of requirements and generation of flowcharts, (2) interface definition, a discussion of hardware available to be operated by the microprocessor, and (3) software definition, the program required to operate the available hardware with maximum reliability, economy, safety, and fairness to all drivers.

SYSTEM DEFINITION   Before a flowchart can be generated, a list of requirements must be established.

1. Traffic lights should be sensitive to traffic conditions on the main and side streets.

2. The main street should always have priority, but the side street must always be serviced when it contains traffic.

3. Underground sensors should be used to detect the presence of vehicles.

4. The signal should normally be green to main street.

5. If a vehicle is detected on the side street, the microprocessor should first check to see that it has been 60 s since it last favored side street.

6. If 60 s has passed since the side street was favored and a side-street vehicle is present, the signal should change to green for side, going through the normal 3-s amber caution interval.

7. The side street should not be serviced more than 60 s regardless of the number of vehicles detected on side.

8. If 10 or more vehicles are detected on main during the side-street routine, the traffic signal should change to favor main even if the 60-s timer for side has not been completed.

The generation of a flowchart is an iterative process. The first flowchart is usually quite general, with many items covered in each block. Each succeeding flowchart produces a finer gradation of instructions until the final flowchart contains only a few software instructions for each block. The first flowchart might appear as shown in Fig. 15.1. Each block showing traffic signals represents a change to the light pattern shown. When the system is first turned ON, all registers must be loaded with quantities which guarantee that the sequence will start and remain in valid states. The sequence begins with main street on green and the side street on red. This is the normal state of the signal with no traffic on the side street. If a vehicle is detected on the side street and it has been at least 60 s since the side street was last favored, the side-street sequence begins. All the remaining blocks in the flowchart belong to the side-street sequence. The amber light first appears to the main-street traffic. After 3 s the main-street signal turns red and the side-street signal changes to green. The side street is favored for

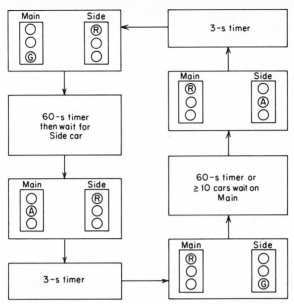

Fig. 15.1   First-cut flowchart for a microprocessor-controlled traffic-light servicing a main street and a side street.

60 s or until 10 or more cars queue up on the main street, whichever comes first.   The side-street light then turns amber for 3 s, after which it turns red and the main-street signal changes back to its normal green.

We now understand enough of the sequence of operations to proceed with a more detailed flowchart.   Figure 15.2 displays what we would judge to be the final flowchart required before writing the software. It is critical that the flowchart be logically correct before proceeding with the writing of program code.   It is much easier to detect flaws in reasoning in a flowchart than in the assembly-language code.   Other people involved in the system design should be consulted at this point to make sure they agree with the flowchart.

Interface definition   After a firm flowchart has been developed, the interface and hardware design can proceed.   Often this phase of the design is done in parallel with the task of writing software.   These two phases of design interact with each other.   That is, progress in each task depends on developments in the other design task.

The traffic-signal controller has four inputs and six outputs.   As shown in Fig. 15.3, some inputs are the ORed result of several sensors in the road near the intersection.   These sensors may be buried variable-reluctance coils or pressure plates on the road surface.   Sensors on the main street should be spread out sufficiently to count 10 cars.   Each sensor produces a pulse each time a car is detected, but when a vehicle passes

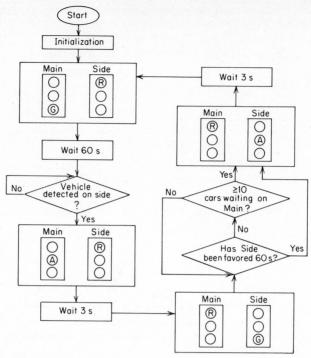

Fig. 15.2 Final flowchart for the microprocessor-controlled traffic-light sequencer.

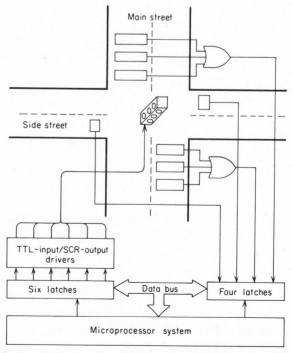

Fig. 15.3 Overall picture of a microprocessor-controlled traffic-light system.

over three sensors on main, it must be counted as only one vehicle. This decision process is handled by the sensor electronics.

The pulses could be counted using software or hardware. A software approach is more economical and will be used here. The car-counter pulses cannot go directly to the microprocessor but must be buffered with a flip-flop having a tristate output. If the car counters provide positive output pulses, the CD4043 quad tristate latch could be used. As shown in Fig. 15.4, each set input is connected to the sensors on one side of each road. The sensors for each side of each road could be ORed together before going to the microprocessor, but since the CD4043 has four inputs, we choose to do the OR function with software. Whenever external logic can be replaced with software logic, it should be because software has lower cost and higher reliability. In some cases, however, if we try to do all logical calculations with software, the interface wiring and interface chips become too numerous. A trade-off is required in each design task.

Once a latch has been set, it remains in that state until the reset line is pulsed to the HIGH state. This allows the software to read the state of the latch on a periodic basis. Each time the microprocessor reads the 4-bit latch, its data are stored and the latch is reset. This is called the *polling technique*. The microprocessor is programmed to read the input latches at those particular times in the execution of the program when their content is of interest. The microprocessor treats the latches as flags, interrogating them under program control and resetting them so that they will be ready for the next input from the sensors.

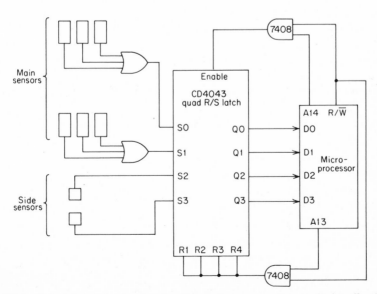

Fig. 15.4 Sensor inputs to the data bus for a microprocessor-controlled traffic signal.

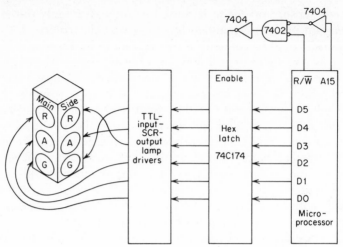

Fig. 15.5 Detail of the traffic-light actuator portion of the microprocessor-controlled traffic-light system.

The output lines to the traffic signal are controlled by the 8-bit bidirectional data bus indicated in Fig. 15.5; six data-bus lines D0 to D5 are utilized for outputs.   Table 15.1 summarizes the lamp status at various times.   Data will be written into the latches when $R/\overline{W} = 0$, that is, when the microprocessor executes a STAA LAMP output instruction. Each time this instruction is given, two of the lamp driver lines are activated and four lines are disabled.   The lamp drivers are commercial TTL-to-SCR modular devices.   The device driving the lamp drivers is the 74LS174 hex D flip-flop chip.   Whenever the clock input to the 74LS174 is low, the information on the data bus is latched into the hex flip-flop.

Since this microprocessor system has only one input port and one output port, addressing the input and output latches is extremely simple. The reset of the CD4043 is simply derived by ANDing $R/\overline{W}$ with A13.

TABLE 15.1   Data-Bus Codes Corresponding to All Four Lamp States

Side-street lamps	Main-street lamps	D7	D6	Side G D5	Side A D4	Side R D3	Main G D2	Main A D1	Main R D0	Hex data word
Red	Green	0	0	0	0	1	1	0	0	OC
Red	Amber	0	0	0	0	1	0	1	0	OA
Green	Red	0	0	1	0	0	0	0	1	21
Amber	Red	0	0	0	1	0	0	0	1	11

The enable for the CD4013 (read the sensors) is obtained by ANDing R/$\overline{W}$ with A14. The clock signal for the output latches is simply the AND of R/$\overline{W}$ in the LOW state and $\overline{A15}$ in the LOW state. The program for this system requires less than 100 words of memory. It utilizes only the lower 8 bits of address (256 words), so that the upper 8 address lines are available for low-cost address decoders, as described above.

Software definition We will describe the software utilizing the 6800 language. Any 8-bit microprocessor can be used for this application. In fact, with a slight redesign of the circuitry driving the six lamps a 4-bit microprocessor could also perform satisfactorily.

As indicated by the program listing in Table 15.2, the routine is initialized by clearing the sensor input latches. The lamp output lines are next latched into the main = green and side = red mode of operation. This is the normal state for the system. A 60-s timer locks out any sensor signals from side traffic when the main = green mode begins. After this 60-s wait the program goes into a three-instruction loop consisting of

```
SIDECK        LDAA        SENSOR
              ANDA        #$0C
              BEQ         SIDECK
```

The system will spend most of its life repeatedly executing this small loop. The ANDA #$0C instruction masks out the main sensors and allows decisions to be based only on the side sensors. If a vehicle is detected on either side sensor, the three-instruction loop terminates and the lamp output is latched into the main = amber and side = red mode.

When the side-street traffic is given the green light, a double decision process begins. Side-street traffic will be terminated if either a 60-s counter times out or if 10 or more cars are waiting on the main street. The first of these two criteria to be satisfied causes the lights to change to main = red and side = amber. This condition lasts for 3 s, after which the program returns to main = green and side = red.

The timer subroutines use double loops to provide the 3- and 60-s delays. Single loops (even using the index register) do not provide sufficient delay. For example, suppose we cleared the A accumulator and decremented it until **0** was detected. The instruction set would appear as

```
              CLRA
TIMER         DECA
              BNE         TIMER
```

Assuming a 1-MHz clock, the DECA and BNE instructions require 2 and 4 μs, respectively. The timer requires 256 loops before it terminates. This provides a delay of 256×6 μs $= 1536$ μs $= 1.536$ ms.

TABLE 15.2 Program Listing for Microprocessor-Controlled Traffic Signal Using the 6800 Assembly Language

Label	Instruction	Operand	Comment
	LDAA	RESET	Reset sensor latches
LOOP	LDAA	#$0C	Main = green, side = red
	STAA	LAMP	
	LDX	#$9736	⎫
	CLRA		⎪ Wait 60 s
SIXTY	DECA		before side sensor
	BNE	SIXTY	is checked
	DEX		⎪
	BNE	SIXTY	⎭
SIDECK	LDAA	SENSOR	Get sensor data
	ANDA	#$0C	Mask out main sensors
	BEQ	SIDECK	Are side sensors activated?
	LDAA	RESET	Yes, reset sensor latches
	LDAA	#$0A	
	STAA	LAMP	Main = amber, side = red
	LDX	#$078F	⎫
	CLRA		⎪ 3-s duration
THREE	DECA		for amber light
	BNE	THREE	on main
	DEX		⎪
	BNE	THREE	⎭
	LDAA	#$21	
	STAA	LAMP	Main = red, side = green
	LDAA	#$0A	
	STAA	COUNTR	Preset car counter to 10
	LDX	#$9736	⎫
	CLRA		⎪ Conditional 60-s
SIXTYC	DECA		timer; go to main if
	BNE	COUNT	60 s passed or
	DEX		≥ 10 cars waiting
	BNE	COUNT	on main
	BRA	MAIN	⎭
COUNT	LDAA	SENSOR	Car on main?
	ANDA	#$03	
	BEQ	SIXTYC	No, go back to 60-s timer;
	LDAA	RESET	Reset sensor latches
	DEC	COUNTR	Yes, countr = Countr-1
	BNE	SIXTYC	Is countr = 0?
MAIN	LDAA	#$11	Yes, let main = red,
	STAA	LAMP	side = amber
	LDX	#$078F	⎫
	CLRA		⎪ 3-s duration
AMBER	DECA		for amber
	BNE	AMBER	light on side
	DEX		⎪
	BNE	AMBER	⎭
	JMP	LOOP	Repeat loop

Using this same technique with the 16-bit index register provides a delay of 65,536 × 8 μs = 0.524 s (DEX requires 4 μs, whereas DECA requires only 2 μs). An 8-bit timer loop can be combined with a 16-bit timer loop in several ways to provide delays up to 134 s. If the A accumulator loop is placed inside the index register loop, the timer can be programmed for any of 65,536 different delays with a time resolution of 1.536 ms. If the index-register loop is placed inside the A accumulator loop, the timer can be programmed for any of 256 different delays with a time resolution of 0.524 s. Since both these techniques require the same number of instructions, we will use the one having higher resolution. A double-loop timer of the first type (highest resolution) having a delay of 60 s is as follows:

	LDX	#$9736	Initialize index register
	CLRA		Clear A accumulator
SIXTY	DECA		Go through inner loop
	BNE	SIXTY	256 (decimal) times
	DEX		Go through outer loop
	BNE	SIXTY	9736 (Hex) times

The program listing in Table 15.2 could be assembled into machine language by hand or by using an assembler program (available in most microprocessor development systems). In either case, numbers will be assigned for each address and instruction. Each instruction consists of a mnemonic (such as LDAA, BEQ, JMP, etc.) and an operand (such as RESET, #$0C, LAMP, SIXTY, etc.). A few of the mnemonics such as CLRA and DEX require no operand. The labels preceding some instructions will also be assigned an address number. The labels do not have any special significance in the final assembled program. They are merely memory aids useful while the program is still written in assembly language.

The operands are assigned different types of numbers. In some cases the operands are addresses of labels within the program listings such as LOOP, SIXTY, SIDECK, or MAIN. Several operands are the addresses of hardware inputs and outputs such as LAMP, SENSOR, or RESET. One operand, COUNTR, is used in a number of places. This is a scratchpad register, i.e., a temporary register made from random-access-memory (RAM) devices. If the microprocessor has a very simple application, such as this traffic signal (and nothing else), perhaps the B accumulator, index register, or stack pointer could be used for this temporary register. This complication is not necessary on most third-generation microprocessors since they have many words of RAM available on the same chip. Although this program was written using the 6800 microprocessor, the 6802 could also be used. This third-generation device has the same language as the 6800, has a built-in clock circuit, and has 128 words of RAM (8-bits wide) on the chip.

REFERENCE

1. Titus, J.: How to Design a μP-based Controller System, *EDN*, Aug. 20, 1974, p. 49.

Chapter 16

Keyboard Scanner Circuits

OVERVIEW OF SCANNING TECHNIQUES

A large number of circuit techniques have been developed for scanning (sampling) switch arrays. The most common type of switch array familiar to everyone is the standard typewriter alphanumeric keyboard. This type of array typically has 60 or more keys which must be rapidly scanned so that the user is not aware of any time delay between a key depression and some appropriate response from the system. All switches must be periodically examined by the scanning circuit either singly or in groups. The most common technique is to examine 8 or 10 of the switches simultaneously. To perform this group-examination scheme the switches are arranged in an *X-Y* array if both terminals of each switch are available. If one terminal of each switch must be at ground or some given voltage, more hardware is required to perform the scan.

In the following pages we look at three scanning techniques. Each method is unique, due to the constraints of the switch array used. The first method, that using an *X-Y* switch array, is by far the most widely used. Many microcircuits have been designed using implementations of this basic technique.

16.1 *X-Y* MATRIX SWITCH ARRAY

The *X-Y* array scanning technique is a common method used by most alphanumeric keyboards. As shown in Fig. 16.1, two buses are required. In this case an output bus goes in the *X* direction and an input bus goes in the *Y* direction. The inset shows that a switch is placed at each intersection of an *X* line and a *Y* line. A number of microcircuits are available which implement this technique. However, we will present

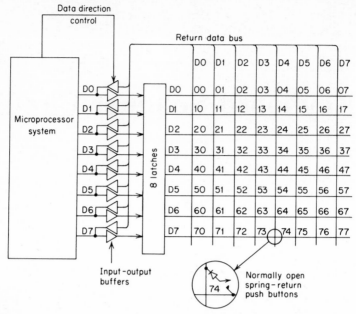

Fig. 16.1 A common switch-scanning technique which uses two buses intersecting in an *X-Y* array.

the concept using a microprocessor. With a microprocessor the technique has much more flexibility, but it ties up most of the microprocessor's time for this one task.

Operation of this circuit is as follows. The microprocessor starts a scanning sequence by first raising line D0 to the HIGH state. Lines D1 to D7 remain LOW while D0 is HIGH. These states are stored in the 8-bit latch. The microprocessor next reads the state of the input data bus by activating the eight input buffers. These lines are vertical in Fig. 16.1. If, for example, switch 02 is depressed, the microprocessor will read 00000100, where D0 is the least significant bit. This word is decoded by microprocessor software into whatever format is needed by the system.

After the read and decode cycle the microprocessor advances the HIGH output line to D1. This new word 00000010 is again stored in the eight latches, the bus is read again, and the cycle repeats itself. The 64-switch keyboard is completely scanned after eight output instructions and eight input instructions have been performed. This complete scan can be performed in less than 100 μs if efficient software is used. After each complete scan the complete state of the switch array is stored in 8 data words of 8 bits each. These data can be processed by the system using software independent of the scanning circuit or its software.

This scanning technique has a potential problem when two or more keys are simultaneously depressed. Suppose switches 12 and 22 are closed. These key closures cause a short circuit between the D1 and D2 output lines. If these two output lines are in opposite states, we may destroy some devices in the latch circuits. Placing a diode in series with each switch avoids this problem.

Some commercial microcircuits which utilize the X-Y scanning technique have a feature called N-key rollover. This means that if a key is depressed before a previously depressed key is returned to the OFF position, the second key will still generate a code. Likewise, a third key will generate a code if the first two keys are still down. With true N-key rollover we could have any number of keys down and any new key depression would still generate a code. However, some manufacturers of these devices caution the user that if any combination of depressed keys forms a right triangle in the X-Y array, the last key down in this set of three will not be encoded. A microprocessor-controlled encoder (Fig. 16.1) does not have this problem. For example, suppose all the switches 02,12,22, . . . , 72 and 03,13,23, . . . , 73 are simultaneously depressed. On the first output code 00000001 the corresponding input will be 00001100. The software will be able to determine that switches 02 and 03 are depressed. On the next output cycle 00000010 the input will again be 00001100. The software now finds that the 12 and 13 switches are depressed. The same mutually exclusive tests are made all the way down to the 72 and 73 switches. No interaction between switches occurs. Since the X-Y switch array is scanned at a high rate, each new key closure merely changes the input code back to the microprocessor. The microprocessor program is written to detect each change in the input code by looking at individual bits in each input word as it arrives.

16.2 KEYBOARD SCANNER USING A MICROPROCESSOR AND PIA

ALTERNATE NAMES True N-key rollover keyboard encoder, software-controlled music keyboard scanner, multiple-closure switch-array encoder.

PRINCIPLES OF OPERATION The circuit shown in Figs. 16.2 and 16.3 is capable of generating switch-closure codes compatible with the music synthesizer discussed in Chap. 17. It is organized in octaves; i.e., circuit sections can be added or subtracted in whole octaves. The software also counts to the base 12 and stores note information in a format directly usable by a music system.

Hardware description This circuit is designed to interface with a multiple-channel synthesizer (such as Fig. 17.1). Data and control lines coming

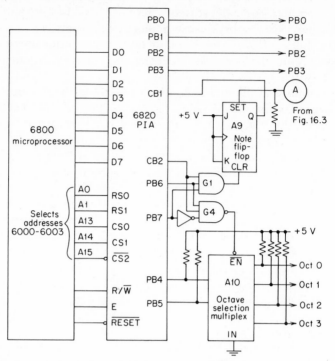

Fig. 16.2 Control circuit for a four-octave musical-keyboard scanner using the 6820 PIA device.

from the microprocessor are organized to handle both the synthesizer and this keyboard scanner simultaneously. Gate circuits to control both circuits must be three-input gates; i.e., G3 and G5 in Fig. 17.1 must also be changed to three-input gates.

The synthesizer circuit has two modes, i.e., select channel 1 or 2 (see Table 16.1). The scanning circuit also has two modes. The first mode clears the note flip-flop (A9), and the second mode samples keys on the keyboard by activating multiplexer A10. These modes are set up by the states of the two most significant bits in the PIA data word (PB6 and PB7). After the correct mode is placed on these lines, a control line (CB2) is momentarily raised to the HIGH state to cause a mode transfer to the appropriate circuit. This two-step mode control utilizing three lines is required since the data bus bits PB6 and PB7 are controlled with different instructions occurring at an earlier time than the control line CB2. The first two modes are described in Chap. 17. The scan-keyboard mode causes the microprocessor to examine each note on the keyboard for a switch closure. If a key is depressed, a signal is generated which sets the note flip-flop. The microprocessor looks at the note flip-flop after each scan cycle to see if a note has

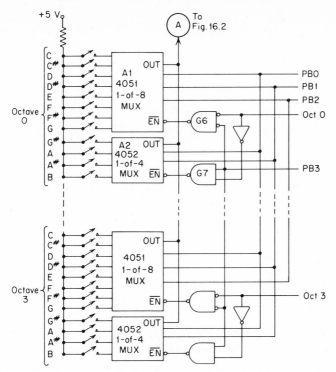

Fig. 16.3 Two octaves of a four-octave musical-keyboard scanner circuit.

been depressed. Once per scan cycle the note flip-flop is cleared to make ready for the next scan.

At this point it will be useful to describe the interface between the microprocessor and the keyboard. The circuit described in this chapter is designed to be controlled by a 6800 microprocessor and its family of peripheral chips. The 6820 peripheral interface adapter (PIA) is used as the parallel interface between the microprocessor and the scanner. This is a dual device, and we will assume the B side is to be used. Quite often the A side is committed to the main alphanumeric keyboard in a typical microcomputer system.

The B side of the PIA has an 8-bit data-word latch which can be used for microprocessor input-output from or to an external device.

TABLE 16.1 Four Modes of Keyboard Scanner and Synthesizer Circuits

PB6	PB7	CB2	Action taken
0	0	⎍	Output T1 to synthesizer 1
0	1	⎍	Output T2 to synthesizer 2
1	0	⎍	Scan keyboard
1	1	⎍	Clear note flip-flop

We will always use these 8 lines as microprocessor outputs for both the synthesizer and the scanner. An input control signal called CB1 is used to sense the status of some external device. In this case it senses the output of the note flip-flop. A PIA output signal called CB2 is often used to control an external device. Here we use CB2 to clear the note flip-flop and enable A10.

The scanner circuit operates as follows. The note flip-flop (A9) is first cleared when the microprocessor sends a PB6 = PB7 = CB2 = 1 command through the PIA. This command is decoded with G1. The microprocessor then moves into the scan mode with PB6 = CB2 = 1 and PB7 in the LOW state. This is decoded by G4, after which a data word already present at PIA output bus B selects one of four octaves using A10. Bits PB4 and PB5 are used for octave selection. The note selection within an octave is performed with bits PB0 through PB3. We can therefore scan four octaves having 12 notes each using this decoding scheme.

Device A10 is a 1-of-8 multiplexer. We use only four of its outputs in this circuit. Since the common input terminal of this device is tied to ground, we will get a LOW level for the selected output. The pullup resistors on each output line make the standby state HIGH. Each output enables two other multiplexers (see Fig. 16.3) in the selected octave circuit. These two multiplexers service 12 keyboard notes. We also could have used a 16-channel multiplexer for note selection within each octave. However, 4- and 8-channel multiplexers are much cheaper (and more readily available) than 16-channel multiplexers.

When a note is depressed, a HIGH level is transferred to the appropriate multiplexer input pin. When the microprocessor scans that particular input, the HIGH level is transferred through one of the note-selection multiplexers, where it sets the note flip-flop. The microprocessor then samples the CB1 line to see if the note flip-flop has been set. If not, the next note is selected for examination. If the note flip-flop has been set, an 8-bit word is stored in memory with a code corresponding to that particular note.

The multiplexer for each octave uses a 1-of-8 multiplexer (a 4051) for the lower 8 notes. The upper 4 notes are sampled with a 1-of-4 multiplexer (half of a 4052). The $\overline{\text{ENABLE}}$ input of the 4051 is LOW (selected) only when PB3 is LOW and the appropriate octave select line is LOW. PB3 is LOW only for the first 8 notes of an octave. When PB3 goes HIGH, notes 9 through 12 are selected and the $\overline{\text{ENABLE}}$ line to the 4052 goes LOW. For example, we note that the output of G6 ($\overline{\text{ENABLE}}$ for A1) goes LOW only when both PB3 and OCT0 are LOW. We also observe that the output of G7 ($\overline{\text{ENABLE}}$ for A2) goes LOW only when PB3 is HIGH and OCT0 is LOW.

Software description   Figure 16.4 shows the flowchart and instructions for a method of scanning the four-octave keyboard described above.

The program begins by clearing all temporary registers in memory locations $0100 through $0106 (the $ means that the number is hexadecimal). Each of these temporary registers is listed in Table 16.2. The PIA is initialized so that the B side is an output bus. We next send out PB6 = PB7 = 1 to clear the note flip-flop. PB6 and PB7 remain latched in the output register while CB2 is momentarily pulsed to the HIGH state. This action causes the output of G1 also to go momentarily HIGH, thereby clearing the note flip-flop. Immediately after the note flip-flop is cleared, the PIA is programmed to look for a CB1 input from the note flip-flop Q output. While the flip-flop is being sensed, the PIA output bus B addresses note 00. If this note is activated (depressed), A1 will transfer a HIGH level over to the SET input of the note flip-flop.

Whenever any note is found in the HIGH state, the program must determine whether this note is of higher or lower pitch than a stored note that was previously evaluated as the highest note activated. At any given instant the note being sampled is N while the highest-pitch note previously stored is T1. If $N > T1$, the new note is higher than the old highest note. The old note is renamed T2 and the new note is named T1. If $N < T1$, the new note is stored as T2 and T1 remains undisturbed.

The program next moves into a debouncing routine. Each time T1 is found in the HIGH state, a counter called T1P is incremented. This counter keeps track of the number of times T1 was present. Since each key is scanned sequentially, the entire keyboard is scanned between samples of any particular note. Another counter called T1M (T1 missing) is cleared whenever T1P is incremented. Counter T1M counts the number of times T1 is found in the LOW state. The number of scans used to detect a valid T1 present or T1 missing is set at 3 in the program. Other numbers can be used depending on the switch-bouncing characteristics of the keys used. When three HIGH values of T1 are detected, the program sends the code corresponding to that particular note to the synthesizer channel 1. When T1M reaches a count of 3, synthesizer channel 1 is turned OFF. Likewise, T2P and T2M control the ON-OFF status of synthesizer channel 2. Neither synthesizer is notified of a T1 or T2 ON-OFF state until that particular note is found in three successive scans of the entire keyboard. This technique debounces all keys and prevents chattering keys from falsely operating the synthesizer.

The program advances N in a base 12 format. Each time N is incremented, its least significant digit is compared to hexadecimal C (decimal 12). If a match is found, N is advanced by 4, so that the most significant digit is incremented by 1. Since we are starting the count at 0, a count of 12 decimal is the first note of the next octave.

If a four-octave synthesizer-keyboard is used, the maximum N is 30

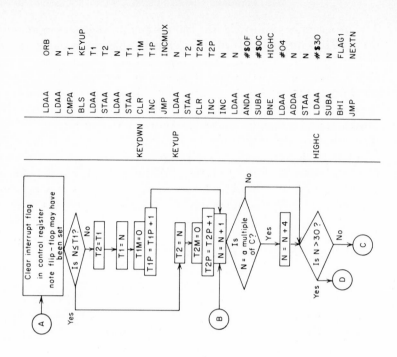

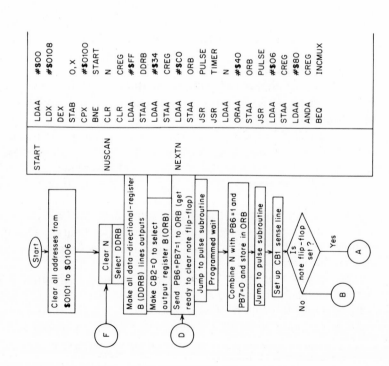

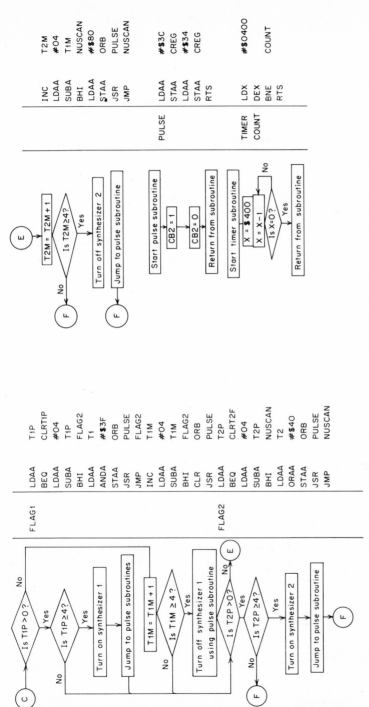

Fig. 16.4 Flowchart of a four-octave musical-keyboard scanning program.

TABLE 16.2   Function of Seven Temporary Registers in Fig. 16.4

Address of register (hex)	Name of register	Register function
0100	N	Note number on keyboard
0101	T1	Higher of two notes activated
0102	T1P	T1-present counter; three scans of activated T1 needed before synthesizer notified
0103	T1M	T1-missing counter; three scans of missing T1 needed before synthesizer turned off
0104	T2	Lower of two notes activated
0105	T2P	T2-present counter, similar to T1P
0106	T2M	T2-missing counter, similar to T1M

(hex). A test is made each time N is advanced to see if the limit 30 has been exceeded. If not, the program counter returns to the clear-note flip-flop routine (NEXTN). If the end of keyboard is detected, a $T1P > 0$ test is performed. If the result is false, the T1M (T1 missing) counter is incremented. When this counter reaches 4, synthesizer 1 is turned OFF. If the result of the $T1P > 0$ test is true, another test to see if $T1P \geq 4$ is performed. If this is true, synthesizer 1 is activated with a clean debounced value of T1. This note will stay latched in the synthesizer tone-generation circuit until T1M advances to 4. The synthesizer is turned OFF by loading an all-zero data word. The T2P test is entered after the T1 synthesizer ON-OFF routine is complete or if $0 < T1P \leq 3$.

The program requires approximately 100 μs for each key and another 50 μs for each activated key. A complete scan of 60 notes therefore takes approximately 6 ms. If four scans are required for debouncing, a 24-ms delay will occur from the time a note is depressed until it is transferred over to the appropriate synthesizer channel. If more debouncing time is required, a timer loop can be inserted after the clear-note flip-flop routine. This timer loop is shown as a subroutine called *timer*. If the index register is loaded with $0400, as shown, the delay per keyboard scan will be approximately 8 ms in addition to the 6 ms for the rest of the program.

16.3 MICROPROCESSOR BCD SWITCH SCANNER

ALTERNATE NAMES   Smart front panel, BCD thumbwheel switch to microprocessor interface, front panel BCD-to-binary interface.

PRINCIPLES OF OPERATION   Instrument front panels commonly utilize BCD switches and BCD displays to simplify human operation and inter-

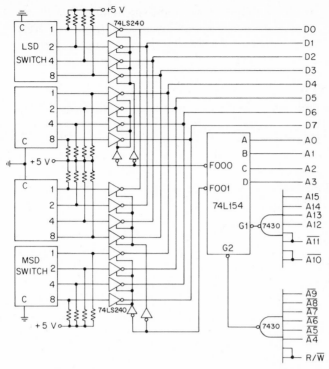

Fig. 16.5 A simple method for interfacing the four BCD thumbwheel switches with the data and address buses of the microprocessor.

pretation.   An array of BCD thumbwheel switches provides a compact method of entering large numbers into an instrumentation system. In many cases this BCD number of N digits must be converted into its binary equivalent.   The following pages describe a technique for scanning four BCD thumbwheel switches and converting the 4 BCD words into a 14-bit binary word.

Figure 16.5 shows one simple method of interfacing the four BCD switches with the microprocessor address and data buses.   Each BCD switch has four outputs with weights 1, 2, 4, and 8.   For example, if the number 9 is selected, the 1 and 8 output lines are internally connected to the common terminal.   Pull-up resistors are needed on each switch output terminal since the various weighted outputs are open-circuited if not internally connected to the grounded common terminal.   The resulting BCD code is inverted at this point, but the 74LS240 buffers again invert the code.   For the 74LS240 devices the maximum pull-up resistor sizes must be

$$R_{\max} = \frac{V_{CC} - V_{IH,\min}}{I_{IH,\max}} = \frac{5 - 2 \text{ V}}{40 \ \mu\text{A}} = 75 \text{ k}\Omega$$

where $V_{IH,\min}$ is the minimum recognizable HIGH input voltage and $I_{IH,\max}$ is the maximum input current with HIGH level applied. Resistors of 10 kΩ would be sufficient. With a worst-case input current of 40 μA the HIGH-level input voltage will be $V_{IH,\min} = RI_{IH,\max} = 10^4 \times 40 \times 10^{-6} = 0.4$ V below V_{CC}.

A 74LS240 is selected when both of its enable lines are driven LOW. When the enable lines are HIGH, the device outputs go into the high-impedance state. The LOW enable signals are generated using a 74L154 decoder. The two lowest addresses decoded by the 74L154 are used in this application. A particular output on this device is selected when G1 and G2 are both LOW and A0 to A3 contain a number corresponding to the desired output. G1 and G2 will be LOW for addresses F000 through F015. A0 through A3 select the exact address within that range. Whenever the computer program issues a read command for address F000, the two least significant thumbwheel digits are transferred over to the data bus. A read command at address F001 moves the two most significant digits to the data bus.

The flowchart of a program to scan the four BCD thumbwheel switches and convert the data into a 14-bit binary word is shown in Fig. 16.6.

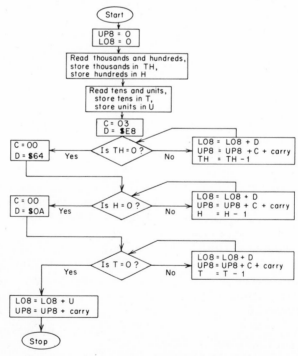

Fig. 16.6 Flowchart for scanning the four BCD thumbwheel switches shown in Fig. 16.5 and converting the data into a 13-bit binary word (see also Fig. 12.8).

An assembly-language listing of the flowchart using the 6800 language is given in Table 16.3. The program starts by clearing UP8 and L08, i.e., the upper and lower 8-bit segments of the 16-bit output register. This register could be part of RAM or some type of output register. The thousands and hundreds switches are read next. This input port, called THH, i.e., thousands and hundreds, is read twice. On the first reading the hundreds bits (D0 to D3) are deleted by shifting the thousands bits (D4 to D7) to the right four places. The LSRA instruction shifts 0s in from the left so that the upper 4 bits are cleared after this operation. This new 8-bit word is stored in register TH (part of RAM) for use later in the program. On the second reading of THH the hundreds data are saved by masking out the upper 4 bits using the ANDA #$0F instruction. These data are saved in register H. The tens and units data are similarly processed and saved in registers T and U.

The program begins the BCD-to-binary conversion routine at START.

TABLE 16.3 Assembly-Language Program to Implement the Flowchart of Fig. 16.6 Using the 6800 Microprocessor Language

BCDBIN	CLR	UP8	Begin subroutine by clearing
	CLR	L08	16-bit output word
	LDAA	THH	Read TH = D7–D4, H = D3–D0
	LSRA		
	LSRA		Move thousands to lower 4 bits and zeros
	LSRA		into upper 4 bits
	LSRA		
	STAA	TH	Store thousands in TH
	LDAA	THH	Read THH again
	ANDA	#$0F	Keep hundreds and mask out thousands;
	STAA	H	store in H
	LDAA	TU	Read tens = D7–D4, U = D3–D0
	LSRA		
	LSRA		Move tens into lower 4 bits; move zeros
	LSRA		into upper 4 bits
	LSRA		
	STAA	T	Store tens in T
	LDAA	TU	Read tens = D7–D4, U = D3–D0
	ANDA	#$0F	Keep units and mask out tens; store in
	STAA	U	U
START	LDAA	#$03	
	STAA	C	C, D = $03E8 (1000 decimal)
	LDAA	#$E8	
	STAA	D	
TEST TH	LDAA	TH	Is TH = 0 ?
	BEQ	HUNDRD	Yes, go to hundreds test
	LDAA	L08	
	ADDA	D	L08 = L08 + D
	STAA	L08	
	LDAA	UP8	
	ADCA	C	UP8 = UP8 + C + carry
	STAA	UP8	
	DEC	TH	TH = TH − 1
	BRA	TEST TH	D0 another 1000s loop

Table 16.3   Assembly-Language Program to Implement the Flowchart of Fig. 16.6 Using the 6800 Microprocessor Language *(Continued)*

HUNDRD	CLR	C	Begin 100s loop
	LDAA	#$64	C, D = $0064 (100 decimal)
	STAA	D	
TESTH	LDAA	H	Is H = 0 ?
	BEQ	TEN	Yes, go to tens test
	LDAA	L08	
	ADDA	D	L08 = L08 + D
	STAA	L08	
	LDAA	UP8	
	ADCA	C	UP8 = UP8 + C + carry
	STAA	UP8	
	DEC	H	H = H − 1
	BRA	TESTH	Do another 100s loop
TEN	LDAA	#$0A	Begin 10s loop
	STAA	D	C, D = $000A (10 decimal)
TESTT	LDAA	T	Is T = 0 ?
	BEQ	UNIT	Yes, to units routine
	LDAA	L08	
	ADDA	D	L08 = L08 + D
	STAA	L08	
	LDAA	UP8	
	ADCA	C	UP8 = UP8 + C + carry
	STAA	UP8	
	DEC	T	T = T − 1
	BRA	TESTT	Do another 10s loop
UNIT	LDAA	L08	
	ADDA	U	L08 = L08 + U
	STAA	L08	
	LDAA	UP8	
	ADCA	#0	UP8 = UP8 + carry
	STAA	UP8	
	RTS		Exit subroutine

A hexadecimal number $03E8, which corresponds to decimal 1000, is loaded into registers C and D.   Next the TH register contents are compared with 0.   If they are equal to 0, the program immediately jumps to the hundreds routine.   If TH > 0, D is added to UP8 and C is added to L08.   The L08 + D operation is performed first, and any resulting carry is added into the UP8 + C operation.   TH is then decremented, and the loop repeats itself until TH becomes 0.   At this point UP8 and L08 will contain a hexadecimal number corresponding to the number of thousands on the BCD thumbwheel switch.

The hundreds and tens routines are identical to the thousands routine. In these cases $0064 is added to UP8, L08 for each hundred and $000A is added for each ten.   The number of units is added directly with no test loop since hex and decimal whole numbers between 0 and 9 are identical.

Chapter 17

Application for the 6800 Microprocessor: A Polyphonic Music Synthesizer

INTRODUCTION

There are numerous applications for microprocessors in the musical field. This chapter explores a method of simultaneously generating multiple tones on the equal-tempered scale using a 6800 microprocessor. Most music synthesizers are capable of producing only one tone at a time. The generator described in this chapter is capable of generating two or more tones simultaneously. Each tone leaves the generator on a separate line. Conceivably, each output could drive separate tone-modifying circuits such as an envelope follower-trigger, an envelope generator, a voltage-controlled filter, and a voltage-controlled amplifier. These four basic circuits can simulate almost any instrumental sound. High-quality music synthesizers also have other circuits available such as *XY* multipliers, ring modulators, frequency multipliers, etc.

In order to keep this chapter reasonably simple, the polyphonic music generator described here has only two outputs, but the technique is expandable to any number of outputs. A practical upper limit would probably be 12 since musicians have only 10 fingers and two feet.

ALTERNATE NAMES Multiplexed signal generator, multiple-output tone generator, programmable tone source.

PRINCIPLES OF OPERATION As shown in Fig. 17.1, the heart of this multiple output generator is the top-octave divider. This class of IC is made by many semiconductor firms. The type used in this example is the Mostec/AMI 50240, which operates from 10 V. We used ±5-V supplies to simplify the interface with the +5-V logic. As shown in the figure, all 50240 outputs are clamped above −0.7 V using diodes D1 to D12. Each diode is buffered with a 10-k Ω resistor to maintain the current from the −5-V supply at a safe level for the 50240.

The block diagram of the 50240 top-octave divider is shown in Fig. 17.2. Although this is a fairly complex chip with several hundred gates, it is simple in concept, consisting merely of 12 frequency dividers with division ratios from 239 to 451. One additional note (the thirteenth) is also provided with a division ratio of 478. This additional note allows organ builders the option of providing a key of C at each end of the keyboard.

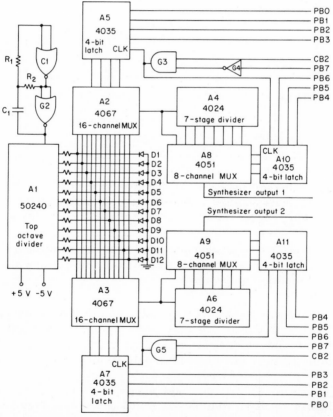

Fig. 17.1 A multiple-output equal-tempered musical-note generator which is controlled by a 6800 microprocessor.

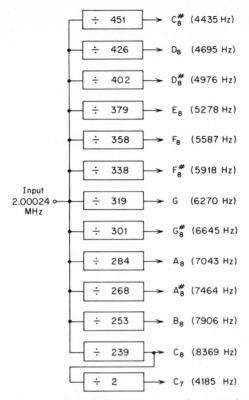

Fig. 17.2   Block diagram of a typical top-octave equal-tempered generator chip.

The generator chip must be driven from a 2.00024-MHz oscillator if it is to provide outputs in tune with the rest of the orchestra. Since the 50240 likes to receive a 10-V input square wave a CMOS multivibrator, as shown in Fig. 17.1, is used (see also Fig. 3.5).

The 16-channel multiplexers (A2 and A3) are used to select a particular note for each synthesizer output. A 10-output polyphonic synthesizer would require 10 multiplexers. Two eight-channel multiplexers could also be used for each channel. If a bipolar 16-channel multiplexer (MUX) such as the 74150 were utilized, the interface would be much more difficult since these devices load the 50240 excessively. Each transmission gate in the 4067 is merely a variable resistor from each input to the common output. Therefore, the real load for the 50240 is A4 and A6. If these are CMOS devices, the 50240 will be very lightly loaded.

Note selection in A2 is made with the data bus lines PB0 to PB3 from the microprocessor system. These data lines must be latched into A5 so that each note will continue to be generated while the data bus does other things. The latching-in of data occurs when pin 6 of A5 (the clock input) sees a positive-edge waveform. The data bits PB0 to PB3 will continue to program A2 until new data are latched into A5 from the microprocessor.

The octave selection of the note coming from A2 is performed with A4 and A8. Device A4 is a seven-stage divider. Each of the A4 outputs, in addition to the original note from A2, is applied to the inputs of an eight-channel multiplexer A8. Selection of the desired octave for the note coming from A2 is performed with PB4 to PB6. These lines are latched into A10 whenever the clock line (pin 6) has a positive transition. This positive transition occurs whenever CB2 goes HIGH and PB7 goes LOW. These events must simultaneously occur after PB0 to PB6 are present at the inputs of A5 and A10.

The interface with the 6800 is implemented with the 6820 peripheral interface adapter (PIA). This device sends the data bus PB0 to PB7 and a control signal CB2 to the synthesizer under computer control. Since we need only 7 bits (PB0 to PB6) to program A5 and A10, PB7 is used as a control signal along with CB2.

If channel 1 is to be turned ON or changed to a new note, the microprocessor first sends PB0 through PB6 to the generator. At the same time PB7 = 0 is sent out. The next few instructions momentarily pulse CB2 HIGH. This latches PB0 through PB6 into A5 and A10 and a note will appear at pin 1 of A8. If this channel must be turned OFF, PB0 to PB3 are programmed to select any of inputs 13 to 16 in the 4067. Since we only use the first 12 inputs of the 4067, no note will be selected. PB7 and CB2 must then change to LOW and HIGH, respectively, to latch in the OFF command.

Operation of channel 2 is identical to that of channel 1 except that data are latched in only when both PB7 and CB2 are HIGH.

PROGRAMMING PROCEDURES How one proceeds to program a microcomputer to run this polyphonic synthesizer depends on the application. A typical application might be to interface this system with the microprocessor-controlled keyboard scanner discussed in Chap. 16. However, it is instructive first to develop simple programs for this circuit and then gradually build up to a full-scale polyphonic synthesizer operated from a keyboard.

Single-channel output After the circuit in Fig. 17.1 is assembled, a test program is written to test each channel independently. Figure 17.3 shows the flowchart of such a program. The program is written to sequentially scan all notes at a rate set by the 16-bit word at addresses 041D to 041E. Before discussing the flow chart we must first provide a brief discussion of the PIA operation. This is a dual device, but in this chapter we will be using the B side. Each side looks like two memory

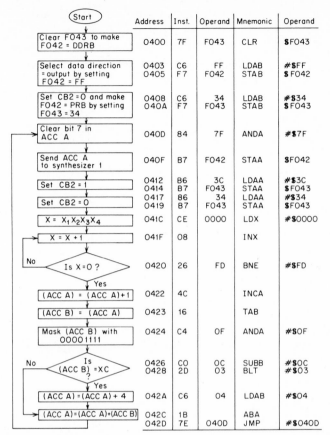

Fig. 17.3 Flowchart and listing of a program to scan through all notes on channel 1 of the polyphonic generator circuit.

addresses. In the author's system the two addresses for the B side are F042 and F043. Location F042 is actually two locations inside the PIA. The choice of these two depends on the content of bit 2 in F043.

Address F043 is called the *PIA control register*. It is always written into at the start of a program to set up the PIA for later use of location F042; 6 of the 8 bits in F043 perform a control function. It is beyond the scope of this chapter to discuss all the things these control functions can do. We will mention only three features of these control functions for our music synthesizer.

Inside the PIA address F042 is routed to an input-output register called *peripheral register B* (PRB) if bit 2 in address F043 is HIGH. If bit 2 in F043 is LOW, address F042 is routed to a circuit called *data direction register B* (DDRB). PRB is used for all input-output data for a given peripheral device. In the present case PRB is always an output port

(PB0 to PB7) which drives the music synthesizer. DDRB is a register which tells PRB whether it is an input register or an output register. For example, if we address F043 and set bit 2 to **0,** we constrain F042 to be the DDRB. When we address F042 and write in all 1s, we fix PRB to be an output register; i.e., the data direction is out. Since this may sound confusing at first, the steps to be followed before PRB can be utilized as an input or an output register are summarized as follows:

    1. Clear address F043. This writes a LOW into bit 2, which sets up F042 so that DDRB is selected.

    2. Address F042 and write in a HIGH for each bit of PRB which is to be an output line and a LOW for each bit which is to be an input line.

    3. Address F043 and write a HIGH into bit 2. This sets up F043 so that PRB is selected.

    4. Address F042 and read or write from or to the peripheral register, as prepared by steps 1 to 3 above.

The reader should review these four steps several times until the process is clear.

As shown in Fig. 17.3, it is convenient for small programs to show the program listing alongside the flowchart. This technique trains one to develop useful and well-organized flowcharts. Each block in the flowchart should be easily convertible into one or more program instructions. If the flowchart is only one block wide, there is room for at least five columns in the program listing.

Starting at the top of the flowchart, we begin by clearing address F043. This selects F042 as data-direction register B (DDRB). Now we address F042 and set all bits HIGH, thus making the PIA an output device. If we had cleared all bits in DDRB, the PIA would have become an input device.

We again set up the PIA by addressing F043 and storing 34 (00110100) in that location. Two things are done with this instruction. The 5 bits on the left (00110) force CB2 to the LOW state. The next HIGH bit transforms F042 into peripheral register B (PRB). We are now ready to begin sending music data from the CPU through PRB to the synthesizer circuit.

Data are latched into A5 and A10 when PB7 is LOW and CB2 is pulsed HIGH. The next two blocks in the flowchart perform this by placing note data on PB0 to PB6 and clearing PB7. The following two blocks send a positive CB2 pulse to G3. This latches PB0 to PB6 into A5 and A10.

The next three instructions are

$$41C: \quad X = X_1 X_2 X_3 X_4$$
$$41F: \quad X = X + 1$$
$$420: \quad \text{If } X \neq 0, \text{go back to } 41F$$

These instructions determine the duration of each note. The X register (index register) is preset to $X_1X_2X_3X_4$. It is incremented and tested to see if it is 0. If not, the X register is continually incremented until 0 is reached. The program is then allowed to advance to line 422.

The size of $X_1X_2X_3X_4$ is inversely proportional to the duration of each note. If X is preset to FFFF, on the next instruction, X = X + 1, we would get a X = 0 result. The duration of each note would only be approximately 40 clock cycles, as found by summing up the clock cycles required for instructions 40D to 42D. If we use a 1.0 MHz clock, these instructions will be executed in 40 μs since each clock cycle takes 1.0 μs. If we preset X to 0000, the instructions 41F and 420 must be executed $2^{16} = 65,536$ times before the program advances to the next note. Since these two instructions take 8 μs, the individual notes will stay on for 8 μs $\times$ 65,536 = 0.52 s.

After the index-register loop is finished, the A accumulator, which holds the note data, is incremented. The A accumulator is then copied into the B accumulator, where it is tested for the end of each octave. Effectively, what we are doing here is forcing the A accumulator to count in base 12. That is, each time accumulator A is incremented, it is tested to see if it is a multiple of 12. If accumulator A is a multiple of 12, that is, A = X_nC, where X_n can be any number, it is advanced to $X_{n+1}0$. The number X_{n+1} is the digit following X_n. Thus, the counter will follow a sequence such as . . . , 33, 34, 35, 36, 37, 38, 39, 3A, 3B, 40, 41, 42, 43, 44, ... The X_nC number will momentarily be in the A and B accumulators, but it will be cleared out before data are sent to the synthesizer (program step 40F).

The octave test is performed by first calculating a logical AND using accumulator B and the word 00001111 = 0F. This deletes the X_n part so it does not cloud the issue. We next subtract 0C from the B accumulator and place the remainder back in the B accumulator. If the remainder is not 0, the program counter advances to the JMP instruction (42D). If B = 0, the octave test has detected a multiple of 12. We load 04 into accumulator B and add this to accumulator A. The program counter advances to JMP (address 42D) and then returns to address 040D, where the next note sequence starts.

Two-channel output   In this programming example we will use a stored list of words (notes). The synthesizers will read pairs of words, one for each channel, and play their respective notes until the next beat in the melody. A timing loop, such as in the last example, will establish the duration of each beat. The flowchart and listing of this dual-note computer-controlled synthesizer are shown in Fig. 17.4.

For this dual-channel circuit we again maintain CB2 = **0** between note changes. If a data word is to be latched into synthesizer 1, PB7 = **0** is placed on the data bus along with the note and octave data. We then momentarily raise CB2 to the HIGH state to latch the 7-bit

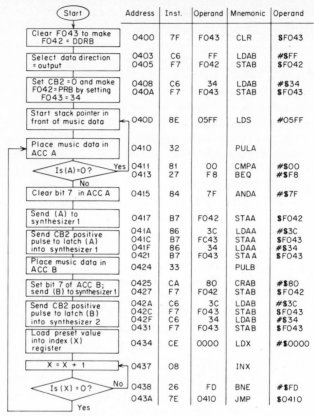

Address	Inst.	Operand	Mnemonic	Operand
Clear F043 to make F042 = DDRB				
0400	7F	F043	CLR	$F043
Select data direction = output				
0403	C6	FF	LDAB	#$FF
0405	F7	F042	STAB	$F042
Set CB2 = 0 and make F042 = PRB by setting F043 = 34				
0408	C6	34	LDAB	#$34
040A	F7	F043	STAB	$F043
Start stack pointer in front of music data				
040D	8E	05FF	LDS	#05FF
Place music data in ACC A				
0410	32		PULA	
Is (A) = 0? Yes				
0411	81	00	CMPA	#$00
0413	27	F8	BEQ	#$F8
Clear bit 7 in ACC A				
0415	84	7F	ANDA	#$7F
Send (A) to synthesizer 1				
0417	B7	F042	STAA	$F042
Send CB2 positive pulse to latch (A) into synthesizer 1				
041A	86	3C	LDAA	#$3C
041C	B7	FC43	STAA	$F043
041F	86	34	LDAA	#$34
0421	B7	F043	STAA	$F043
Place music data in ACC B				
0424	33		PULB	
Set bit 7 of ACC B; send (B) to synthesizer 1				
0425	CA	80	CRAB	#$80
0427	F7	F042	STAB	$F042
Send CB2 positive pulse to latch (B) into synthesizer 2				
042A	C6	3C	LDAB	#$3C
042C	F7	F043	STAB	$F043
042F	C6	34	LDAB	#$34
0431	F7	F043	STAB	$F043
Load preset value into index (X) register				
0434	CE	0000	LDX	#$0000
X = X + 1				
0437	08		INX	
Is (X) = 0? No				
0438	26	FD	BNE	#$FD
043A	7E	0410	JMP	$0410
Yes				

Fig. 17.4 Flowchart and program listing for a dual-note synthesizer controlled by a 6800 microprocessor and 6820 PIA.

word into A5 and A10. This word will remain in A5 and A10 until the next beat, when PB7 = **0** and CB2 = **1** are simultaneously present. If the same note is to be played, the A5 and A10 data will not change. The note will be heard as one uninterrupted frequency for both beats.

Programming of synthesizer 2 is identical to synthesizer 1 except that PB7 must be HIGH when CB2 is momentarily brought to the HIGH state. In this case, whatever information is on the data bus (PB0 to PB6) will be latched into A7 and A11. The selected note will play continuously until PB7 and CB2 again both go the HIGH state.

Referring to Fig. 17.4, we start by clearing address F043, which makes F042 become data-direction register B (DDRB). We next set all bits of DDRB HIGH so that peripheral register B (PRB) is programmed as an output register. Address F043 is again accessed to select F042 = PRB and also to drive CB2 to the LOW state.

The melody is stored as pairs of data words starting at address 600.

All even words will be notes for synthesizer 1 and odd words for synthesizer 2. We use the stack pointer to pull data automatically from this section of memory. The instruction at 040D loads address 599 into the stack pointer to begin the melody-lookup routine. Data are then pulled from location 600 and placed in accumulator A. Since these are data for synthesizer 1, bit 7 (PB7) is cleared before the word is sent to the PIA. CB2 is then pulsed HIGH to latch the word into A5 and A10.

Accumulator B is loaded from the next address using the stack instruction PULB. This instruction increments the stack pointer, then automatically loads accumulator B with the data in the address at which the stack pointer is positioned. This is why we originally initialized the stack pointer at 5FF when the music data actually start at 600. We set bit 7 HIGH and then pulse CB2 HIGH to send this word to synthesizer 2.

As with the single-channel synthesizer, the note duration is determined by preloading the index register with a 16-bit number. We then increment the index register until it becomes 0. After reaching 0, the program returns to address 410, where the next synthesizer 1 note is pulled from memory.

Chapter 18

Special-Purpose LSI Circuits

THE EVOLUTION OF LARGE-SCALE INTEGRATION

In the 20 years since the first microcircuit was introduced there has been an increase by a factor of 2 in capability per chip each year. Beginning with the single flip-flop, or four gates per chip, in the early 1960s, we have witnessed an increase to 10^5 gates per chip by the late 1970s. Figure 18.1 shows this past growth in circuit density and indicates what we can expect in the 1980s. If the trend continues, the chip density may well be more than 100 million gates per chip by the 1990s. New processes are continually being developed to facilitate this growth as each older process reaches the end of its capabilities.

Many of the LSI chips now available do not fit into any of the categories described in Chaps. 3 to 17. In this chapter we will briefly describe several LSI devices which are breaking new ground for microcircuits. Those of us actively following the development of LSI devices are painfully aware of the many new types of circuits and the new challenges they present. We will briefly discuss the following in this chapter:

Direct-memory-access controller
Number cruncher
Manchester encoder-decoder

18.1   DIRECT-MEMORY-ACCESS CONTROLLER

ALTERNATE NAMES   DMA controller, DMAC, DMA supervisor, block-data transfer device.

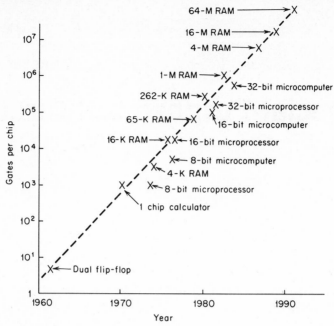

Fig. 18.1 Doubling chip density approximately every 1 or 2 years will provide devices with hundreds of millions of gates by the 1990s.

PRINCIPLES OF OPERATION We will describe the DMA function utilizing a specific device, the Motorola MC6844 direct-memory-access controller. All DMA control chips perform the same basic function, i.e., transfer data between memory locations and peripheral equipment without going through the microprocessor. These chips are typically used to transfer an array of data to or from a peripheral device outside the microcomputer. Applications might include block-data transfers between a floppy-disk system or a CRT display and the microcomputer system memory. If the CRT must be refreshed every $\frac{1}{30}$ s, the DMAC could be used to transfer the block of data representing a screen full of characters without disturbing the microcomputer. The microcomputer would be utilized only for an occasional change of data on the CRT screen.

Figure 18.2 shows a simplified block diagram of the MC6844 DMAC, a four-channel device in which four different peripherals can read or write into the memory without utilizing the microprocessor. These four peripherals can be serviced on either a fixed or a rotating-priority basis. Most peripherals are designed for block-data transfers at rates much slower than the normal data-transfer rate of the microprocessor. If four peripherals were serviced without using a device like the DMAC, the microprocessor would need to operate in the interrupt mode. Otherwise, excessive time would be wasted by the microprocessor as

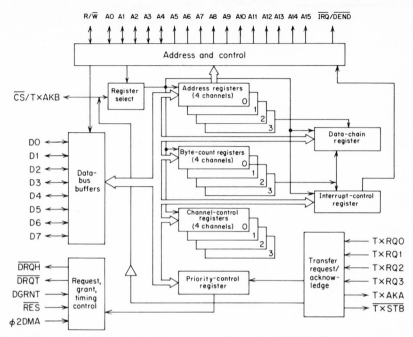

Fig. 18.2 A simplified block diagram of the Motorola MC6844 direct-memory-access controller. (*Motorola Semiconductor Products, Inc.*)

it waits for the peripherals to complete each segment of each block transfer. The DMAC chip does all this work for the microprocessor with minimal software and very little sacrifice of normal execution time of the main program.

The MC6844 is a programmable device with 15 addressable registers. Eight of these registers are 16 bits wide. Each channel has a separate 16-bit *address register* and a 16-bit *byte-count register*. Four *channel-control registers* (8 bits wide) configure the operating conditions for each channel. The three remaining 8-bit addressable registers control operations common to all four input-output channels.

Before any DMA operation the address registers must be loaded with the starting memory addresses, and the byte and count registers are each loaded with a number equal to the number of bytes to be transferred for that channel. Each channel-control register sets up the direction of transfer, mode, and address increment or decrement after each cycle for its particular channel. Three modes are available: (1) three-state control steal, (2) halt steal, and (3) halt burst. The channel-control registers can also be read by the microprocessor. Two bits in each of these registers indicate whether that channel is busy transferring data or the DMA transfer is complete.

The *priority-control register* enables the data-transfer requests from all peripheral controllers, i.e., the electronics which operate each peripheral device. This register establishes either a rotating- or fixed-priority scheme of servicing these requests. The fixed-priority scheme sets channel 0 with highest priority, channel 1 with the next highest priority, etc. At power turn-on or reset the rotating-priority scheme is the same as the fixed-priority scheme, but once a channel has been serviced, it moves to the lowest priority and those below it advance to the next higher priority. The priority-control register also has 4 bits to enable or disable any channel.

When a DMA transfer for any channel is complete, i.e., its *byte-count register* = 0, a DMA END signal is directed to that particular peripheral and an interrupt request ($\overline{\text{IRQ}}$) is simultaneously sent to the microprocessor. Each of these interrupts is separately enabled using bits in the *interrupt-control register.* Bit 7 in this register, called DEND/$\overline{\text{IRQ}}$, is a flag indicating that this chip has requested an interrupt on the $\overline{\text{IRQ}}$/ $\overline{\text{DEND}}$ output line.

The *data-chain register* controls chaining of data transfers. Repetitive reading or writing of a block of data can be performed by using this register. The process is possible only for channels 0, 1, and 2 because the channel 3 registers are utilized for temporary storage in this mode. Before a chained DMA operation the address register and byte-count register of both channel 3 and channel N (channel 0, 1, or 2) are loaded with identical numbers. After the first block of data has been transferred, the address register of channel N is pointing to the last address of the block of data and the byte-count register of channel N is 0. At this point the DMAC transfers the channel 3 address and byte-count registers to their equivalent register in channel N. The data-transfer process then starts over again immediately after these two 16-bit transfers have been performed.

In the DMA mode, the MC6844 has complete control of the address bus, the data bus, R/$\overline{\text{W}}$, and the valid memory address line. When a peripheral device needs a DMA transfer, it issues a transfer request (TxRQ0 to TxRQ3). If that particular channel is enabled in the priority-control register and also meets the test of highest priority, the DMAC will issue a DMA request using the $\overline{\text{DRQT}}$ line. When the DMAC receives a DMA grant on the DGRNT line, it provides a transfer acknowledge (TxAKA) to the peripheral device and the data transfer begins. When the byte counter for that channel goes to 0, the block transfer is complete and an $\overline{\text{IRQ}}$/$\overline{\text{DEND}}$ signal is sent to the peripheral device and to the microprocessor.

18.2 NUMBER CRUNCHER

ALTERNATE NAMES Number-crunching unit, NCU, number-crunching processor, number-oriented microprocessor, number processor, numeric processing computer, digit handler, arithmetic processing unit, APU.

PRINCIPLES OF OPERATION   General-purpose microprocessors can handle complex mathematical calculations only if lengthy software algorithms are developed.   Many microprocessor-based systems could utilize the algebraic and trigonometric calculations and conversions readily available in scientific calculator chips, but this is difficult because these chips are not designed to interface with a microprocessor.   The designer of a microprocessor system with high-level math requirements could interface with a BCD device such as the National 57109 number cruncher or the Advanced Micro Devices AM9511 arithmetic processing unit. We will briefly discuss the 57109 number cruncher, a bit-parallel–digit-serial BCD number-oriented microprocessor.   As shown in Fig. 18.3, all BCD data are entered via D1 to D4 and exit via D01 to D04.   The data formats at the input or the output may be in floating-point or

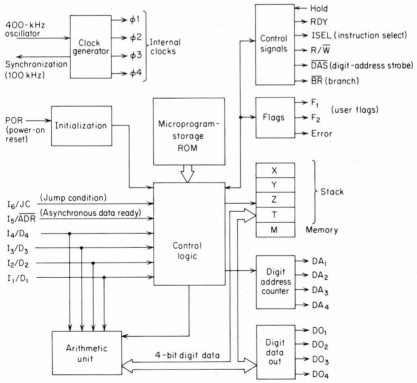

Fig. 18.3   Simplified block diagram of the National 57109 number-cruncher unit. (*National Semiconductor Corp.*)

scientific notation. The mantissa may have up to 8 BCD digits with a 2-digit exponent having a range of ±99.

The 57109 combines many features of a scientific calculator chip with the features of a microprocessor. For example, the NCU input-output functions are much more flexible than those of a calculator chip, which is normally limited to keyboard inputs and 7-segment display outputs. This device is also several orders of magnitude easier to program for complex mathematical routines than a general-purpose microprocessor.

The NCU receives data and instructions via the six input lines I1 to I6. Data utilize only the I1 to I4 lines. The timing of a data or instruction single-word input is shown in Fig. 18.4. If the outside circuitry wants to send an input to the NCU, it raises the HOLD line. When the NCU is ready to accept this input word, it issues RDY = 1 to the outside circuitry. The input data word is then changed to a new value, and the HOLD line is lowered. The NCU then lowers the RDY line and simultaneously reads the I1 to I6 input lines. The RDY line rises every time an instruction has been executed or every 10 μS, whichever is larger. If the outside circuitry has no instruction ready to execute, it keeps the HOLD line HIGH. This places the NCU in a wait state which inhibits RDY from falling until a new input is ready from the outside. This is called the *handshaking mode* since RDY and HOLD both control each other's actions.

There are two other modes of entering BCD data. The first, which utilizes the IN instruction shown in Table 18.1, is called the *synchronous-digit-input mode.* With this method the NCU must first be informed of the exact number of digits to be read in by using the set-mantissa-digit-

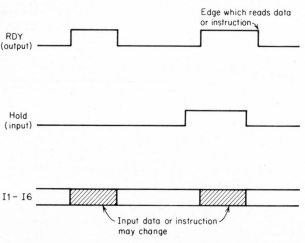

ring diagram of an NCU fetch cycle for input data or an instruction. n word is entered only when both RDY and HOLD are HIGH.

TABLE 18.1 NCU Input-Output Mode Instructions, Input-Output Data, and Their Octal Codes

Mode instruction	Digit entry	Octal code	Description
IN		27	Multidigit synchronous digit input to follow on lines D1–D4
AIN		16	Single-digit asynchronous digit input to follow on lines D1–D4
OUT		26	Multidigit synchronous digit output to follow on lines DO1–DO4
TOGM		42	Change mode from floating-point to scientific mode or vice versa
SMDC		30	Mantissa-digit count (1 to 8) is set
MCLR		57	Clear all internal registers and memory
DP		12	Enter decimal point
EE		13	Digits that follow will be exponent
CS		14	Change sign of mantissa or exponent
EN		41	Terminates digit entry
	0	00	⎫
	1	01	
	2	02	
	3	03	
	4	04	Mantissa or
	5	05	exponent digits
	6	06	
	7	07	
	8	10	
	9	11	⎭
SF1		47	Set flag 1 output HIGH
PF1		50	Toggle flag 1 to opposite state
SF2		51	Set flag 2 output HIGH
PF2		52	Toggle flag 2 to opposite state
PRW1		75	Pulses R/$\overline{\text{W}}$ output to an active LOW state
PRW2		76	Same as PRW1

count (SMDC) instruction. SMDC is a 2-word instruction in which the second word specifies the number of mantissa digits (1 to 8) in the sequence to be loaded. The mantissa and exponent digits and signs must follow a fixed-entry format. In this mode, all information is entered at a fixed rate without handshaking, as mentioned for the entry mode using RDY and HOLD. The synchronous-input mode has the advantage of speed. A number containing an 8-digit signed mantissa with a signed 2-digit exponent requires approximately 6 ms for entry. The handshaking method for a number this complex would require approximately 30 ms.

The second nonhandshaking input technique uses the asynchronous-digit-input instruction (AIN). This mode is useful in NCU stand-alone applications, where sequencing is done without the aid of a microprocessor. One to eight digits can be entered into any positions of the NCU X register. This method must use the following NCU outputs: RDY,

TABLE 18.2 Description of All Signal Inputs or Outputs of the 57109 Number Cruncher

Name	Description
POR	Power-on reset; must be HIGH at least 8 clock cycles after power is applied
OSC	Single-phase clock input; typically 400 kHz
SYNC	Sync output; one-fourth clock frequency
RDY	NCU ready output; rising edge indicates NCU is ready to execute next instruction or get second word of 2-word instruction
HOLD	Input used to stop NCU when HIGH; should go LOW only while RDY is HIGH
$\overline{\text{BR}}$	Active low output for four sync pulses to indicate a program branch
ISEL	A HIGH input requests 6-bit instruction mode for NCU; a LOW input places NCU in 4-bit instruction mode
R/$\overline{\text{W}}$	Active LOW output during OUT instruction to write data digits into external RAM or register
I6, JC	Most significant instruction bit input if ISEL = 1; jump condition for TJC instruction when ISEL = 0
I5, $\overline{\text{ADR}}$	Second most significant instruction bit input if ISEL = 1; AIN data ready for AIN instruction when ISEL = 0
I1–I4, D1–D4	Four least significant bits of 6-bit instruction if ISEL = 1 or all 4 bits of regular instruction when ISEL = 0
DA1–DA4	Digit address of input or output digits using AIN, IN, or OUT instructions
$\overline{\text{DAS}}$	Digit address strobe; LOW if digit address is changing; new address is valid on second positive going edge
DO1–DO4	Output digits; BCD output digit for each OUT instruction
F1, F2	User-controlled output flags

TABLE 18.3 Arithmetic Instructions for the 57109 NCU

Instruction	Octal code	Description
+	71	$Y + X \rightarrow X$
−	72	$Y - X \rightarrow X$
X	73	$X \cdot Y \rightarrow X$
/	74	$Y \div X \rightarrow X$
YX	70	$Y^x \rightarrow X$
INV+	40,71	$M + X \rightarrow M$
INV−	40,72	$M - X \rightarrow M$
INVx	40,73	$M \cdot X \rightarrow M$
INV/	40,74	$M \div X \rightarrow M$
1/X	67	$1 \div X \rightarrow X$
SQRT	64	$\sqrt{X} \rightarrow X$
SQ	63	$X^2 \rightarrow X$
10X	62	$10^x \rightarrow X$
EX	61	$e^x \rightarrow X$
LN	65	$\ln X \rightarrow X$
LOG	66	$\log X \rightarrow X$
SIN	44	$\text{Sin } X \rightarrow X$
COS	45	$\text{Cos } X \rightarrow X$
TAN	46	$\text{Tan } X \rightarrow X$
INV SIN	40,44	$\text{Sin}^{-1} X \rightarrow X$
INV COS	40,45	$\text{Cos}^{-1} X \rightarrow X$
INV TAN	40,46	$\text{Tan}^{-1} X \rightarrow X$
DTR	55	Convert X from degrees to radians
RTD	54	Convert X from radians to degrees
Clear	57	Clear all internal registers and memory

TABLE 18.4 Data-Movement Instructions of the 57109 NCU

Instruction	Octal code	Description
ROLL	43	Roll stack; $X \to T$, $T \to Z$, $Z \to Y$, $Y \to X$
POP	56	Pop stack; $Y \to X$, $Z \to T$, $T \to Z$, $0 \to T$
XEY	60	Exchange X and Y
XEM	33	Exchange X and M
MS	34	Store X in M
MR	35	Recall M into X
LSH	36	Shift X mantissa digits left
RSH	37	Shift X mantissa digits right

ISEL, DA1 to DA4, F2, and $\overline{\text{DAS}}$. The $\overline{\text{ADR}}$ input must also be used in addition to the 4-bit input digit. Table 18.2 describes the function of all these input-output signal lines.

The NCU has three classes of instructions which operate on input data. The main class is the arithmetic instructions shown in Table 18.3. These instructions utilize the X, Y, Z, and T registers, which are arranged in a stack configuration. This allows calculations in Polish notation, which avoids the need for parentheses. The next class of instructions, shown in Table 18.4, moves data between the X, Y, Z, T, and M registers. As indicated in Table 18.5, the final class of instructions performs jump and branch operations. This NCU is able to execute most instructions contained in general-purpose 8-bit microprocessors. Since the instruction bus (I1 to I6) is only 6 bits wide, a maximum of 64 instructions is possible. The 8-bit word of a standard microprocessor can accommodate 256 instructions although most microprocessors do not use all this capability. This NCU utilizes all 64 of its possible instruction codes.

TABLE 18.5 Branch Instructions of the 57109 Number Cruncher

Instruction	Octal code	Description
JMP	25	Unconditional branch to address specified by second instruction word
TJC	20	Branch to address specified in second word if $JC = 1$; otherwise skip second word
TERR	24	Branch to address specified in second word if error flag = 1; otherwise skip second word
TX = 0	21	Branch to address specified in second word if $X = 0$; otherwise skip second word
TXF	23	Branch to address specified in second word if $|X| < 1$; otherwise skip second word
TXLTO	22	Branch to address specified in second word if $X < 0$; otherwise skip second word
IBNZ	31	Increment memory (M); if $M \neq 0$, branch to address specified in second word; skip second word if $M = 0$
DBNZ	32	Decrement memory (M); if $M \neq 0$, branch to address specified in second word; if $M = 0$, skip second word

18.3   MANCHESTER ENCODER-DECODER

ALTERNATE NAMES   MIL-STD-1553A interface device, Manchester II serial interface.

PRINCIPLES OF OPERATION   The Manchester serial data format is useful for interfaces requiring no transmission of a dc component. The format also facilitates easy extraction of a clock from the data waveform. These features allow transformer coupling and excellent isolation between various systems. Note in Fig. 18.5 that a Manchester code has a transition in the middle of each bit period. To read the code one must observe the level during the beginning of each bit period. A HIGH level during this first half bit period implies a 1 transmitted.

To the casual observer systems using the Manchester code appear to have a steady stream of binary data. However, synchronization must be implemented at the beginning of each command or data word. If the line is first HIGH for 1½ bit periods then LOW for 1½ bit periods (according to MIL-STD-1553), the following 17 bits represent a command word (16 bits of command and 1 bit for parity). If the line is first LOW for 1½ bit periods then HIGH for 1½ bit periods, the following 17 bits represent a data word (16 data bits plus 1 bit for parity).

The Harris HD-15530 CMOS Manchester encoder-decoder is a typical device able to implement the requirements of MIL-STD-1553. Block diagrams of the encoder and decoder sections of this device are shown in Fig. 18.6. The encoder requires a send clock with a frequency twice the desired data rate. This clock chops each bit in the middle of each bit period, as shown in Fig. 18.5. If the encoder is to transmit at the

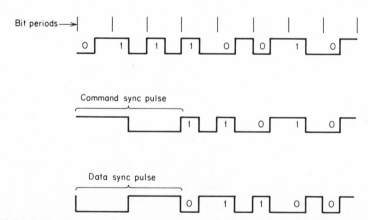

Fig. 18.5   A Manchester code has a transition in the middle of each bit period. A 1 is HIGH at the beginning of the period, and a 0 is LOW at the beginning of the period. A command sync pulse is HIGH for 1½ bit periods, then LOW for 1½ bit periods. A data sync pulse is just the opposite.

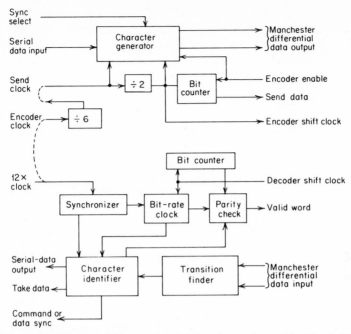

Fig. 18.6 Block diagrams of the encoder and decoder sections of the Harris HD-15530 Manchester interface device. *(Harris Semiconductor.)*

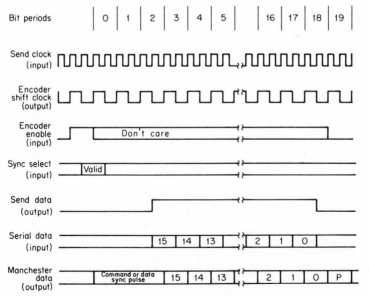

Fig. 18.7 Timing diagram of the Harris HD-15530 Manchester encoder inputs and outputs.

same bit rate as the decoder, the 12× decoder clock can also be applied at the auxiliary ÷6 input and no separate 2× send clock is needed.

As indicated in Fig. 18.7, the encoder's cycle begins when encoder enable is HIGH during the falling edge of encoder shift clock (ESC). The encoder cycle lasts for 1 word length, that is, 20 ESC periods. At the next positive transition of ESC, a HIGH at the sync-select input

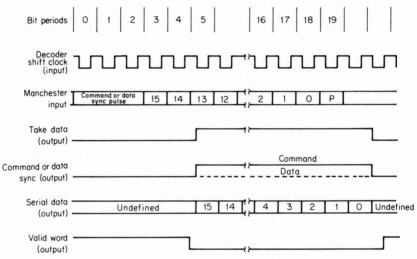

Fig. 18.8 Timing diagram of the Harris HD-15530 Manchester decoder inputs and outputs.

starts a command sync pulse (see Fig. 18.5). Likewise, a LOW sync-select input starts a data sync pulse. When the sync pulse is nearly finished, the send-data output rises for 16 ESC periods. During the 16 periods the data should be clocked into the serial-data input at each positive transition of ESC. The Manchester output lines begin transmitting the data ½ bit period behind the serial-data input. After 16 bits have been transmitted, the circuitry adds a parity bit.

The decoder timing in Fig. 18.8 requires a single decoder-clock input at 12 times the desired data rate. The decoder is free-running. It continuously samples the Manchester input lines for a valid sync pulse and 2 valid data bits. When these are detected, the take-data output goes HIGH, allowing external circuitry to begin receiving the serial-data output. After all 16 bits have been received, the parity is checked. If no error is detected, the valid-word output line goes LOW.

REFERENCES

1. Nelson, P.: The Number Crunching Processor, *Byte,* August 1978, p. 64.
2. Weissberger, A. J., and T. Toal: Tough Mathematical Tasks Are Child's Play for Number Cruncher, *Electronics,* Feb. 17, 1977, p. 102.
3. Parker, R. D., and J. H. Kroeger: Algorithm Details for the AM9511 Arithmetic Processing Unit, *Advanced Micro Devices, Inc. Publ.* AM-PUB072, 1978.
4. Kim, S. N.: Number Cruncher (MM57109) Interface to Microprocessor, *National Semiconductor, Inc. Appl. Note* AN-186, July 1977.

Chapter 19

Operational Amplifiers

19.1 OVERVIEW OF OPERATIONAL AMPLIFIERS

The operational amplifier (op amp) is a direct-coupled high-gain amplifier which uses feedback to control its performance characteristics. Internally, it consists of several series-connected transistor amplifiers. Externally, it is represented by the symbol shown in Fig. 19.1. The op amp is widely popular with analog-circuit designers because of its nearly ideal characteristics.

The op amp is capable of amplifying, controlling, or generating any sinusoidal or nonsinusoidal waveform over frequencies from dc to many megahertz. All classical computational functions are possible such as addition, subtraction, multiplication, division, integration, differentiation, etc. It is useful for innumerable applications in control systems, regulating systems, signal processing, instrumentation, and analog computation.

Functionally, as shown in Fig. 19.1, the op amp contains one output terminal which is controlled by two input terminals. If a positive voltage is applied to the positive (+) input, the op amp output will go positive. Likewise, a positive voltage on the negative (−) input will cause the output to go negative. An alternate symbol for the op amp (Fig. 19.2) shows that an op amp can be represented by a voltage source which is controlled by two "floating" terminals. The op amp has high voltage gain for differential signals effective between the two inputs. The op amp also has very low gain for signals applied to both inputs simultaneously (common-mode input signals).

The two inputs are labeled positive and negative or noninverting and inverting. The positive input is in phase with the output, and the nega-

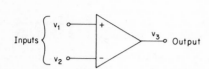

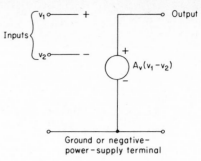

Fig. 19.1 The basic electrical symbol for the op amp.

Fig. 19.2 An alternate symbol for the op amp.

tive input is 180° out of phase with the output. If two resistors are connected to the op amp as shown in Fig. 19.3, we have the basic noninverting amplifier circuit. The voltage gain from v_1 to v_2 is determined only by the resistors R_1 and R_2:

$$A_{vc} = \frac{v_2}{v_1} = 1 + \frac{R_2}{R_1}$$

If R_2 is replaced by a short circuit and R_1 is removed, we have a popular circuit known as the *voltage follower.* It has a gain of exactly 1. The

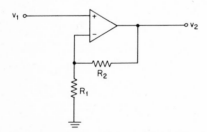

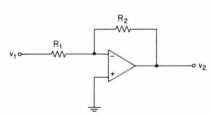

Fig. 19.3 The basic noninverting amplifier.

Fig. 19.4 The basic inverting-amplifier circuit.

basic inverting amplifier is shown in Fig. 19.4. Again, its gain is determined only by R_1 and R_2:

$$A_{vc} = \frac{v_2}{v_1} = \frac{-R_2}{R_1}$$

Both inputs to the op amp can be used simultaneously for differential amplifier circuits, as shown in Fig. 19.5. The output voltage is proportional to the difference in the voltages applied to the two inputs.

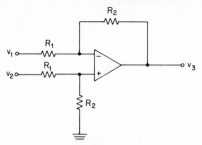

Fig. 19.5 Basic differential-amplifier circuit.

The constant of proportionality depends only on the size of the resistors R_1 and R_2:

$$\frac{v_3}{v_2 - v_1} = A_{vd} = \frac{R_2}{R_1}$$

The op amp is quite versatile in simulating mathematical functions.

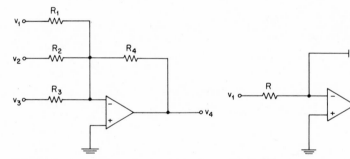

Fig. 19.6 An op-amp circuit which simulates the mathematical function of summation.

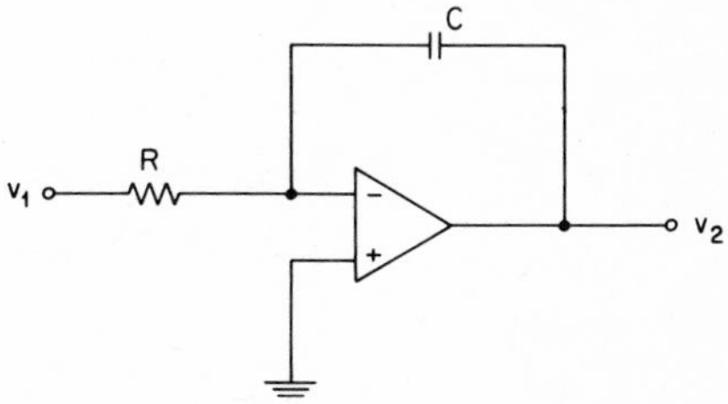

Fig. 19.7 An op-amp circuit which performs integration.

Figure 19.6 shows a circuit which produces the sum of the inputs. The input-output relation for this circuit is

$$v_4 = -R_4 \left(\frac{v_1}{R_1} + \frac{v_2}{R_2} + \frac{v_3}{R_3} \right)$$

Figure 19.7 produces an output voltage which is proportional to the integral of the input voltage (this is an integrator):

$$v_2 = -\frac{1}{RC} \int v_1 \ dt$$

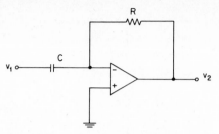

Fig. 19.8 The mathematical function of differentiation performed with an op-amp circuit.

Likewise, Fig. 19.8 produces an output voltage which is proportional to the derivative of the input voltage (this is a differentiator):

$$v_2 = - RC \frac{dv_1}{dt}$$

Op amps can be used to convert voltage into voltage, as in Figs. 19.3 and 19.4; voltage into current, as in Fig. 19.9; current into voltage, as in Fig. 19.10; or current into current, as in Fig. 19.11.

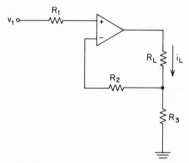

Fig. 19.9 A circuit which converts voltage into current according to $i_L = v_1/R_3$.

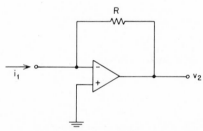

Fig. 19.10 An op amp used to implement a current-to-voltage converter; that is, $v_2 = -i_1 R$.

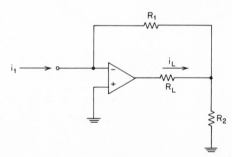

Fig. 19.11 An op-amp circuit which converts current into current according to $i_L = i_1(1 + R_1/R_2)$.

PARAMETERS OF THE IDEAL OP AMP   An op-amp circuit will follow its ideal design equations most satisfactorily if the op amp has the following ideal properties (see Sec. 19.4 for a discussion of each parameter):

1. Differential voltage gain = ∞
2. Common-mode voltage gain = 0
3. Bandwidth = ∞
4. Input impedance = ∞
5. Output impedance = 0
6. Output voltage = 0 when input voltage = 0
7. Parameter drift with temperature = 0
8. Equivalent input noise = 0

None of these ideal parameters is achieved by any op amp, nor will they ever be achieved, but manufacturers of op amps are continually improving their products, and some of these parameters are now so close to the ideal that the difference is hardly discernible.

When selecting an op amp for a specific application the designer usually will be looking for optimum performance of two or three parameters. After finding an op amp with the best possible values for these parameters, one often discovers that many of the other parameters are much less than optimum. This is the trade-off that every designer must go through during the design phase of a circuit. It is not possible to find an op amp which has state-of-the-art performance of every parameter. For example, low-noise op amps are not usually wide-bandwidth devices. Likewise, wide-bandwidth op amps may not have a very high input impedance.

Many specialized op amps have been developed with highly superior input characteristics. The 108 superbeta op amp is probably the best known device of this type. Other op amps with MOS inputs rival the performance of the superbeta devices. Monolithic chopper-stabilized op amps are also commercially available.

19.2 APPLICATIONS OF OP AMPS

Op amps can be used in nearly every area of linear and nonlinear electronics and also in a few digital applications.

LINEAR AMPLIFIERS   The primary use of op amps is for linear amplifiers where highly stable gain is required. The high gain of the op amp combined with heavy feedback results in amplification which is almost independent of temperature, time, and op-amp gain changes. For example, assume that an op amp has a gain of 10^5 and feedback is used to reduce the circuit gain to 10. The ratio of op-amp gain to circuit gain is $10^5/10 = 10^4$. Factors which affect the op-amp gain will have 10^4 times less effect on the circuit gain. If the op-amp gain changes from

10^5 to 2×10^5 (a 100 percent change), the circuit gain will change (100 percent)$/10^4 = 0.01$ percent. This is the major benefit of high-gain op amps. The circuit can also be designed so that gain is traded for gain stability. If higher gains are desired, one automatically gets less stability.

In addition to stability, the heavy feedback mentioned above has several other benefits. Nonlinearity is reduced, bandwidth is increased, input and output impedances are changed, and the op amp can be replaced without affecting circuit gain.

Linear amplifiers using op amps can be constructed with either positive or negative gain. The magnitude of gain can range from less than 1 to several million, the upper limit depending on the particular op amp used. As mentioned above, however, the very high gain circuit will have poor gain stability. Its bandwidth will also be very narrow.

NONLINEAR AMPLIFIERS See also Chap. 22. Many types of nonlinear amplifiers are designed using op amps, one of the most common being an op-amp circuit which precisely amplifies signals of one polarity but not the other. This is known as a *precision rectifier*. Rectification in either direction is possible. Adding a capacitor across the feedback elements converts the circuit into a filtered precision rectifier. Slight rearrangements of the diode and capacitor change the circuit into a peak detector, peak-to-peak detector, or average-value detector. Another slight modification converts this circuit into an absolute-value amplifier (full-wave rectifier).

COMPARATORS Most op amps can be utilized as a comparator. Because of their high gain, the output terminal changes from plus saturation to minus saturation, or vice versa, with a change of 1 mV or less across the input terminals. By definition, this is a zero-cross detector: the output changes polarity whenever the input passes through zero voltage. With a small bias on one input or the other, the circuit becomes a level detector. In this case the output changes state only when the signal input passes through the value of the bias on the other input.

Comparators have other more complex capabilities, e.g., double-ended level detectors, level detectors with prescribed hysteresis, window detectors, and pulse-height analyzers. Most types of A/D converters require a comparator. The comparator is such a widely required device that many op amps have been developed with high-speed switching characteristics, open-collector output, and the open-loop stability required for this specialized application. These op amps are marketed as comparators, not as specialized op amps.

FILTERS See also Chap. 20. Filter design has been revolutionized by the op amp. One of the disadvantages with conventional filter design was its reliance on inductors. A simple op-amp circuit is able to act as a very stable, highly linear inductor. Modern filters, accordingly, use resistors, capacitors, and op amps as long as the application is within the frequency limitations of the op amps. These simulated inductors avoid the disadvantages of real inductors: nonlinearity, hysteresis, core loss, radiation, unwanted coupling, large size, and difficult fabrication.

All the popular filter types are realizable with op-amp filters (usually called *active filters*). A few of these types include:

Low-pass filters
High-pass filters
Bandpass filters
Bandstop filters

The ripple, phase, and rolloff of these filters can be tightly controlled with Chebyshef, Butterworth, or Bessel characteristics. Any combination of these classical characteristics is also possible using active filters.

Several other important advantages are offered by active filters. First, their output impedance (for most configurations) in less than 1 Ω, and their input impedance is usually many kilohms. This relieves the designer of the task of impedance matching. Second, active filters can be designed to provide gain or supply large amounts of power or both.

Passive RLC filters require impedance matching at both the input and output and have less than unity gain at all frequencies. Although they can supply large amounts of power by appropriate choice of components, they cannot provide power gain as active filters can.

LOGARITHMIC APPLICATIONS See also Chap. 22. Using nonlinear elements (diodes or transistors) in the feedback circuit makes a host of logarithmic circuits possible. The basic circuit provides an output voltage proportional to the logarithm of the input voltage. Either dc or ac signals can be converted in this manner. A simple part rearrangement turns the circuit into an antilog amplifier.

Log and antilog circuits are sometimes known as *compressors* and *expanders,* respectively, when utilized for audio and video signals. By use of appropriate combinations of log and antilog circuits the following types of functions are possible:

Multiplier
Divider
Squaring circuit
$Y = X^n$ circuit
Square-root circuit
Square root of the sum of squares

All these functions can be performed on either ac or dc signals.

MULTIVIBRATORS See also Chaps. 4 and 9. All three basic types of multivibrators are possible with the op amp: the bistable (flip-flop), monostable (one-shot or single-shot), and the astable (rectangular-waveform generator). It is true that these are really digital functions and are easily implemented with digital microcircuits, but more flexibility is offered with the op-amp approach. For example, most digital microcircuits operate between specific voltages such as +5V and ground or −10V and ground. The maximum power level of these devices is also restricted. A multivibrator built with an op amp can operate between a wide range of minimum voltages (−1 down to −50 V is available) and a wide range of maximum voltages (+1 up to +50 V is available). Thus, a multivibrator with a 100-V (±50-V) pulse output or a ±1-V output could be designed. The power level could be increased any desired amount by inserting buffer transistors after the op amp but inside the feedback loop.

OSCILLATOR-GENERATORS See also Chaps. 9 and 23. Waveforms of many shapes, sizes, and frequencies are realizable using op amps. Sine-wave oscillators are implemented by using phase-shift feedback, inductor-capacitor feedback, or twin-tee feedback. The oscillator frequency and/or amplitude can be voltage-controlled. Negative feedback can be simultaneously incorporated to provide a highly stable output amplitude.

Waveform generators have been designed which provide all the commonly required wave shapes. Rectangular shapes with either fixed or independently adjustable width and period adjustments are possible. Sawtooth generators with either fixed or independently adjustable rise and fall times are also common.

REGULATORS Tight control of some parameter such as voltage, current, temperature, etc., is easily performed with op-amp circuits. The op amp is such a useful device for voltage regulation that a whole class of specialized op amps has been developed for this application. These devices, called *monolithic voltage regulators,* are discussed in Chap. 21. Monolithic voltage regulators use the same theory and electronic parts as regulators using op amps and discrete parts, but the former contain most of the other electronic parts on the same monolithic chip as the op amp.

Voltage regulators of the following types can be designed around the ordinary op amp:

    Series-pass regulation
    Shunt regulation
    Positive output
    Negative output
    Switching
    Foldback

Current-limited
Floating
High-voltage
Precision

Current regulators can be designed to handle many specialized applications. They are often seen supplying current to floating loads, grounded loads, or even active complex loads.

SAMPLING CIRCUITS   See also Chaps. 24 and 25.   In this age of computerized control of analog processes, sampling circuits are indispensable.   Typical applications using op amps are precision analog gates, sample-and-hold circuits, and A/D converters.   The sampling portion of these circuits requires one or two FETs in conjunction with an op amp.   A/D converters often require two or three op amps.

19.3   BASIC OP-AMP CIRCUITS

Two simple op-amp circuits with wide use, the basic inverting and noninverting op-amp circuits, will be discussed in this section.   A large number of the more complex circuits utilizing op amps are merely extensions of these two basic circuits.

THE INVERTING OP-AMP CIRCUIT   Figure 19.12 shows the basic inverting op-amp circuit.   The voltage gain at dc and low frequencies, called A_{vco}, is equal to

$$A_{vco} = \frac{v_2}{v_1} = -\frac{R_2}{R_1} \tag{19.1}$$

The circuit voltage gain as a function of frequency is called A_{vc}.   Since R_1 and R_2 represent resistances, the amplifier gain is somewhat independent of frequency, as shown in Fig. 19.13.   At higher frequencies, where the op-amp gain has fallen off to the point where it equals A_{vco}, the circuit gain falls off at the same rate.

The circuit of Fig. 19.12 has an input impedance of Z_1 (or R_1). The output impedance is very small, typically less than 1 Ω.

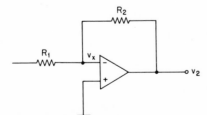

Fig. 19.12.   The basic inverting op-amp circuit.

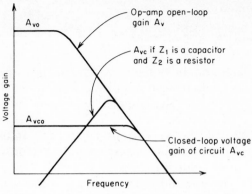

Fig. 19.13. The open-loop voltage gain of a typical op amp plotted as a function of frequency. When R_1 and R_2 in Fig. 19.12 are installed, the voltage gain becomes A_{vc}.

The resistors R_1 and R_2 can be replaced with any arbitrary impedances. In this case the closed-loop gain is

$$A_{vc} = \frac{v_2}{v_1} = -\frac{Z_2}{Z_1} \tag{19.2}$$

Simple filters can be constructed by using appropriate reactive components for Z_1 and Z_2. For instance, if Z_1 is a capacitor $(Z_1 = 1/j\,2\pi fC)$ and Z_2 is a resistor $(Z_2 = R)$, then

$$A_{vc} = -\frac{Z_2}{Z_1} = \frac{-R}{1/j\,2\pi fC} = -j\,2\pi fRC$$

The circuit gain therefore increases with frequency until the op-amp gain is reached. Then the circuit gain will fall with frequency at the same rate as the op amp.

THE NONINVERTING OP-AMP CIRCUIT   If the input voltage v_1 is applied to the positive op-amp input, as shown in Fig. 19.14, we have a noninverting amplifier. Negative feedback is still required, however, to stabilize

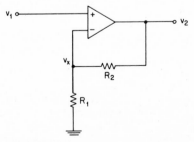

Fig. 19.14 The basic noninverting op-amp circuit.

the circuit and to set the gain. For this circuit the closed-loop voltage gain at dc and low frequencies is

$$A_{vco} = \frac{v_2}{v_1} = 1 + \frac{R_2}{R_1} \tag{19.3}$$

Note that no minus sign is used. If impedances are used instead of resistances, the closed-loop gain as a function of frequency is

$$A_{vc} = \frac{v_2}{v_1} = 1 + \frac{Z_2}{Z_1} \tag{19.4}$$

Z_2 and Z_1 may be either linear or nonlinear.

The input impedance of a noninverting op-amp circuit is very high, approximately equal to the op-amp input impedance times the ratio A_v/A_{vc}. The output impedance is very low, usually below 1 Ω.

TWO RULES WHICH SIMPLIFY OP-AMP-CIRCUIT DESIGN By using the ideal properties of op amps listed in Sec. 19.1 we arrive at two basic rules [1] which greatly simplify op-amp-circuit design:

1. Op-amp input terminals draw no current.
2. Voltage across input terminals is zero.

These rules are adequate for most design work. They are also useful for an initial design in cases where op-amp parameter drifts must be considered later. We now present two examples using the two design rules.

Inverting amplifier The most fundamental version of an inverting op-amp circuit is shown in Fig. 19.12. Let us consider each of the two basic rules individually to see how easily they are applied.

Basic Rule 1. No current goes into positive or negative input terminals. This means that the current passing through R_1 must be identical with the current through R_2. Two equations can be developed using Ohm's law:

$$v_1 - v_x = iR_1 \tag{19.5}$$

$$v_x - v_2 = iR_2 \tag{19.6}$$

Basic Rule 2. The voltage across input terminals is zero. Since the positive input terminal is grounded (at zero voltage), the negative input terminal must also be at zero voltage. Thus, $v_x = 0$ and Eqs. (19.5) and (19.6) become

$$v_1 = iR_1 \quad \text{and} \quad -v_2 = iR_2$$

These two equations can be rearranged into

$$i = \frac{v_1}{R_1} \quad \text{and} \quad i = -\frac{v_2}{R_2}$$

Setting these two equations equal to each other gives

$$i = \frac{v_1}{R_1} = -\frac{v_2}{R_2}$$

or
$$\frac{v_2}{v_1} = -\frac{R_2}{R_1} \tag{19.7}$$

Equation (19.7) is the fundamental gain equation for inverting amplifiers. It should be committed to memory.

Noninverting amplifier   The fundamental noninverting amplifier circuit is shown in Fig. 19.14.   We can use the same approach as that in the first example.

Basic Rule 1.   No current goes into positive or negative input terminals; the same current i therefore flows in R_1 and R_2.   We can develop two equations using Ohm's law:

$$v_2 - v_x = iR_2 \tag{19.8}$$

$$v_x - 0 = iR_1 \tag{19.9}$$

Basic Rule 2.   Voltage across input terminals is zero.   This means that $v_1 = v_x$.   Incorporating this into Eqs. (19.8) and (19.9), we get

$$v_2 - v_1 = iR_2 \quad \text{and} \quad v_1 - 0 = iR_1$$

Solving for i in each case gives

$$i = \frac{v_2 - v_1}{R_2} = \frac{v_2}{R_2} - \frac{v_1}{R_2} \quad \text{and} \quad i = \frac{v_1}{R_1}$$

Setting these equal to each other, we get

$$i = \frac{v_2}{R_2} - \frac{v_1}{R_2} = \frac{v_1}{R_1}$$

Solving for v_2/v_1 gives the final result

$$\frac{v_2}{v_1} = 1 + \frac{R_2}{R_1} \tag{19.10}$$

This is the fundamental equation of noninverting amplifiers.   It should also be committed to memory.

19.4   OP-AMP PARAMETERS

Most design engineers must eventually consider the adverse effects of nonideal op-amp parameters on their circuits.   In this section we summarize the common op-amp parameters and in each case discuss a method for minimizing the circuit degradation caused by these parameters.

OPEN-LOOP VOLTAGE GAIN

Definitions

$$A_{vo} = \text{differential voltage gain at dc}$$

$$= \frac{v_3}{v_1 - v_2} \quad \text{see Fig. 19.1}$$

$$A_v = \text{differential voltage gain at all frequencies}$$

$$= \frac{V_3(s)}{V_1(s) - V_2(s)}$$

The relationship between A_v (open-loop gain) and A_{vc} (closed-loop gain) is (see Figs. 19.13 and 19.15)

$$A_{vc, \text{ inverting}} = \frac{-Z_2/Z_1}{1 + 1/A_v + Z_2/A_v Z_1} \tag{19.11}$$

and

$$A_{vc, \text{ noninverting}} = \frac{1 + Z_2/Z_1}{1 + 1/A_v + Z_2/A_v Z_1} \tag{19.12}$$

The numerators of both above equations are simply the ideal closed-loop gains. The degradation in the ideal closed-loop gains comes about from the terms in the denominator. Ideally, the term $A_v = \infty$. In this ideal case we have $1/A_v = 0$ and $Z_2/A_v Z_1 = 0$. The denominator then becomes $1 + 0 + 0$, and the effect of the open-loop gain vanishes.

If $A_v \neq \infty$, the effect of A_v on closed-loop gain A_{vc} becomes larger as A_v becomes smaller. We first examine the significance of this statement for the dc case (using A_{vo} and A_{vco}). Later we will examine the ac case (using A_v and A_{vc}). As a specific example, assume $Z_2/Z_1 = 100$ and $A_{vo} = 1000$. Equation (19.11) becomes

$$A_{vco, \text{ inverting}} = \frac{-100}{1 + 1/1000 + 100/1000}$$

$$= \frac{-100}{1.101} = -90.83$$

This indicates a 9.2 percent reduction of dc gain from the 100 desired. It also means (and this is probably more important) that a 100 percent change in A_{vo} will result in a change in A_{vco} of approximately 9 percent. The exact change is determined by using Eq. (19.11) twice, once for each value of A_{vo}.

If A_{vo} is increased to 10,000, the A_{vco} is reduced only to

$$A_{vco} = \frac{-100}{1 + 1/10000 + 100/10000} = -99.00$$

Thus, the gain error is only 1 percent if the op amp has a gain of 10,000. By similar calculations we find that if the required A_{vco} is only 10 and A_{vo} is still 10,000, the error is reduced to 0.1 percent. We conclude that the gain error depends on the ratio of open- and closed-loop gains. Table 19.1 summarizes this conclusion.

TABLE 19.1   Gain Error at DC Caused by Finite Open-Loop DC Gain

$\dfrac{A_{vo}}{A_{vco}}$	Gain error (dc), %
1	−50
10	−9
10^2	−1
10^3	−0.1
10^4	-10^{-2}
10^5	-10^{-3}
10^6	-10^{-4}

We can also make a table of gain errors for the ac case. As shown in Fig. 19.15, the op-amp open-loop gain has a $-90°$ phase shift over

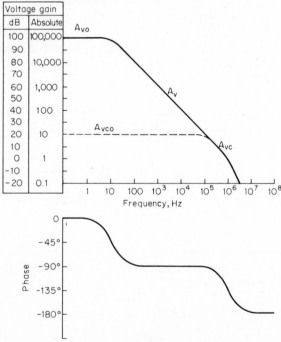

Fig. 19.15 The open-loop gain and phase shift of a typical op amp. The plot of a typical ×10 amplifier is also shown.

much of the usable range of frequencies. This causes a phase difference of 90° between A_v and A_{vc} for these frequencies. As a result, the degradation of A_{vc} by A_v is much less than shown for the dc case. Between the first-pole frequency (10 Hz) and the second-pole frequency (10^6 Hz) the true circuit gain is

$$A_{vc, \text{ inverting}} = \frac{-A_v R_2/R_1}{[A_v^2 + (1 + R_2/R_1)^2]^{1/2}} \tag{19.13}$$

and
$$A_{vc, \text{ noninverting}} = \frac{+A_v R_2/(R_1 + R_2)}{\{A_v^2 + [1 + R_2/(R_1 + R_2)]^2\}^{1/2}} \tag{19.14}$$

These equations are the basis for Table 19.2, which applies only for the region between the first pole of A_v and the first pole of A_{vc}, that is, from 10 to 10^5 Hz in Fig. 19.15.

TABLE 19.2   AC Gain Error
Caused by Finite Open-Loop
AC Gain

$\dfrac{A_v}{A_{vc}}$	Gain error (ac), %
1	−33
10	−0.6
10^2	−0.006
10^3	$−5 \times 10^{-5}$
10^4	$−5 \times 10^{-7}$
10^5	$−5 \times 10^{-9}$
10^6	$−5 \times 10^{-11}$

As an example, suppose we wish to find the maximum frequency at which we can expect 1 percent accuracy for a ×10 amplifier using the 741 op amp. We first refer to the data sheet and determine the minimum open-loop gains and minimum unity-gain bandwidth at +25°C. As shown in Fig. 19.16, we use the above data to make a plot of A_v as a function of frequency. The minimum value of A_v at dc, that is, A_{vo}, is given as 50,000. We draw a horizontal line at that value. The unity-gain crossover frequency (called the bandwidth in this data sheet) is 0.44 MHz. We then draw a line having a slope of −20 dB/decade such that it passes through $A_v = 1$ at 0.44 MHz. This line is extended up and to the left until it intersects the $A_v = 50,000$ line. We define the resulting plot as the worst-case minimum curve of A_v at +25°C.

Now, to determine the worst-case A_v over temperature we must assume (1) that the first pole at 8.5 Hz does not appreciably change frequency with temperature and (2) that the slope of A_v above the pole frequency remains at −20 dB/decade. We next examine the data sheet to find the minimum A_{vo} over temperature. The test conditions for this A_{vo}

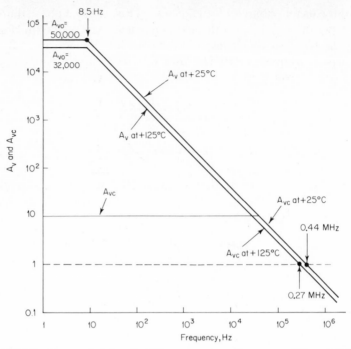

Fig. 19.16   Curves of A_v for the 741 op amp at temperatures of 25 and 125°C.

(and the A_{vo} at +25°C) must not be too different from the planned use of the amplifier. Assuming the power supplies to be used are ±15 V, the minimum A_{vo} over temperature is 32,000 at 125°C. We draw a horizontal line at 32,000 on Fig. 19.16. At 8.5 Hz we change the slope to −20 dB/decade and note that the $A_v = 1$ intersection occurs at 0.27 MHz.

The closed-loop-gain A_{vc} curve is drawn as a straight horizontal line at $A_{vc} = 10$ until it intersects the two A_v curves. We next determine the gain errors due to the lower A_v curve. According to Table 19.1, the error will be less than −0.1 percent at dc since $A_{vo}/A_{vco} = 3200$ at dc. At 28 Hz the phase shift of A_v will be almost −90° relative to A_{vc}. At this frequency $A_v/A_{vc} = 1000$, so that the gain error according to Table 19.2 is $−5 \times 10^{-5}$ percent. Likewise, at 280 Hz the gain error is −0.006 percent, at 2800 Hz the gain error is −0.6 percent, and at 28 kHz the gain error is −33 percent. We can reasonably assume this circuit will be better than 1 percent stable up to about 3000 Hz and over the temperature range −55 to +125°C.

Our suggested methods for reducing the errors caused by changes in A_v and A_{vo} are:

1. Choose an op amp with a high dc gain and/or wide bandwidth.
2. Keep the temperature variations to a minimum.

3. Make sure the ratios A_v/A_{vc} and A_{vo}/A_{vco} are high over all frequencies at which precision gain is required.

BANDWIDTH

Definitions

f_u = frequency where op-amp voltage gain is unity
f_{cp} = frequency of first pole in closed-loop voltage gain
f_f = maximum frequency of full-power response

The bandwidth f_u can also be determined by measuring the op-amp rise time when the op amp is connected as a unity-gain noninverting amplifier. The bandwidth is then computed from

$$f_u = \frac{0.35}{t_r} \tag{19.15}$$

where t_r is the 10 to 90 percent rise time.

We are usually more interested in the closed-loop bandwidth than in the open-loop bandwidth. Throughout this chapter we will use f_{cp} as the frequency at which the closed-loop gain is down 3 dB. This is the dominant (first) pole frequency of the closed-loop circuit. Referring back to Fig. 19.16, we see that the closed-loop bandwidth is the frequency where the open- and closed-loop curves intersect. In that case the closed-loop bandwidth is 44 kHz at +25°C and 27 kHz at +125°C.

When the circuit must supply large output-voltage swings, the closed-loop bandwidth is much less than f_u. Depending on the size of the peak output voltage, the bandwidth may only be 0.1 f_u or $0.01f_u$. This high-level bandwidth is called f_f since it is the maximum frequency at which full-power output response can be expected. Many data sheets provide curves showing the maximum peak-to-peak output voltage as a function of frequency. Figure 19.17 is a curve of this type for the 101 op amp. The curve was obtained by noting the voltage at each frequency where about 5 percent distortion occurs. If more than 5 percent distortion is acceptable, slightly higher voltages may be allowed.

The strong relationship between full-power response and the compensation capacitor is immediately obvious. Also, it should be noted that the curves flatten off rather abruptly at the top because the data were taken using ±15-V power supplies. One would expect abrupt limiting as soon as the peak-to-peak amplitude approaches 30 V.

Bandwidth can be extended by:

1. Keeping output amplitude low so that the full-power response curves are not approached

2. Using the minimum compensation on the widest-bandwidth device

3. Allowing slightly more distortion in the output signal

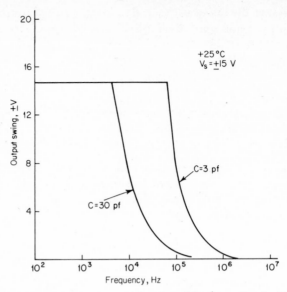

Fig. 19.17 Full-power response as a function of frequency for the 101 op amp.

SLEW RATE

Definition

$$S = \text{maximum rate at which op-amp output voltage can change}$$
$$= \frac{\Delta V}{\Delta t}$$

Maximum slew rate is sometimes called *slew-rate limiting*. This limiting action does not take place suddenly. It is observed to begin at one location on a sine wave as the frequency or amplitude is increased and then broaden out to include most of the sine wave. The exact nature of this complicated phenomenon depends on the type of op amp, the compensation used, and the load capacitance.

Slew-rate limiting is characterized by a definite flattening on one portion of the sine wave. This flat portion is due to a constant-current source charging a capacitance. Precision ramp generators are built with the same principle. If the constant current is I, the slope of this slew-rate limit is

$$\frac{\Delta v}{\Delta t} = \frac{I}{C} \tag{19.16}$$

The capacitances and current generators causing the slew-rate limiting may be in several locations.

1. Often a compensation capacitor C_c is placed between the collec-

tors of the input differential stage. The slew-rate limit at this point is then

$$\frac{\Delta v}{\Delta t} = \frac{2\,I_c}{C_c}$$

where I_c is the quiescent collector current of either input transistor.

2. If a large load capacitor C_L is connected to the op amp, the maximum slew rate is

$$\frac{\Delta v}{\Delta t} = \frac{I_o}{C_L}$$

where I_o is the maximum available op-amp output current. The ultimate slew-rate limitation of the circuit will be the smallest I/C ratio in the op amp.

If slew-rate limiting is a problem, the designer may consider the following suggestions:

1. Slew rate is higher for high-gain circuits. Perhaps the input signal can be reduced and the circuit gain increased.

2. The compensation capacitor size may be too large. Some op amps have several methods for compensation from which to choose. Use of input lead-lag compensation is one possible suggestion for increasing slew rate (a series RC circuit placed across the op-amp input terminals).

3. If a large load capacitance is the cause of slew-rate limiting, perhaps an emitter-follower buffer will help isolate the op amp from C_L.

Maximum slew rate S is related to full power response by

$$S = 2\pi f_f V_{pp} \tag{19.17}$$

where f_f and V_{pp} are the coordinates of a point on one of the curves in Fig. 19.17. Suppose we wish to find the maximum slew rate of the 101 op amp if $V_{pp} = 10$ V and $C_f = 30$ pF. From Fig. 19.17 we note that $f_f = 8$ kHz at $V_{pp} = 10$ V and $C_f = 30$ pF. Thus,

$$S = 2\pi f_f V_{pp} = (6.28)(8 \times 10^3 \text{ Hz})(10 \text{ V})$$
$$= 5 \times 10^5 \text{ V/s} = 0.5 \text{ V/}\mu\text{s}$$

INPUT OFFSET VOLTAGE

Definition

$V_{io} =$ voltage required across input terminals to drive op-amp output voltage to zero

All op amps have a slight mismatch of the emitter-base forward-bias voltages of the two input transistors. This results in a voltage offset at the op-amp output. The input offset voltage V_{io} (between the bases of the two input transistors) is related to the output offset V_o by

$$V_o = \pm V_{io}\left(1 + \frac{R_f}{R_1}\right) \tag{19.18}$$

This equation holds true for both the inverting and noninverting configurations. If, for example, the circuit has a voltage gain of -1000 (an inverting amplifier), we divide the output offset voltage by 1001 to obtain the input offset voltage. Op-amp data sheets always specify the offset voltage at the input terminals since the magnitude of output offset depends on the circuit gain. The input offset is independent of the circuit. By definition, the input offset voltage is that voltage required across the input terminals which nulls the output. The input offset voltage V_{io} is typically in the range of a fraction of a millivolt to several millivolts. In high-gain circuits the output offset voltage may therefore be several volts. This offset will vary with temperature and could cause problems in a dc-coupled system. It is often neglected in ac circuits unless the output offset voltage and peak ac voltage add up to a voltage approaching either power-supply voltage. Clipping of the output signal would then begin to occur.

Many op amps have offset-adjustment terminals. A potentiometer is placed between these two terminals with the wiper connected to the plus or minus power-supply terminal. Adjustments of ± 15 mV equivalent input offset voltage is possible with this method. This adjustment merely places the output offset at the desired value. The temperature effect on offset voltage is still present. The temperature coefficient of input offset voltage is typically 5 to 10 $\mu V/°C$ for bipolar monolithic op amps. For chopper-stabilized op amps this coefficient may be only 0.1 to 1.0 $\mu V/°C$.

INPUT BIAS CURRENT

Definition

I_b = average value of input current required to forward-bias op-amp
input transistors

The input transistors in the first differential-amplifier stage of the op amp must be forward-biased. This requires a small current into each of their bases. The input bias current I_b is defined as one-half the sum of these two currents, in other words, the average of the two currents. This definition applies only if the output terminal is balanced or nulled to 0 V.

The typical input bias current for bipolar monolithic op amps ranges from 10 nA to several thousand nanoamperes. High-quality chopper-

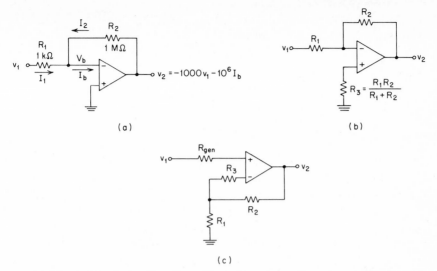

(a)   (b)

(c)

Fig. 19.18 (a) Circuit showing effect of input bias current (inverting amplifier); (b) the most popular method of reducing the error caused by input bias current; (c) solution to problem for noninverting current.

stabilized or parametric op amps may have input bias currents under 10 pA. The input bias current varies dramatically with temperature, and the op-amp circuit design must account for this error source.

As with input offset voltage, the input bias current is a dc parameter and may not affect the design of an ac amplifier. If the circuit must amplify ac and dc, the bias current must be considered. If the amplifier is for ac only, one must determine whether the resulting output offset plus the peak ac signal approaches the saturation region of the op amp. If this occurs, the peaks of the ac signal will be clipped.

By reference to Fig. 19.18a we will show how the input bias current is a potential error source. Suppose we require an inverting amplifier with a gain of -1000 and an input resistance of 1000 Ω. The input resistor R_1 must be 1000 Ω to satisfy the input-resistance requirement. R_2 must be 1000 times greater to satisfy the gain requirement:

$$A_{vco} = -\frac{R_2}{R_1} = -1000$$

Thus, R_2 must equal 1 MΩ. Assume the input bias current is 100 nA. This current will pass through R_1 and R_2. We then have $I_b = I_1 + I_2$

$$I_b = \frac{V_1 - V_b}{R_1} + \frac{V_2 - V_b}{R_2}$$

We assume $V_1 = 0$ so that the bias current can be isolated from the signal. The term V_b will be less than V_2 by a factor A_v (10,000 or more). We can therefore drop V_b and solve for V_2:

$$V_2 = I_b R_2 = 10^{-7} \times 10^6 = 0.1 \text{ V} \tag{19.19}$$

The above error caused by input bias current can be almost totally canceled if an extra resistor with a resistance

$$R_3 = \frac{R_1 R_2}{R_1 + R_2} \tag{19.20}$$

is added to the circuit as shown in Fig. 19.18b. This resistor is often made adjustable with a maximum value 2 or 3 times the computed R_3. The I_b into the op-amp positive input will develop a voltage across R_3 which will cancel the effects of I_b into the negative input. Perfect cancellation does not take place, however, because the two bias currents are not exactly equal. We will look into this problem in the next section.

INPUT OFFSET CURRENT

Definition

I_{io} = difference between two input bias currents of an op amp

The two input transistors of any op amp will always require slightly different bias currents. The difference between these two currents is defined as the input offset current I_{io}. As might be expected, I_{io} also varies with temperature. Since the offset current may flow into either op-amp input terminal, the resulting output voltage error is

$$V_o = \pm I_{io} R_2 \tag{19.21}$$

The bias-current-compensation scheme presented in the last section, i.e., using R_3, does nothing to cancel the effects of I_{io}. We can compensate for any given I_{io} by choosing an R_3 slightly larger or smaller than $R_1 R_2/(R_1 + R_2)$. However, at another temperature I_{io} is different, and consequently the op-amp output voltage will shift. Several circuits which cancel the effects of I_{io} changes with temperature are shown in Ref. 7.

INPUT RESISTANCE

Definitions

R_{id} = differential input resistance, i.e., between op-amp input terminals

R_{ic} = common-mode input resistance, i.e., between either op-amp input and negative power-supply terminal

These parameters lower the closed-loop gain slightly and also limit the high input impedance expected with the noninverting amplifier. The equations for closed-loop gain considering input resistance are

$$A_{vc,\text{ inverting}} = \frac{-R_2/R_1}{1 + 1/A_v + R_2/A_vR_1 + R_2/A_vR_{id} + R_2/A_vR_{ic}} \qquad (19.22)$$

and

$$A_{vc,\text{ noninverting}} = \frac{1 + R_2/R_1}{1 + 1/A_v + R_2/A_vR_1 + R_2/A_vR_{id} + R_2/A_vR_{ic}} \qquad (19.23)$$

Equations (19.22) and (19.23) are identical to Eq. (19.11) and (19.12) except that the factor $R_2/A_vR_{id} + R_2/A_vR_{ic}$ has been added to the denominators. If the three terms A_v, R_{id}, and R_{ic} are all very large, this entire factor is very small and does not affect gain. This is one case where one op-amp parameter may help another. That is, if for some reason R_{id} is low but A_v is very large, the net result is a large product A_vR_{id}. The effect of R_{id} on closed-loop gain would then be negligible.

Table 19.3 indicates the percentage effect on gain due to finite R_{id} and R_{ic}. A_v is also included since it is a dominant variable in Eqs. (19.22) and (19.23). For purposes of illustration, assume $R_{ic} \gg R_{id}$ (as is usually the case). We can therefore drop R_2/A_vR_{ic} for this example. We can now make a table of percentage error of A_{vc} as a function of A_v and R_{id}. If A_v is high and R_{id} is low, A_{vc} will have little degradation.

TABLE 19.3 The Effect of Input Resistance on Closed-Loop Gain*

A_{vo}	R_{id}	Gain error, %	A_{vo}	R_{id}	Gain error, %
1	10^4	-92.3	10^4	10^4	-0.12
1	10^6	-91.7	10^4	10^6	-0.11
1	10^8	-91.7	10^4	10^8	-0.11
10^2	10^4	-10.7	10^6	10^4	-0.0012
10^2	10^6	-9.92	10^6	10^6	-0.0011
10^2	10^8	-9.91	10^6	10^8	-0.0011

* This table assumes $R_f = 10$ kΩ, $R_1 = 1$ kΩ, and $A_{vco} = 10$.

If both A_v and R_{id} are high, A_{vc} will be essentially undisturbed. However, if both A_v and R_{id} are low, severe degradation of A_{vc} will occur. For simplicity, the table is computed only at dc, so A_{vo} is used in place of A_v. Table 19.3 provides the interesting result that gain error is much more sensitive to A_{vo} than it is to R_{id}. Changes in R_{id} are only 0.01 to 0.1 times as influential on gain error as A_{vo}.

INPUT CAPACITANCE

Definitions

C_{id} = capacitance between op-amp input terminals
C_{ic} = capacitance between either input terminal and negative-power-
supply terminal

These two parameters are seldom stated on am-amp data sheets. Their effect on the closed-loop gain of the inverting-amplifier configuration is negligible. Both C_{id} and C_{ic} have typical values of 1 to 2 pF and maximum values of 3 pF for monolithic op amps. The common-mode input capacitance C_{ic}, however, does have some deleterious effect on the noninverting amplifier at high frequencies. Since this type of amplifier is often driven by a high-impedance source, a capacitance to ground at the op-amp positive input will attenuate high frequencies. The only way around this problem is to use careful layout procedures and to choose an op amp with a low C_{ic}.

These capacitances also reduce feedback stability in applications requiring a large feedback resistance.

OUTPUT RESISTANCE

Definition

$$R_o = \text{op-amp open-loop output resistance}$$

The output resistance of an op amp affects the circuit output resistance, circuit gain, and feedback stability. The heavy feedback usually incorporated in an op-amp circuit makes the circuit output resistance R_{out} effectively very low. The relationship between the op-amp output resistance and the circuit output resistance is

$$R_{\text{out}} = \frac{R_o(R_1 + R_2)}{R_1 A_v} = \frac{R_o(A_{vc} + 1)}{A_v} \tag{19.24}$$

For a dc calculation of R_{out} we use A_{vo} in place of A_v.

R_o is commonly 100 Ω or less. Since A_{vo} is quite large in most op amps, typically 5×10^4 to 10^6, a circuit with a gain of 10 will have a dc output resistance of

$$R_{\text{out}} = \frac{100(10 + 1)}{5 \times 10^4} = 0.022 \ \Omega$$

At high frequencies the situation is not so good. R_o often goes up to several hundred ohms near unity open-loop gain A_v. Thus, A_v is small and R_o is large at the same time. At these frequencies R_{out} may be nearly as large as R_o.

As shown in Fig. 19.19, we can represent R_{out} as a resistance in series

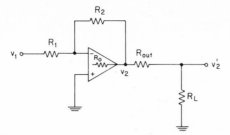

Fig. 19.19 Effective position of the circuit output resistance R_{out}.

with the load R_L. The actual circuit output voltage v_2' will be slightly lower than v_2 due to the voltage divider formed by R_{out} and R_L. Thus,

$$v_2' = \frac{R_L}{R_L + R_{out}} v_2 \qquad (19.25)$$

At low frequencies, where A_v is large and R_0 is small, R_{out} will be small. As an example, consider the 748 op amp. At frequencies up to 50 kHz R_0 is approximately 70 Ω. At 50 kHz the open-loop gain is 200. If $A_{vc} = 4$, the output resistance of the circuit at this frequency is

$$R_{out} = \frac{R_0(A_{vc} + 1)}{A_v} = \frac{70(4 + 1)}{200} = 1.4 \ \Omega$$

If $R_L = 1$ kΩ and $v_2 = 10$ V, the output voltage reduction due to R_{out} is [using Eq. (19.25)]

$$v_2 - v_2' = \left(1 - \frac{R_L}{R_L + R_{out}}\right) v_2$$

$$= \left(1 - \frac{1000}{1000 + 1.4}\right) 10 = 14 \text{ mV reduction}$$

At lower frequencies the voltage reduction will be even smaller. This error would probably go unnoticed. Now consider what happens at 1 MHz where at the same time R_0 has increased to 280 Ω:

$$R_{out, \ 100 \ kHz} = \frac{R_0(A_{vc} + 1)}{A_v} = \frac{280(4 + 1)}{7} = 200 \ \Omega$$

The output voltage reduction will be

$$v_2 - v_2' = \left(1 - \frac{R_L}{R_L + R_{out}}\right) v_2$$

$$= \left(1 - \frac{1000}{1000 + 200}\right) 10 = 1.67 \text{ V}$$

The error in this case is more than 16 percent.

How do we reduce these errors caused by R_o? The most obvious suggestions are: (1) do not require operation at frequencies where $A_v = 10$ or less, (2) keep R_L large, and (3) place an emitter-follower between the op amp and R_L. This last suggestion will make the load resistance R_L seen by the op amp very large. Thus, the voltage-divider action between R_{out} and R_L will be very small.

COMMON-MODE REJECTION RATIO

Definition

$$\text{CMRR} = \frac{A_{vo}}{A_{cmo}} \quad \text{at dc} \tag{19.26}$$

or
$$\text{CMRR} = \frac{A_v}{A_{cm}} \quad \text{at any frequency} \tag{19.27}$$

where A_{vo} = op-amp differential gain at dc

A_v = op-amp differential gain as a function of frequency

A_{cmo} = op-amp common-mode gain at dc

A_{cm} = op-amp common-mode gain as a function of frequency

Nearly all op amps have differential inputs. Many applications of op amps require both these differential inputs for proper operation. Some of these same applications require that any common voltage simultaneously applied to both inputs should not be amplified. This is not entirely possible in real op amps. This "common-mode" voltage always arrives at the output at some finite level. The common-mode rejection ratio (CMRR) is a measure of how much this common-mode signal is rejected relative to the desired differential-mode signal.

Figure 19.20 clarifies the definition of A_v and A_{cm} (or A_{vo} and A_{cmo}, which are merely the dc components of A_v and A_{cm}). Data sheets provide CMRR data in several forms. The most common form is merely one or two numbers stating the minimum and/or typical CMRR at dc. More useful data sheets provide curves showing minimum CMRR vs. frequency.

Figure 19.21a shows a typical CMRR curve for the 101A op amp. The data sheet also states that the minimum CMRR at dc is 80 dB over the military temperature range (-55 to $+ 125°C$). The curve of

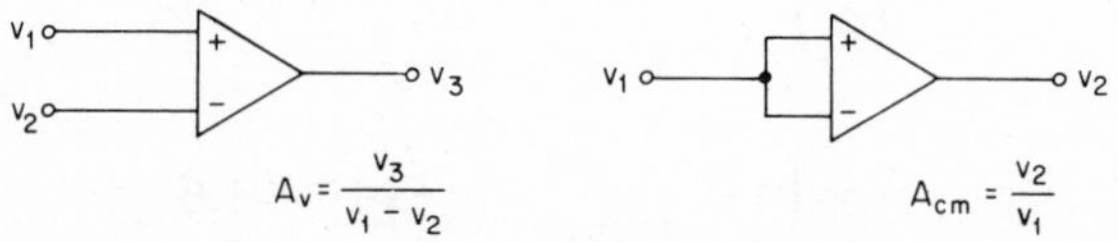

Fig. 19.20 Definitions of op-amp differential gain A_v (usually called the open-loop gain) and common-mode gain A_{cm}.

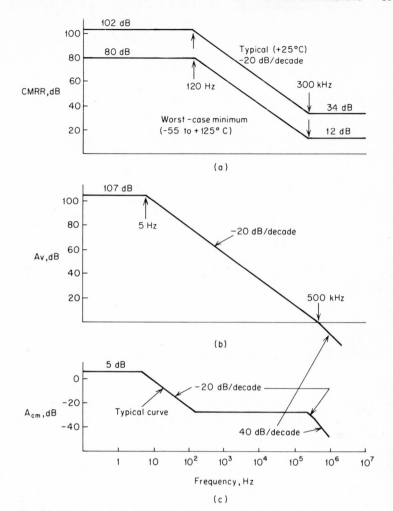

Fig. 19.21 Curves of *(a)* CMRR, *(b)* A_v, and *(c)* A_{cm} for the 101A op amp.

CMRR vs. frequency (Fig. 19.21*a*) must accordingly be lowered by 22 dB at all frequencies if worst-case performance calculations are to be made.

Assume that the designer knows the nature of the common-mode input voltage over frequency. A reasonable question to ask is: What will the output voltage be over frequency due to the common-mode input? To answer this question we need plots of CMRR and A_v both obtained under similar operating conditions. Figure 19.21*a* and *b* will be used. If both plots are in decibels vs. frequency, the calculations will be simplified. Since CMRR $= A_v/A_{cm}$, $A_{cm} = A_v/$CMRR. Therefore

$$20 \log A_{cm} = 20 \log A_v - 20 \log \text{CMRR}$$

The three terms are all expressed in decibels.   To obtain 20 log A_{cm} (in decibels) at a given frequency, we merely subtract 20 log CMRR (in decibels) from 20 log A_v (in decibels) at that frequency.   The resulting plot of 20 log A_{cm} is shown in Fig. 19.21c.   To obtain the plot of output voltage vs. frequency due to a common-mode voltage we must multiply A_{cm} by the input common-mode voltage.   Again, this is most easily done if both are in decibels.

The circuit surrounding the op amp also has a CMRR of its own. In Ref. 7 it is shown how to incorporate the op-amp CMRR into the circuit CMRR to obtain the total CMRR.   As it turns out, the total CMRR is always less than the circuit or op-amp CMRR.   The op-amp CMRR can be optimized in the following ways:

1. Choose an op amp with a large minimum CMRR at dc.

2. Choose an op amp with the largest possible CMRR values over the same frequency range to be used in circuit.

3. Measure each op amp for maximum CMRR before installation in circuit.

4. Make sure circuit CMRR is at least 20 times larger than op-amp CMRR.

POWER-SUPPLY REJECTION RATIO

Definition

$$\text{PSRR} = \Delta V_{io}/\Delta V_{CC}$$

where ΔV_{io} is a change in the effective input offset voltage and ΔV_{CC} is a change in the power-supply voltage.

The test for PSRR is usually performed at some given frequency (60 Hz, 1 kHz, etc.), but data sheets seldom state the frequency.   The resulting number is expressed in microvolts per volt or decibels.   When stated in decibels, the number is actually negative since the input offset change is much smaller than the power-supply change.   A negative decibel value means that the ratio is less than 1; that is, -20 dB = 0.1.

To determine the effect of PSRR on a given circuit we can use the same type of calculations used to determine the effects of input offset voltage.   As we recall, the op-amp output voltage V_2 is related to the input offset voltage V_{io} by

$$V_2 = \left(1 + \frac{R_2}{R_1}\right)V_{io} \tag{19.28}$$

This expression holds for both inverting and noninverting amplifiers. The definition of PSRR is also valid for the ac equation

$$\text{PSRR} = \frac{V_{io}}{V_s} \tag{19.29}$$

where V_{io} is an equivalent ac rms voltage across the op-amp input terminals and V_s is an ac rms voltage on both power-supply terminals (in phase). We can find the op-amp output ripple voltage by combining Eqs. (19.28) and (19.29). The result is

$$V_2 = \left(1 + \frac{R_2}{R_1}\right) V_s \times \text{PSRR}$$

Suppose the power supplies have a 0.1 V rms ripple and the op amp has a PSRR of 20 μV/V (-94 dB). If the op amp is connected as a $\times1000$ inverting amplifier, its output ripple will be

$$V_{2,\text{rms}} = (1 + 1000)(0.1 \text{ V rms})(20 \times 10^{-6} \text{ V/V})$$
$$= 0.02 \text{ V rms}$$

Whether or not this disturbs the circuit function depends on the size of the real output signal and its required signal-to-noise ratio.

The effects of op-amp PSRR can be minimized several ways:

1. Choose an op amp with a small PSRR.
2. Reduce the power-supply ripple with additional filtering.
3. Increase the input signal.
4. Decrease the circuit gain.

EQUIVALENT INPUT NOISE

Definitions   (See Fig. 19.22)

$V_n =$ equivalent input noise voltage
$I_n =$ equivalent input noise current

These parameters affect both the ac and dc characteristics of op-amp circuits. Op-amp spec sheets tabulate data on equivalent input noise from 0.01 Hz to over 1 MHz. Op-amp noise is specified by using an equivalent input-noise voltage and an equivalent input noise current. The actual noise is generated in a number of places in the first few stages in the op amp. To simplify noise calculations, all these noise sources are assumed to be lumped into a single equivalent input noise current and/or voltage, as shown in Fig. 19.22.

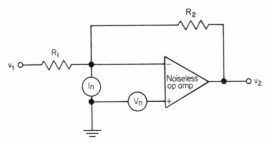

Fig. 19.22 Placement of equivalent input-noise voltage and equivalent input-noise current generators in front of a noiseless op amp.

At least four different types of units are used to specify input noise:

$V/(Hz)^{1/2}$ or $A/(Hz)^{1/2}$

V^2/Hz or A^2/Hz

Volts peak to peak or amperes peak to peak

Noise figure in decibels

This book uses $V/(Hz)^{1/2}$, which are rms numbers quite easily correlated with simple measurements. Before we get into methods of optimizing op-amp circuits for low noise, we first work out methods to make conversions from the other three types of noise specifications to the $V/(Hz)^{1/2}$ and $A/(Hz)^{1/2}$ terminology. These units allow the easiest computation of circuit parameters from which the designer can minimize noise. When we say $V/(Hz)^{1/2}$, we mean the rms voltage over a 1-Hz bandwidth. This number must also be specified at some center frequency. Some texts call this the *spot noise*. It is not usually measured with instruments having a 1-Hz bandwidth as this is difficult. Instead a more convenient bandwidth such as one-tenth or one-hundredth the center operating frequency is used. The number obtained in this manner is then divided by the square root of the bandwidth. This converts the number into $V/(Hz)^{1/2}$.

As an example, suppose we measure the spot noise of some device at 10 kHz using an rms voltmeter having a bandwidth of 100 Hz. The equivalent input noise voltage measures 300 nV. The spot noise is therefore

$$V_n = \frac{300 \text{ nV}}{(100 \text{ Hz})^{1/2}} = \frac{30 \text{ nV}}{(Hz)^{1/2}} \qquad \text{at 100 kHz}$$

The same type of calculation could be performed using data from a noise-current measurement.

Some data sheets provide noise data in units of V^2/Hz or A^2/Hz. To obtain $V/(Hz)^{1/2}$ and $A/(Hz)^{1/2}$ one need merely take the square root of these numbers. For example, the 741 op amp has a noise voltage of 4×10^{-16} V^2/Hz at 10 kHz. The spot noise at 10 kHz is therefore 2×10^{-8} $V/(Hz)^{1/2}$.

Very low frequency noise, such as that in the region from 0.01 to 1 Hz, is difficult to measure with an rms meter. One approach to this problem is to pass the signal through a low-pass filter having a 1-Hz upper cutoff frequency. The filter output is applied to an oscilloscope and the peak-to-peak excursions estimated. The rms voltage is then $0.707 \times$ one-half the measured peak-to-peak amplitude. The numbers obtained with these peak-to-peak measurements are only a rough estimate since the amplitude of low-frequency noise in most devices increases as the frequency goes down. As a result, if a peak-to-peak estimate is once made, it will always be exceeded by a larger signal if one waits long enough. The correct peak-to-peak reading is that amplitude which is exceeded only 10 to 15 percent of the time [3].

The fourth type of noise information often shown in data sheets is the *noise figure* expressed in decibels. The noise figure is a measure of additional noise contributed by the amplifier above the noise already in the input signal. If the input signal is only that due to source resistor noise, its equivalent rms voltage is

$$V_R = (4kTR_s)^{1/2} \text{ V/(Hz)}^{1/2}$$

where k = Boltzmann's constant = 1.374×10^{-23} J/K

T = temperature, K

R_s = source resistance, Ω

At room temperature the resistor noise is

$$V_R = 0.13(R_s)^{1/2} \qquad \text{nV/(Hz)}^{1/2}$$

If the amplifier contributes an equivalent input noise voltage of V_n and an equivalent input noise current of I_n, the noise figure is defined as

$$\text{NF} = 10 \log \frac{V_n^2 + I_n^2 R_s^2 + 4kTR_s}{4kTR_s} \qquad \text{dB} \qquad (19.30)$$

To find V_n and I_n at any given frequency we must use data-sheet curves, such as shown in Fig. 19.23, which relate NF to R_s and frequency. The equation must be solved for two NFs at the given frequency since both V_n and I_n are unknowns.

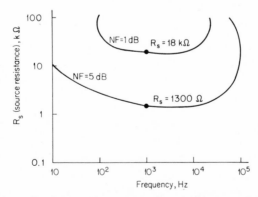

Fig. 19.23 Narrow-band spot-noise contours for a μA725. *(Fairchild Semiconductor.)*

An example of the above procedure is worthwhile at this point. Suppose we want to find V_n and I_n of the Fairchild μA725 at 1 kHz. Using Fig. 19.23, we note that NF = 1dB if f = 1 kHz and R_s = 18 kΩ. Likewise, NF = 5 dB if f = 1 kHz and R_s = 1300 Ω. These numbers are substituted into Eq. (19.30), giving

$$V_n^2 + I_n^2(18{,}000)^2 = 4kT(18{,}000)\left[\left(\text{antilog }\frac{1\text{ dB}}{10}\right) - 1\right] \quad (19.31)$$

$$V_n^2 + I_n^2(1300)^2 = 4kT(1300)\left[\left(\text{antilog }\frac{5\text{ dB}}{10}\right) - 1\right] \quad (19.32)$$

If the second equation is subtracted from the first, we get

$$I_n^2\left[(18{,}000)^2 - (1300)^2\right] = 4kT\left\{18{,}000\left[\left(\text{antilog }\frac{1}{10}\right) - 1\right]\right.$$
$$\left. - 1300\left[\left(\text{antilog }\frac{5}{10}\right) - 1\right]\right\}$$

Now we only need to substitute in k and T, then solve for I_n. The final result is

$$I_n = 0.307 \text{ pA/(Hz)}^{1/2} \quad \text{and} \quad V_n = 6.77 \text{ nV/(Hz)}^{1/2}$$

We will now show how to minimize the op-amp output noise once we have curves of V_n and I_n vs. frequency. The output noise voltage due to V_n in the circuit of Fig. 19.22 is

$$V_{onv} = \frac{R_1 + R_2}{R_1} V_n \qquad (19.33)$$

or

$$V_{onv} = \frac{Z_1 + Z_2}{Z_1} V_n \qquad (19.34)$$

The relation $(Z_1 + Z_2)/Z_1$ is called the *noise voltage gain* of the circuit. Note that the noise voltage gain is larger than the gain for normal input voltages. For a noninverting amplifier the gains will be identical. The op-amp output noise voltage due to I_n is

$$V_{oni} = R_2 I_n \qquad (19.35)$$

or

$$V_{oni} = Z_2 I_n \qquad (19.36)$$

The total output noise is the rss sum of V_{onv} and V_{oni}:

$$V_{on} = (V_{onv}^2 + V_{oni}^2)^{1/2}$$

Therefore, the minimum value of V_{on} is achieved when $V_{onv} = V_{oni}$. To satisfy this we must set Eq. (19.33) equal to Eq. (19.35) [or Eq. (19.34) equal to Eq. (19.36)]. This results in

$$\frac{R_1 + R_2}{R_1} V_n = R_2 I_n$$

Rearranging this, we get

$$\frac{V_n}{I_n} = \frac{R_1 R_2}{R_1 + R_2}$$

Minimum noise is therefore achieved when the parallel resistance of R_1 and R_2 is made equal to the ratio V_n/I_n. This latter ratio is appropriately called the *noise resistance* of the op amp.

19.5 OP-AMP APPLICATION: BASIC INVERTING AMPLIFIER

ALTERNATE NAMES   Inverting-mode amplifier, inverting configuration, phase inverter, inverter.

PRINCIPLES OF OPERATION   In Fig. 19.24a assume that v_1 starts to drive v_2, that is, $v_2 - v_3$, in the same direction as v_1. The op-amp output $A_{vo}(v_3 - v_2)$ will then be driven in a direction opposite to that of v_1. This will cause a feedback current through R_2 which attempts to drive $v_2 - v_3$ back toward zero. If A_{vo} is very large, $v_2 - v_3$ is driven to nearly zero and the current into the op-amp negative input terminal also approaches zero. All the input current through R_1 must therefore flow through R_2. The current through R_3 also approaches zero as A_{vo} becomes very large. The voltage at v_2 can thus be assumed equal to zero, which makes computation of the circuit gain quite easy. Since the same current i_1 flows in R_1 and R_2,

$$i_1 = \frac{v_1}{R_1} = -\frac{v_4}{R_2}$$

The closed-loop circuit voltage gain at dc becomes

$$A_{vco} = \frac{v_4}{v_1} = -\frac{R_2}{R_1}$$

In the design equations and design procedure to follow we show the effect of various op-amp parameters on the accuracy and stability of A_{vco}. Likewise, we examine the closed-loop voltage gain as a function

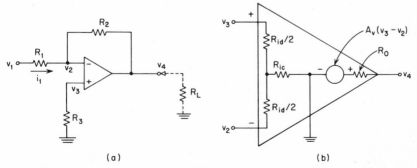

(a)   (b)

Fig. 19.24   (a) The basic inverting amplifier and (b) the definition of input-output resistances and voltage gain in an op amp.

of frequency A_{vc}. The effects of nonideal op-amp parameters on A_{vc} will also be summarized. Figure 19.24b defines some of these nonideal parameters.

DESIGN PARAMETERS

Parameter	Description
A_{vc}	Closed-loop voltage gain of circuit as a function of frequency
A_{vco}	Closed-loop dc voltage gain of circuit
A_v	Open-loop voltage gain of op amp as a function of frequency
A_{vo}	Open-loop dc voltage gain of op amp (A_{vo} may be substituted for A_v in any equation if dc characteristics are wanted)
f_{cp}	First pole frequency of circuit, i.e., the -3-dB bandwidth
f_{op}	First pole frequency of op amp
I_b	Input bias current of op amp
I_{io}	Input offset current of op amp
I_n	Equivalent input noise current of op amp
R_1	Input resistor
R_2	Feedback resistor
R_3	Resistor used to cancel the effects of I_b
R_{ic}	Common-mode input resistance of op amp
R_{id}	Differential input resistance of op amp
R_{in}	Input resistance of circuit
R_L	Load resistance of circuit
R_o	Output resistance of op amp
R_{out}	Output resistance of circuit
t_r	Rise time of circuit (10 to 90%)
v_1	Input voltage to circuit
v_4	Output voltage of circuit
V_{io}	Input offset voltage of op amp
V_n	Equivalent input-noise voltage of op amp
V_{on}	Output noise voltage of circuit
β	Voltage feedback ratio of R_1 and R_2, that is, $\beta = R_1/(R_1 + R_2)$

DESIGN EQUATIONS

Description	Equation	
Closed-loop voltage gain assuming ideal op-amp parameters	$A_{vc} = -\dfrac{R_2}{R_1}$	(1)
Closed-loop voltage gain if finite op-amp gain A_v is included	$A_{vc} = \dfrac{-R_2/R_1}{1 + 1/\beta A_v}$	(2)
	where $\beta = \dfrac{R_1}{R_1 + R_2}$	
Closed-loop voltage gain if differential input resistance R_{id} is included (A_v must also be included)	$A_{vc} = \dfrac{-R_2/R_1}{1 + 1/\beta A_v + R_2/A_v R_{id}}$	(3)
Closed-loop voltage gain if op-amp output resistance R_o is included (A_v must also be included)	$A_{vc} = \dfrac{(-R_2/R_1)}{1 + (R_2 + R_o)/\beta A_v R_2}$	(4)

Size of R_2 for minimum gain error due to A_v, R_{id}, and R_o	$R_{2,\text{opt}} = \left(\dfrac{R_{id}R_o}{2\beta}\right)^{1/2}$	(5)
Input resistance of circuit assuming ideal op-amp parameters	$R_{\text{in}} = R_1$	(6)
Input resistance of circuit assuming finite A_{vo}	$R_{\text{in}} = R_1\left(1 + \dfrac{R_2}{A_{vo}R_1}\right)$	(7)
Output resistance of circuit assuming ideal op-amp parameters	$R_{\text{out}} = 0$	(8)
Output resistance of circuit assuming finite op-amp output resistance R_o and finite A_v	$R_{\text{out}} = \dfrac{R_o}{1 + \beta A_v}$	(9)
Bandwidth of circuit assuming bandwidth (-3 dB) of op amp is at f_{op} (f_{op} = first pole of op amp)	$f_{cp} = \dfrac{f_{op}A_{vo}R_1}{R_2}$	(10)
Small-signal rise time of circuit (10 to 90%)	$t_r = \dfrac{0.35R_2}{f_{op}A_{vo}R_1}$	(11)
Output offset voltage due to input offset voltage of op amp (assuming I_b and $I_{io} = 0$)	$V_o = \pm V_{io}\dfrac{R_1 + R_2}{R_1}$	(12)
Output dc voltage due to input bias current of op amp assuming $R_3 = 0$ and $V_{io} = 0$	$V_o = I_b R_2$	(13)
Output offset voltage due to input offset current of op amp assuming	$V_o = \pm I_{io}R_2$	(14)
$R_3 = \dfrac{R_1 R_2}{R_1 + R_2}$ and $V_{io} = 0$		
Output-noise voltage due to an equivalent op-amp input-noise voltage, V/(Hz)$^{1/2}$	$V_{on} = V_n\left(1 + \dfrac{R_2}{R_1}\right)$ V/(Hz)$^{1/2}$	(15)
Output-noise voltage due to both equivalent op-amp input noise voltage and current [V/(Hz)$^{1/2}$ and A/(Hz)$^{1/2}$ or V^2/Hz and A^2/Hz]	$V_{on} = \left[V_n^2\left(1 + \dfrac{R_2}{R_1}\right)^2 + I_n^2 R_2^2\right]^{1/2}$	(16)
Optimum value for R_3 to minimize output voltage offset due to I_b	$R_3 = \dfrac{R_1 R_2}{R_1 + R_2}$	(17)

DESIGN PROCEDURE

The approach one takes in designing an inverting amplifier depends on the application of the circuit. Many designs will merely require use of Eqs. (1), (6), and (9) (voltage gain and input-output resistances). High-frequency applications may require a trade-off between Eqs. (1) (voltage gain) and (10) (closed-loop bandwidth). Low-level precision dc amplifiers may require compromises between Eqs. (2) to (5), (7), (9), (12) to (14), (16), and (17). This last category would be the most difficult since there are many conflicting requirements in the above list of equations.   For example, Eq. (5) indicates that an optimum R_2 can be chosen which reduces the effect of changes in A_v, R_{id}, and R_o on the circuit.   Contrary to this, Eqs. (7), (12) to (14), and (16) imply that R_2 should be as small as practical.   The final choice for R_2 in this case would probably be simplified if a few extra dollars were invested in a good-quality instrumentation op amp.

In the design steps to follow, we assume that the op-amp type is already chosen and that both the closed-loop dc gain A_{vco} and the minimum input resistance R_{in} are specified.

The variation of dc closed-loop gain A_{vco} as a result of variations in open-loop dc gain A_{vo}, differential input resistance R_{id}, R_1, R_2, and output resistance R_o will be calculated.   The drift of the circuit output dc voltage as a result of drifts in input offset voltage V_{io} and input offset current I_{io} will be determined.   The small-signal rise time, bandwidth, and output noise will also be computed.

DESIGN STEPS

Step 1.   Compute the optimum R_2 with Eq. (5):

$$R_{2,\ \text{opt}} = \left(\frac{R_{id}R_o}{2\beta}\right)^{1/2} \qquad \text{where } \beta = \frac{1}{1 - A_{vco}}$$

Step 2.   Determine the size of R_1 using Eq. (1) and the results of step 1 above:

$$R_1 = \frac{-R_2}{A_{vco}}$$

If this computed R_1 is less than the minimum specified circuit input resistance, a compromise between Eqs. (1), (5), and (6) may be required. Assuming that Eqs. (1) and (6) are more important, we calculate the errors caused by a nonoptimum R_2 in the following steps.   This approach means that we let $R_1 = R_{in}$ and $R_2 = -A_{vco}R_1$.

Step 3.   Given the range of temperatures within which the circuit is to be operated, the variations in A_{vo}, R_{id}, R_1, R_2, V_{io}, and I_{io} are determined from the op-amp and resistor data sheets.

Step 4.   Assuming that a given range of frequencies must be amplified with minimum error and noise, the following parameters as a function of frequency are also found from the data sheet: A_v, R_o, V_n, and I_n.

Step 5.   Compute the variations in closed-loop dc voltage gain A_{vco} using data from step 3 and Eqs. (2) to (4).   Repeat this step using A_v at selected frequencies of interest.

Step 6.   Determine the circuit output resistance R_{out} at selected frequencies using Eq. (9).   Compute the reduction of A_{vc} using R_{out} and R_L with voltage-divider theory.   The effective voltage gain A_{vc} with R_L attached is

$$A_{vc} \text{ (with } R_L) = \frac{A_{vc}(R_L = \infty) R_L}{R_L + R_{\text{out}}}$$

Step 7.   Determine the true input resistance by using Eq. (7). If the input resistance is critical and this calculation reveals a design deficiency, R_1 may need to be increased.   Steps 2 and 5 to 7 will then have to be repeated.

Step 8.   Compute the variation in A_{vc} due to resistance changes in R_1 and R_2.   Since $A_{vc} = -R_2/R_1$, a ± 1 percent change in either R_2 or R_1 will result in a ± 1 percent change in A_{vc}.   If R_1 increases 1 percent as R_2 decreases 1 percent, A_{vc} will decrease 2 percent.   Since resistor variations are usually unpredictable, the $\pm$ sign must be used.

Step 9.   Compute R_3 according to Eq. (17).

Step 10.   If R_3 has been determined according to Eq. (17), Eq. (13) will not need to be calculated.   The output offset drift voltage will be controlled only by ΔV_{io} and ΔI_{io}.   Determine the total output offset voltage change from Eq. (12) and (14).   If the temperature-dependent drifts of V_{io} and I_{io} are known in both magnitude and direction, the actual ΔV_o can be determined.   However, in most cases $\pm$ signs should be used since either input terminal of the op amp can require more bias voltage or more bias current than the other.   The output dc voltage error caused by the value of V_{io} at $+25°C$ can be nulled out using special terminals on most op amps.

Step 11.   Use Eq. (10) to compute the small-signal bandwidth and Eq. (11) for the small-signal rise time.

Step 12.   If the equivalent input noise current and voltage is available as a function of frequency, the output noise voltage as a function of frequency can be determined.   Equation (16) requires that V_n be in units of $V/(Hz)^{1/2}$ or V^2/Hz and I_n be in units of $A/(Hz)^{1/2}$ or A^2/Hz. This equation can be solved at several frequencies using V_n and I_n data at these same frequencies.

EXAMPLE OF INVERTING-AMPLIFIER DESIGN   An actual design of an inverting amplifier with a gain of -100 will now be presented.

Design Requirements

$$A_{vco} = -100 \qquad R_{in} \geq 1000 \; \Omega$$

$$\text{Op amp} = \mu\text{A741A (Fairchild)} \qquad R_L \geq 2000 \; \Omega$$

Device Data

For -55 to $+125°C$ and ±20-V supply voltages

$$V_{io} = \begin{cases} \pm0.8 \text{ mV} & \text{typical} \\ \pm4.0 \text{ mV} & \text{worst case } (-55 \text{ or } +125°C) \end{cases}$$

$\Delta V_{io} = \pm15 \ \mu V/°C$ maximum

$$I_{io} = \begin{cases} \pm3.0 \text{ nA} & \text{typical} \\ \pm70 \text{ nA} & \text{worst case } (-55°C) \end{cases}$$

$\Delta I_{io} = \pm0.5$ nA/°C max

$$R_{id} = \begin{cases} 6 \text{ M}\Omega & \text{typical} \\ 0.5 \text{ M}\Omega & \text{worst case } (-55°C) \end{cases}$$

$$A_{vo} = \begin{cases} 5 \times 10^4 \text{ min} & \text{at } +25°C \\ 3.2 \times 10^4 \text{ min} & \text{over temperature } (-55 \text{ to } +125°C) \end{cases}$$

$$f_{op} = \begin{cases} 8 \text{ Hz} & \text{typical} \\ 6 \text{ Hz} & \text{at } +125°C \end{cases}$$

$$R_o = \begin{cases} 70 \ \Omega & \text{dc to 10 kHz} \\ 90 \ \Omega & \text{at 100 kHz} \\ 280 \ \Omega & \text{at 1 MHz} \end{cases}$$

$$V_n = \begin{cases} 5 \times 10^{-15} \text{ V}^2/\text{Hz} & \text{at 10 Hz} \\ 10^{-15} \text{ V}^2/\text{Hz} & \text{at 100 Hz} \\ 5 \times 10^{-16} \text{ V}^2/\text{Hz} & \text{from 1 to 100 kHz} \end{cases}$$

$$I_n = \begin{cases} 5 \times 10^{-23} \text{ A}^2/\text{Hz} & \text{at 10 Hz} \\ 5 \times 10^{-24} \text{ A}^2/\text{Hz} & \text{at 100 Hz} \\ 8 \times 10^{-25} \text{ A}^2/\text{Hz} & \text{at 1 kHz} \\ 3 \times 10^{-25} \text{ A}^2/\text{Hz} & \text{from 10 to 100 kHz} \end{cases}$$

$\Delta R_1 = \pm100$ ppm/°C $= \pm10^{-4}$ change in 1°C

$\Delta R_2 = \pm100$ ppm/°C

Step 1. The value of β, the voltage transfer ratio of the feedback network, is

$$\beta = \frac{1}{1 - A_{vco}} = \frac{1}{1 + 100} = 0.0099$$

The optimum R_2 can now be computed from

$$R_{2,\text{opt}} = \left(\frac{R_{id} R_o}{2\beta}\right)^{1/2} = \left[\frac{(6 \times 10^6)(70)}{2 \times 0.0099}\right]^{1/2} = 145{,}600 \ \Omega$$

Step 2. The size of R_1 becomes

$$R_1 = \frac{-R_{2,\text{opt}}}{A_{vco}} = \frac{-1.456 \times 10^5}{-100} = 1456 \ \Omega$$

This resistance also satisfies the 1000-Ω minimum-input-resistance requirement.

Step 3. The variations of the parameters over the temperature range of -55 to $+125°C$ (using data at $+25°C$ as a reference) are shown in Table 19.4.

TABLE 19.4

Parameter*	$-55°C$	$+25°C$	$+125°C$
A_{vo}	. . . .	5×10^4	3.2×10^4
R_{id}	0.5×10^6	6×10^6	. . . .
ΔV_{io}	$\pm 2 \text{ mV}$	$\pm 0.8 \text{ mV}$	$\pm 2.3 \text{ mV}$
ΔI_{io}	$\pm 43 \text{ nA}$	$\pm 3 \text{ nA}$	$\pm 53 \text{ nA}$
R_1	$1456 \pm 11.6 \ \Omega$	$1456 \ \Omega$	$1456 \pm 14.6 \ \Omega$
R_2	$145,600 \pm 1160 \ \Omega$	$145,600 \ \Omega$	$145,600 \pm 1456 \ \Omega$

* Values from the Fairchild data sheet.

Step 4. The variations of the parameters over the frequency range of dc to 1 MHz are given in Table 19.5.

TABLE 19.5

Parameter*	dc	10 Hz	100 Hz	1 kHz	10 kHz	100 kHz	1 MHz
A_v	5×10^4	4×10^4	4000	400	40	4	0.4
R_o, Ω	70	70	70	70	70	90	280
V_n^2, V^2/Hz	. . . . .	5×10^{-15}	10^{-15}	5×10^{-16}	5×10^{-16}	5×10^{-16}	. . . .
I_n^2, A^2/Hz	. . . . .	5×10^{-23}	5×10^{-24}	8×10^{-25}	3×10^{-25}	3×10^{-25}	. . . .

* Values from the Fairchild data sheet.

Step 5. The ideal gain should be

$$A_{vco,\text{ideal}} = -\frac{R_2}{R_1} = -\frac{145,600}{1456} = -100$$

Since the open-loop dc gain A_{vo} is only 5×10^4 (the worst-case minimum at $+25°C$), the actual gain (assuming ideal values for R_1 and R_2) will be

$$A_{vco} \ (A_{vo} \neq \infty) = \frac{-R_2/R_1}{1 + 1/\beta A_{vo}} = \frac{-145,600/1456}{1 + 1/0.0099(5 \times 10^4)} = -99.798387$$

If A_{vo} decreases to 3.2×10^4 at $125°C$, the dc closed-loop gain is reduced to

$$A_{vco} \ (A_{vo} = 3.2 \times 10^4) = \frac{-145,600/1456}{1 + 1/0.0099(3.2 \times 10^4)} = -99.685336$$

This gain is approximately 0.1 percent lower than the closed-loop gain at room temperature. The degradation caused by the typical input resistance R_{id} at $+25°C$ is

$$A_{vco}(R_{id} = 6\ M\Omega) = \frac{-R_2/R_1}{1 + 1/\beta A_{vo} + R_2/A_{vo} R_{id}}$$

$$= \frac{-145,600/1456}{1 + 1/0.0099(5 \times 10^4) + 145,600/(5 \times 10^4)(6 \times 10^6)}$$

$$= -99.798339$$

An input resistance of 6 MΩ is obviously not an error source of concern. If the worst-case minimum R_{id} of 0.5 MΩ is used, we get

$$A_{vco}(R_{id} = 0.5\ M\Omega)$$

$$= \frac{-145,600/1456}{1 + 1/0.0099(5 \times 10^4) + 145,600/(5 \times 10^4)(0.5 \times 10^6)}$$

$$= -99.797807$$

This is still a very small error compared with the effects of a finite A_{vo}. The change of R_{id} over the worst-case temperature range will affect the closed-loop gain by approximately 0.001 percent.

The effects of a finite R_0 in conjunction with R_2 are determined by Eq. (4):

$$A_{vco}(R_0 = 70\ \Omega) = \frac{-R_2/R_1}{1 + (R_2 + R_0)/\beta A_{vo} R_2}$$

$$= \frac{-145,600/1456}{1 + (145,600 + 70)/0.0099(5 \times 10^4)(145,600)}$$

$$= -99.79829$$

This gain is approximately the same as that computed when only the finite A_{vo} was considered.

The effects of output resistance at frequencies above the first op-amp pole (≈ 8 Hz) must be determined with caution. Since A_v lags A_{vc} 90° for these frequencies, Eq. (4) must take this into account. This is done as follows (at 1 kHz where $A_v = -j400$):

$$A_{vc}(f = 1\ kHz,\ R_0 = 70\ \Omega)$$

$$= \frac{-100}{1 + (145,600 + 70)/0.0099(-j400)(145,600)}$$

$$= \frac{-100}{1 + j0.2526} = \frac{-100}{1.0314 \angle 14°}$$

$$= 96.95 \angle -14°$$

The closed-loop gain has a slight phase lag at 1 kHz. If we had not considered the phase shift of A_v, we would have mistakenly computed A_{vc} as

$$A_{vc} = \frac{-100}{1 + (145{,}600 + 70)/0.0099(400)(145{,}600)} = 79.83$$

Step 6. The output resistance of the circuit is calculated. The phase shift of A_v must be considered if the op amp first pole frequency is exceeded

$$R_{\text{out,dc}} = \frac{R_o}{1 + \beta\,A_v} = \frac{70}{1 + 0.0099(5\times 10^4)} = 0.1411\ \Omega$$

$$R_{\text{out,1 kHz}} = \frac{70}{1 + 0.0099(-j400)} = 17.14\ \angle 76°\ \Omega$$

$$R_{\text{out,100 kHz}} = \frac{90}{1 + 0.0099(-j4)} = 89.93\ \angle 0°\ \Omega$$

The reduction in closed-loop gain A_{vc} due to interaction between R_{out} and R_L is computed by use of voltage-divider theory. If $R_{\text{out}} = 17.14$ Ω (1 kHz) and $R_L = 2$ kΩ, the effective closed-loop gain A_{vc} is

$$A_{vc}\ (\text{with } R_L) = \frac{A_{vco}\ (R_L = \infty)\ R_L}{R_L + R_{\text{out}}}$$

$$= \frac{-96.95 \times 2000}{2000 + 17.14} = -96.12$$

Step 7. The true input resistance at dc is

$$R_{\text{in}} = R_1\left(1 + \frac{R_2}{A_{vo}\,R_1}\right) = 1456\left[1 + \frac{145{,}600}{(5\times 10^4)(1456)}\right] = 1459\,\Omega$$

Step 8. Using $+25°C$ as a reference, we compute the change in closed-loop gain for a temperature change up to $+125°C$:

$$A_{vco} = -\frac{R_2 \pm \Delta R_2}{R_1 \pm \Delta R_1} = -\frac{145{,}600 \pm (145{,}600 \times 10^{-4})(100)}{1456 \pm (1456 \times 10^{-4})(100)}$$

$$= -\frac{145{,}600 \pm 1456}{1456 \pm 14.56} = -98.02 \text{ to } -102.02$$

Step 9. The required value of R_3 is

$$R_3 = \frac{R_1\,R_2}{R_1 + R_2} = \frac{1456(145{,}600)}{1456 + 145{,}600} = 1442\,\Omega$$

Step 10. The change in dc output voltage due to the change of V_{io} over $100°C$ is $(+25$ to $+125°C)$

$$\Delta V_o = \pm V_{io} \frac{R_1 + R_2}{R_1}$$

$$= \pm \frac{(15 \times 10^{-6})(100°C)}{°C} \frac{1456 + 145,600}{1456} = \pm 0.1515 \text{ V}$$

The change in dc output voltage due to the change of I_{io} over $100°C$ is

$$\Delta V_o = \pm \Delta I_{io} R_2 = \pm \frac{(0.5 \times 10^{-9})(100°C)(145,600)}{°C} = 7.28 \text{ mV}$$

Step 11. The small-signal bandwidth is

$$f_{cp} = \frac{f_{op} A_{vo} R_1}{R_2} = \frac{(8 \text{ Hz})(5 \times 10^4)(1456)}{145,600} = 4 \text{ kHz}$$

The small-signal rise time is

$$t_r = \frac{0.35 R_2}{f_{op} A_{vo} R_1} = \frac{0.35(145,600)}{8(5 \times 10^4)(1456)} = 87.5 \text{ }\mu\text{s}$$

Step 12. The output noise at 10 Hz is computed from

$$V_{on,10 \text{ Hz}} = \left[V_n^2 \left(1 + \frac{R_2}{R_1} \right)^2 + I_n^2 R_2^2 \right]^{1/2}$$

$$= \left[(5 \times 10^{-15}) \left(1 + \frac{145,600}{1456} \right)^2 + (5 \times 10^{-23})(145,600)^2 \right]^{1/2}$$

$$= 7.2 \text{ }\mu\text{V rms} \quad \text{at } 10 \text{ Hz}$$

At 100 Hz, 1 kHz, 10 kHz, and 100 kHz the output noise is found to be 3.2, 2.3, 2.3, and 2.3 μV, respectively.

REFERENCES

1. Giles, J. N.: "Fairchild Semiconductor Linear Integrated Circuits Applications Handbook," Fairchild Semiconductor, Mountain View, Calif., 1967.
2. Millman, J., and C. C. Halkias: "Integrated Electronics: Analog and Digital Circuits and Systems," McGraw-Hill, New York, 1972.
3. Smith, L., and D. H. Sheingold: Noise and Operational Amplifier Circuits, *Analog Dialogue* (Analog Devices, Inc.), vol. 3, no. 1, March 1969.
4. Niu, G.: Gain-Error Nomograms for Op Amps, *EEE,* February 1967, p. 104.
5. Moschytz, G. S.: The Operational Amplifier in Linear Active Networks, *IEEE Spectrum,* January 1970, p. 42.
6. Graeme, J. G., G. E. Tobey, and L. P. Huelsman: "Operational Amplifiers Design and Applications," McGraw-Hill, New York, 1971, pp. 427–436.
7. Stout, D. F., and M. Kaufman: "Handbook of Operational Amplifier Circuit Design," McGraw-Hill, New York, 1976, p. 9-1.

Chapter 20

Active Filters

INTRODUCTION

Nearly all active filters use op amps as the active element Packaged active filters are no exception. Most of these packages contain three or four op amps and about a dozen resistors and capacitors. We therefore confine our discussion in this chapter to active filters which use one to four op amps. We do not discuss which resistors and capacitors are contained inside the filter package since each manufacturer has a slightly different approach. The theory of operation, design equations, and a design procedure are presented for each of the more popular active-filter configurations. The reader is left with the task of determining which manufacturer can place most of a given circuit in one package or on one chip. In most cases designers will probably use a dual or quad op amp and tailor the external resistors and capacitors to their needs.

20.1 SECOND-ORDER LOW-PASS FILTER

ALTERNATE NAMES Unity-gain low-pass filter, active RC low-pass filter, active inductorless low-pass filter.

PRINCIPLES OF OPERATION This circuit provides two complex poles with adjustable damping. By proper choice of R_1, R_2, C_1, and C_2 of Fig. 20.1 the transfer function can be made to exhibit the range of characteristics shown in Fig. 20.2. The curves of Fig. 20.2 were obtained with an actual circuit using a 741 op amp and a pole frequency f_{cp} of 1000 Hz. Using the design steps for this circuit (which will be listed later), we calculate the following component values:

Bessel: $R_1 = R_2 = 10,800 \ \Omega$
            $C_1 = 0.0133 \ \mu F$ $C_2 = 0.01 \ \mu F$

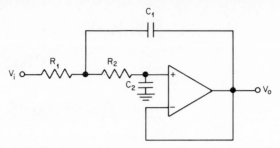

Fig. 20.1   A single-feedback second-order low-pass filter utilizing a unity-gain amplifier.

Butterworth: $R_1 = R_2 = 10{,}800\ \Omega$
$\qquad\qquad\qquad C_1 = 0.02\ \mu\text{F} \quad C_2 = 0.01\ \mu\text{F}$

3-dB Chebyshev: $R_1 = R_2 = 49{,}400\ \Omega$
$\qquad\qquad\qquad C_1 = 0.01\ \mu\text{F} \quad C_2 = 1470\ \text{pF}$

The transfer function of Fig. 20.1 is

$$A_{vc} = \frac{V_o}{V_i} = \frac{1}{s^2(C_1 C_2 R_1 R_2) + sC_2(R_1 + R_2) + 1}$$

The damping factor ζ determines the shape of A_{vc} in the frequency

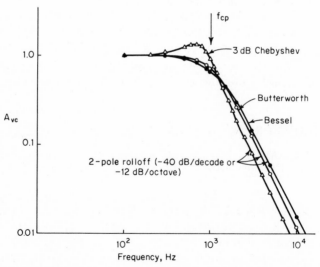

Fig. 20.2   Test data of three typical 2-pole low-pass filters using the circuit of Fig. 20.1.

region near f_{cp}.   Low values of ζ cause the frequency-response curve to have more peaking near the pole frequency.   This term is related to the familiar circuit Q by

$$\zeta = \frac{1}{2Q}$$

The circuit transfer function can be put in the classical form to help us find ζ:

$$A_{vc} = \frac{1}{s^2 + 2\zeta\omega_n s + \omega_n^2}$$

where $\omega_n = 2\pi f_{cp}$ is the natural resonant radian frequency of the circuit. We now can determine ζ to be

$$\zeta = \frac{R_1 + R_2}{2}\left(\frac{C_2}{R_1 R_2 C_1}\right)^{1/2}$$

The formula for ζ gives a value of 0.383 for the Chebyshev filter plotted in Fig. 20.2. Similarly, values of $\zeta = 0.866$ and 0.707 are obtained for these particular Bessel and Butterworth filters, respectively.

TABLE 20.1   Unscaled Capacitor Values for Fig. 20.1

Type of 2-pole low-pass filter	ζ	C_1'',F	C_2'',F
Bessel	0.8659	0.9066	0.6799
Butterworth	0.7072	1.414	0.7071
Chebyshev,			
0.1 dB peak	0.6516	1.638	0.6955
0.25 dB peak	0.6179	1.778	0.6789
0.5 dB peak	0.5789	1.949	0.6533
1 dB peak	0.5228	2.218	0.6061
2 dB peak	0.4431	2.672	0.5246
3 dB peak	0.3833	3.103	0.4558

DESIGN PARAMETERS

Parameter	Description
A_v	Op-amp voltage gain as a function of frequency
A_{vc}	Voltage gain of circuit as a function of frequency
C_1, C_2	Final values for capacitors after both impedance and frequency scaling; determine f_{cp} and ζ
C_1', C_2'	Intermediate values for C_1 and C_2 after frequency scaling
C_1'', C_2''	Unscaled capacitor values from Table 20.1
f_{cp}	Pole (or corner) frequency of circuit
Q	Determines height of peak in frequency response
R	Common value of R_1 and R_2 (both $= R$)
R_1, R_2	Determine f_{cp} and ζ
S_A^B	Sensitivity of B to variations in A (applies to all sensitivity functions listed in design equations)
V_i	Circuit input voltage
V_o	Circuit output voltage
ζ	Damping factor
ω_n	Natural radian frequency of poles of circuit

DESIGN EQUATIONS

Description	Equation
Transfer function (voltage gain) of circuit	$A_{vc} = \dfrac{V_o}{V_i} = \dfrac{1}{s^2(C_1 C_2 R_1 R_2) + s C_2(R_1 + R_2) + 1}$ (1a)
	or
	$A_{vc} = \dfrac{1}{s^2 + 2\zeta \omega_n s + \omega_n^2}$ (1b)
Location of two complex poles of Eqs. (1) in s domain	$s_1, s_2 = \dfrac{-C_2(R_1 + R_2) \pm [C_2^2(R_1 + R_2)^2 - 4 C_1 C_2 R_1 R_2]^{1/2}}{2 C_1 C_2 R_1 R_2}$ (2)
Relationship between initial and final capacitor values	$C_1 = \dfrac{C_1''}{2\pi f_{cp} R} \qquad C_2 = \dfrac{C_2''}{2\pi f_{cp} R}$ (3)
Damping factor of circuit	$\zeta = \dfrac{R_1 + R_2}{2}\left(\dfrac{C_2}{R_1 R_2 C_1}\right)^{1/2}$
	$= \left(\dfrac{C_2}{C_1}\right)^{1/2} \quad \text{if } R_1 = R_2$ (4)
Relationship between circuit Q and damping factor	$Q = \dfrac{1}{2\zeta}$ (5)
Relationship between pole frequency and natural radian frequency	$\omega_n = 2\pi f_{cp}$ (6)
Sensitivity* of f_{cp} to variations in R_1, R_2, C_1, or C_2	$S_{R_1}^{f_{cp}} = S_{R_2}^{f_{cp}} = S_{C_1}^{f_{cp}} = S_{C_2}^{f_{cp}} = -\tfrac{1}{2}$ (7)
Sensitivity of ζ to variations in R_1, R_2, C_1, or C_2	$S_{R_1}^{\zeta} = \dfrac{1}{2} - \dfrac{1}{4\pi\zeta f_{cp} R_1 C_1}$ (8a)
	$S_{R_2}^{\zeta} = \dfrac{1}{2} - \dfrac{1}{4\pi\zeta f_{cp} R_1 C_1}$ (8b)
	$S_{C_1}^{\zeta} = \dfrac{1}{2} - \left(\dfrac{1}{R_1} + \dfrac{1}{R_2}\right)\dfrac{1}{4\pi\zeta f_{cp} C_1}$ (8c)
	$S_{C_2}^{\zeta} = \tfrac{1}{2}$ (8d)
Required op-amp open-loop gain to assure accuracy	$A_v \gg \dfrac{C_1}{2 C_2} \qquad$ at f_{cp} and lower frequencies (9)

* $S_{R_1}^{f_{cp}} = -\tfrac{1}{2}$ means that if R_1 increases in value by 1%, f_{cp} will decrease in frequency by $\tfrac{1}{2}\%$. After all sensitivities of a given parameter are computed they are algebraically added to determine the total result.

DESIGN PROCEDURE

There are several approaches given in the literature for designing this circuit. The one we have chosen to present here [1] requires only very simple calculations. Its only disadvantage is that the capacitor values are different whereas some design approaches result in identical capacitors and different resistors.

DESIGN STEPS

Step 1. Choose C_1'' and C_2'' from Table 20.1 according to the type of filter required.

Step 2. Using the required corner frequency f_{cp}, perform the frequency scaling

$$C_1' = \frac{C_1''}{2\pi f_{cp}} \quad C_2' = \frac{C_2''}{2\pi f_{cp}}$$

Step 3. Choose a value $R = R_1 = R_2$ which will produce practical sizes for C_1 and C_2 according to

$$C_1 = \frac{C_1'}{R} \quad C_2 = \frac{C_2'}{R}$$

This procedure is called *impedance scaling.*

Note: The remaining steps are required only if the designer wants to gain further insight into error sources, etc.

Step 4. Compute the damping factor ζ using Eq. (4). Compare the result with data in Table 20.1 to verify that the correct filter has been designed.

Step 5. If required, use Eq. (7) to compute the sensitivity of f_{cp} to variations in R_1, R_2, C_1, and C_2. Likewise, Eqs. (8) can be used to determine the sensitivity of ζ to changes in R_1, R_2, C_1, and C_2.

Step 6. From the op-amp data sheet determine A_v at f_{cp}. This value of A_v must satisfy Eq. (9) by at least a factor of 100 in order to keep the actual frequency response (Fig. 20.2) less than 0.2 dB from the ideal frequency response.

EXAMPLE OF SECOND-ORDER LOW-PASS FILTER DESIGN The eight design steps will be numerically illustrated through an example. The results of tests on a circuit designed using these steps were previously shown in Fig. 20.2 (3-dB Chebyshev).

Design Requirements

$$f_{cp} = 1000 \text{ Hz} \qquad \text{peaking} \approx 3 \text{ dB (Chebyshev)}$$

$$\text{Maximum capacitor size} \approx 0.01 \ \mu\text{F}$$

Device Data

$$A_v \ (1000 \text{ Hz}) = 1000$$

At -55 to $+125°C$

$$\Delta R_1 = 0.018 \, R_1 \quad \Delta R_2 = 0.018 \, R_2$$
$$\Delta C_1 = 0.01 \, C_1 \quad \Delta C_2 = 0.01 \, C_2$$

Step 1. From Table 20.1 we obtain $C_1'' = 3.103$ F and $C_2'' = 0.4558$ F.

Step 2. Frequency scaling:

$$C_1' = \frac{C_1''}{2\pi f_{cp}} = \frac{3.103}{2\pi \times 1000} = 4.94 \times 10^{-4} \text{ F}$$

$$C_2' = \frac{C_2''}{2\pi f_{cp}} = \frac{0.4558}{2\pi \times 1000} = 7.25 \times 10^{-5} \text{ F}$$

Step 3. Since C_1 is always the largest capacitor in this design approach, we scale R so that $C_1 = 0.01 \ \mu\text{F}$.

$$R = \frac{C_1'}{C_1} = \frac{4.94 \times 10^{-4}}{10^{-8}} = 49,400 \ \Omega$$

Also

$$C_2 = \frac{C_2'}{R} = \frac{7.25 \times 10^{-5}}{49,400} = 1470 \text{ pF}$$

Step 4.

$$\zeta = \left(\frac{C_2}{C_1}\right)^{1/2} = \left(\frac{1.47 \times 10^{-9}}{10^{-8}}\right)^{1/2} = 0.383$$

Step 5. The sensitivity functions are

$$S_{R_1}^{f_{cp}} = S_{R_2}^{f_{cp}} = S_{C_1}^{f_{cp}} = S_{C_2}^{f_{cp}} = -\tfrac{1}{2}$$

The fractional variations of R_1 and R_2 over temperature are

$$\frac{\Delta R_1}{R_1} = \frac{\Delta R_2}{R_2} = 0.018 \quad -55 \text{ to } +125°\text{C}$$

Thus if these resistances increase as temperature increases, f_{cp} will decrease by $\tfrac{1}{2}(0.018)f_{cp} = 9$ Hz as the temperature increases from -55 to $+125°$C.

The fractional variations of C_1 and C_2 over temperature are

$$\frac{\Delta C_1}{C_1} = \frac{\Delta C_2}{C_2} = 0.01 \quad -55 \text{ to } +125°\text{C}$$

If the capacitances increase as temperature increases, f_{cp} will decrease by $\tfrac{1}{2}(0.01)(1000) = 5$ Hz as the temperature increases from -55 to $+125°$C.

Changes to ζ as temperature varies are

$$S_{R_1}^{\zeta} = \frac{1}{2} - \frac{1}{4\pi\zeta f_{cp} R_1 C_1}$$

$$= \frac{1}{2} - \frac{1}{4\pi(0.383)(1000)(49,400 \times 10^{-8})} = 0.0794$$

If $\Delta R_1/R_1$ increases by 0.018 as the temperature increases from -55

to +125°C, ζ will correspondingly increase by $0.0794(0.383)(0.018) =$ 5.5×10^{-4}. It should be realized the generator driving R_1 has a finite resistance. This resistance (and its variations) must be incorporated into all R_1 calculations.

Continuing the calculations, we have

$$S_{R2} = \frac{1}{2} - \frac{1}{4\pi\zeta f_{cp} R_2 C_1}$$

$$= \frac{1}{2} - \frac{1}{4\pi(0.383)(1000)(49,400 \times 10^{-8})} = 0.0794$$

Also $\quad S_{C_1}^{\zeta} = \frac{1}{2} - \left(\frac{1}{R_1} + \frac{1}{R_2}\right)\frac{1}{4\pi\zeta f_{cp} C_1}$

$$= \frac{1}{2} - \left(\frac{1}{49,400} + \frac{1}{49,400}\right)\frac{1}{4\pi(0.383)(1000 \times 10^{-8})}$$

$$= -0.341$$

$$S_{C_2}^{\zeta} = \frac{1}{2}$$

Step 6. We must satisfy

$$A_v \gg \frac{C_1}{2C_2} = \frac{10^{-8}}{2(1.49 \times 10^{-9})} = 3.4$$

at 1000 Hz. This is assured by more than 100 since $A_v(1000 \text{ Hz}) \gtrsim$ 1000 in most monolithic op amps.

20.2 THIRD-ORDER LOW-PASS FILTER

ALTERNATE NAMES Unity-gain low-pass filter, active low-pass filter, active inductorless low-pass filter, active *RC* filter.

PRINCIPLES OF OPERATION The circuit shown in Fig. 20.3 is nearly identical to the circuit shown in Fig. 20.1 except for the additional *RC*

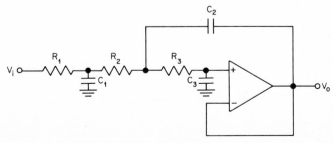

Fig. 20.3 A single-feedback third-order low-pass filter utilizing a unity-gain amplifier.

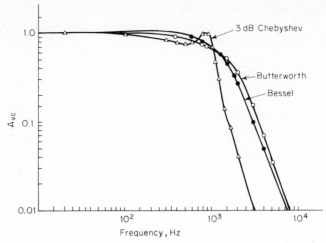

Fig. 20.4 Measured frequency response of three typical third-order filters using the circuit of Fig. 20.3.

input stage. These two additional passive parts add a pole on the negative real axis. The resulting 3-pole (third-order) filter rolls off at -60 dB/decade or -18 dB/octave at frequencies above f_{cp}. This is shown in Fig. 20.4, where the actual response curves of three circuits with different peaking characteristics are plotted. A 741 op amp was used with a selected pole frequency of 1000 Hz.

The transfer function of Fig. 20.3 is

$$A_{vc} = \frac{V_o}{V_i} = \frac{1}{s^3 A + s^2 B + sC + 1}$$

where $A = C_1 C_2 C_3 R_1 R_2 R_3$
$B = C_1 C_2 C_3 (R_1 + R_2) + C_1 C_3 R_1 (R_2 + R_3)$
$C = C_1 R_1 + C_3 (R_1 + R_2 + R_3)$

TABLE 20.2 Unscaled Capacitor Values for Fig. 20.3

Type of 3-pole low-pass filter	C_1'', F	C_2'', F	C_3'', F
Bessel	0.9880	1.423	0.2538
Butterworth	1.392	3.546	0.2024
Chebyshev, 0.1-dB peak	1.825	6.653	0.1345
0.25-dB peak	2.018	8.551	0.1109
0.5-dB peak	2.250	11.23	0.08950
1-dB peak	2.567	16.18	0.06428
2-dB peak	3.113	27.82	0.03892
3-dB peak	3.629	43.42	0.02533

DESIGN PARAMETERS

Parameter	Description
A_v	Op-amp voltage gain as function of frequency
A_{vc}	Voltage gain of circuit as function of frequency
C_1, C_2, C_3	Final values for capacitors after both impedance and frequency scaling; these capacitors determine f_{cp} and the magnitude of peaking near f_{cp}
C_1', C_2', C_3'	Intermediate values for C_1, C_2, and C_3 after frequency scaling
C_1'', C_2'', C_3''	Unscaled capacitor values from Table 20.2
f_{cp}	Pole or corner frequency of circuit
R	Common value of R_1, R_2, and R_3
R_1, R_2, R_3	Determine f_{cp} along with C_1, C_2, and C_3
V_i	Input voltage to circuit
V_o	Output voltage of circuit

DESIGN EQUATIONS

Description	Equation	
Transfer function (voltage gain) of circuit	$A_{vc} = \dfrac{V_o}{V_i} = \dfrac{1}{s^3 A + s^2 B + sC + 1}$	(1)
	where $A = C_1 C_2 C_3 R_1 R_2 R_3$ $\quad\quad B = C_1 C_2 C_3 (R_1 + R_2) + C_1 C_3 R_1 (R_2 + R_3)$ $\quad\quad C = C_1 R_1 + C_3 (R_1 + R_2 + R_3)$	
Relationship between initial and final capacitor values	$C_1 = \dfrac{C_1''}{2\pi f_{cp} R}$	(2a)
	$C_2 = \dfrac{C_2''}{2\pi f_{cp} R}$	(2b)
	$C_3 = \dfrac{C_3''}{2\pi f_{cp} R}$	(2c)
Recommended minimum A_v at f_{cp}	$A_v(f_{cp}) \geqq 100$	(3)

DESIGN PROCEDURE

At least two simplified design approaches are possible for this circuit. One could assume $R = R_1 = R_2 = R_3$ and solve for R and the capacitor values. Conversely, we could let $C = C_1 = C_2 = C_3$ and solve for C and the resistor values. The design steps will utilize the first method in conjunction with Table 20.2.

Step 1. Choose C_1'', C_2'', and C_3'' from Table 20.2 according to the type of filter required.

Step 2. Using the required corner frequency f_{cp}, perform the frequency scaling

$$C_1' = \frac{C_1''}{2\pi f_{cp}} \quad\quad C_2' = \frac{C_2''}{2\pi f_{cp}} \quad\quad C_3' = \frac{C_3''}{2\pi f_{cp}}$$

Step 3. Choose a value $R = R_1 = R_2 = R_3$ which will produce convenient sizes for C_1, C_2, and C_3 according to

$$C_1 = \frac{C_1'}{R} \qquad C_2 = \frac{C_2'}{R} \qquad C_3 = \frac{C_3'}{R}$$

This procedure is called *impedance scaling.*

Step 4. To minimize errors due to the op amp, verify that the following is satisfied

$$A_v \text{ (at } f_{cp}) \geqq 100$$

20.3 SECOND-ORDER HIGH-PASS FILTER

ALTERNATE NAMES   Unity-gain high-pass filter, active high-pass filter, active inductorless high-pass filter, ac-coupled voltage follower, active RC filter.

PRINCIPLES OF OPERATION   The circuit shown in Fig. 20.5 provides zero response at dc and unity gain from f_{cp} up to the frequency where the op-amp gain crosses unity.   By proper choice of R_1, R_2, C_1, and C_2 the transfer function can be made to exhibit the range of characteris-

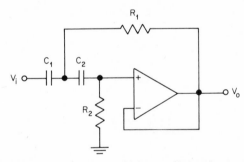

Fig. 20.5 A single-feedback second-order high-pass filter using a unity-gain amplifier.

tics shown in Fig. 20.6.   These curves were obtained from an actual circuit using a 741 op amp.   The f_{cp} was chosen to be 100 Hz.   It should be remembered that this circuit is a high-pass filter only for frequencies between f_{cp} and f_u of the op amp.   Above f_u the gain rolls off at 20 or 40 dB/decade.   Also, the filter will have progressively more error as f_u is approached.

The transfer function of Fig. 20.5 is

$$A_{vc} = \frac{V_o}{V_i} = \frac{s^2}{s^2 + s(1/R_2 C_1 + 1/R_2 C_2) + 1/R_1 R_2 C_1 C_2}$$

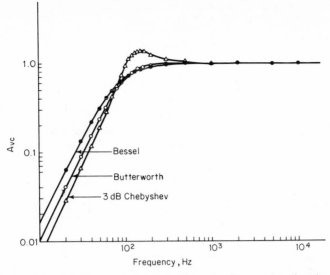

Fig. 20.6   Test data of three typical 2-pole high-pass filters using the circuit of Fig. 20.5.

The damping factor ζ determines the shape of A_{vc} in the frequency region near f_{cp}. Low values of ζ cause the frequency-response curve to have more peaking near the pole (corner) frequency. This term is related to the familiar circuit Q by

$$\zeta = \frac{1}{2Q}$$

DESIGN PARAMETERS

Parameter	Description
A_v	Op-amp voltage gain as function of frequency
A_{vc}	Voltage gain of circuit as function of frequency
C, C_1, C_2	Common value of C_1 and C_2; determine both f_{cp} and ζ
f_{cp}	Poles, or corner frequency, of circuit
f_u	Op-amp unity-gain crossover frequency
Q	Determines height of peak in frequency response
R_1, R_2	Final values for resistors after both impedance and frequency scaling
R_1', R_2'	Unscaled resistor values from Table 20.3
s_1, s_2	Locations of complex poles in s domain
S_A^B	Sensitivity of parameter B to a change in parameter A
V_i	Circuit input voltage
V_o	Circuit output voltage
ζ	Damping factor
ω_n	Natural radian frequency of poles of circuit

TABLE 20.3   Unscaled Resistor Values for Fig. 20.5

Type of 2-pole high-pass filter	ζ	R_1', Ω	R_2', Ω
Bessel	0.8659	1.103	1.471
Butterworth	0.7072	0.7072	1.414
Chebyshev, 0.1-dB peak	0.6516	0.6105	1.438
0.25-dB peak	0.6179	0.5624	1.473
0.5-dB peak	0.5789	0.5131	1.531
1-dB peak	0.5228	0.4509	1.650
2-dB peak	0.4431	0.3743	1.906
3-dB peak	0.3833	0.3223	2.194

DESIGN EQUATIONS

Description	Equation	
Transfer function (voltage gain) of circuit	$$A_{vc} = \frac{V_o}{V_i} = \frac{s^2}{s^2 + s\left(\dfrac{1}{R_2 C_1} + \dfrac{1}{R_2 C_2}\right) + \dfrac{1}{R_1 R_2 C_1 C_2}}$$	(1)
Location of two complex poles of Eq. (1)	$$s_1,\, s_2 = -\frac{1}{2 R_2}\frac{C_1 + C_2}{C_1 C_2} \pm \left[\left(\frac{C_1 + C_2}{2 R_2 C_1 C_2}\right)^2 - \frac{1}{R_1 R_2 C_1 C_2}\right]^{1/2}$$	(2)
Damping factor of circuit	$$\zeta = \frac{1}{2}\left(\frac{R_1 C_1}{R_2 C_2}\right)^{1/2} + \frac{1}{2}\left(\frac{R_1 C_2}{R_2 C_1}\right)^{1/2}$$	(3)
Sensitivity* of f_{cp} to variations in R_1, R_2, C_1, or C_2	$$S_{R_1}^{f_{cp}} = S_{R_2}^{f_{cp}} = S_{C_1}^{f_{cp}} = S_{C_2}^{f_{cp}} = -\tfrac{1}{2}$$	(4)
Sensitivity of ζ to variations in R_1, R_2, C_1, or C_2	$$S_{R_1}^{\zeta} = \tfrac{1}{2}$$	(5a)
	$$S_{R_2}^{\zeta} = \frac{1}{2} - \frac{1}{2\zeta\omega_n R_2}\left(\frac{1}{C_1} + \frac{1}{C_2}\right)$$	(5b)
	$$S_{C_1}^{\zeta} = \frac{1}{2} - \frac{1}{2\zeta\omega_n R_2 C_1}$$	(5c)
	$$S_{C_2}^{\zeta} = \frac{1}{2} - \frac{1}{2\zeta\omega_n R_2 C_2}$$	(5d)
Relationship between initial and final resistor values	$$R_1 = K R_1' \qquad R_2 = K R_2'$$	(6)
	where $K = \dfrac{1}{2\pi f_{cp} C_1} = \dfrac{1}{2\pi f_{cp} C_2}$	
Required op-amp open-loop gain to assure accuracy	$A_v \geq 100$ at all frequencies where high-pass operation is required	(7)

* $S_{R_1}^{f_{cp}} = -\tfrac{1}{2}$ means that if R_1 increases in value by 1%, f_{cp} will decrease in frequency by ½%. After all sensitivites of a given parameter are computed, they are algebraically added to determine the total result.

DESIGN PROCEDURE

The design approach to be presented requires very simple calculations even though Eqs. (1) to (7) may appear quite difficult. The end result is capacitors of equal value but resistors of different values. Other approaches may give equal-sized resistors and different-sized capacitors. Setting two variables equal at the outset is the key to simplified design.

Step 1. Choose R_1' and R_2' from Table 20.3 according to the type of filter required.

Step 2. Using the chosen corner frequency f_{cp}, perform the frequency scaling

$$C = 1/2\pi f_{cp}$$

Step 3. Choose a constant K which will provide practical capacitor sizes for C_1 and C_2 according to

$$C_1 = C_2 = C/K$$

Step 4. Calculate values for resistors with

$$R_1 = KR_1' \qquad R_2 = KR_2'$$

These last two steps are called impedance scaling.

Note: Steps 1 to 4 cover the basic design of the filter. The remaining steps are provided only for the designer who wishes to obtain more insight into the filter operation, error sources, etc.

Step 5. Compute the damping factors ζ using Eq. (3). Compare the result with data in Table 20.3 to verify that the correct filter has been designed.

Step 6. If required, use Eq. (4) to compute the sensitivity of f_{cp} to variations in R_1, R_2, C_1, and C_2. Likewise, Eqs. (5) can be used to determine the sensitivity of ζ to changes in R_1, R_2, C_1, and C_2.

Step 7. Determine the range of frequencies where $A_v \geq 100$ from the op-amp data sheet. This value of A_v must be maintained in order to keep the actual frequency response within 0.1 dB of the theoretical response. This is even more important as temperature changes since A_v has a strong dependence on temperature in most op amps.

EXAMPLE OF SECOND-ORDER HIGH-PASS FILTER   A Butterworth high-pass filter will be designed to illustrate the seven design steps. The response of an actual filter built with this procedure was shown in Fig. 20.6 along with Chebyshev and Bessel filters.

Design Requirements

$$f_{cp} = 100 \text{ Hz} \qquad \text{response} = \text{Butterworth } (\zeta = 0.707)$$
$$\text{Maximum capacitor size} = 0.1 \ \mu\text{F}$$

Device Data

$$A_v \cong 100 \qquad \text{up to 10 kHz}$$

Step 1.   From Table 20.3 we get

$$R_1' = 0.7072 \ \Omega \qquad R_2' = 1.414 \ \Omega$$

Step 2.   Frequency scaling gives

$$C = \frac{1}{2\pi f_{cp}} = \frac{1}{2\pi \times 100} = 1.592 \times 10^{-3} \ \text{F}$$

Step 3.   If we want the capacitor sizes of C1 and C2 to be 0.1 μF,

$$K = \frac{C}{C_1} = \frac{C}{C_2} = \frac{1.592 \times 10^{-3}}{10^{-7}} = 1.592 \times 10^4$$

Step 4.   The final resistor values become

$$R_1 = KR_1' = (1.592 \times 10^4)(0.7072) = 11,255 \ \Omega$$
$$R_2 = KR_2' = (1.592 \times 10^4)(1.414) = 22,505 \ \Omega$$

Step 5.   The damping factor is checked at this point:

$$\zeta = \frac{1}{2}\left(\frac{R_1 C_1}{R_2 C_2}\right)^{1/2} + \frac{1}{2}\left(\frac{R_1 C_2}{R_2 C_1}\right)^{1/2}$$

$$= \frac{1}{2}\left(\frac{R_1}{R_2}\right)^{1/2} + \frac{1}{2}\left(\frac{R_1}{R_2}\right)^{1/2} = \left(\frac{R_1}{R_2}\right)^{1/2}$$

Thus
$$\zeta = \left(\frac{11,255}{22,505}\right)^{1/2} = 0.7072$$

This checks out with Table 20.3.

Step 6.   The sensitivity of f_{cp} to component variations is

$$S_{R_1}^{f_{cp}} = S_{R_2}^{f_{cp}} = S_{C_1}^{f_{cp}} = S_{C_2}^{f_{cp}} = -\tfrac{1}{2}$$

Therefore, if any of these passive components increase in value by 1 percent, f_{cp} will decrease in frequency by 0.5%.

The sensitivity of ζ to component variations is

$$S_{R_1}^{\zeta} = \tfrac{1}{2}$$

$$S_{R_2}^{\zeta} = \frac{1}{2} - \frac{1}{2\zeta\omega_n R_2}\left(\frac{1}{C_1} + \frac{1}{C_2}\right)$$

$$= \frac{1}{2} - \frac{1/10^{-7} + 1/10^{-7}}{2(0.7072)(2\pi)(100)(22,505)} = -0.50$$

$$S^{\zeta}_{C_1} = \frac{1}{2} - \frac{1}{2\zeta\omega_n R_2 C_1}$$

$$= \frac{1}{2} - \frac{1}{2(0.7072)(2\pi)(100)(22{,}505 \times 10^{-7})}$$

$$= 1.310 \times 10^{-6}$$

$$S^{\zeta}_{C_2} = \frac{1}{2} - \frac{1}{2\zeta\omega_n R_2 C_2}$$

$$= \frac{1}{2} - \frac{1}{2(0.7072)(2\pi)(100)(22{,}505 \times 10^{-7})}$$

$$= 1.310 \times 10^{-6}$$

Step 7. Since $A_v > 100$ from dc to 10 kHz, the response curve should be stable from f_{cp} to 10 kHz. A 100 percent variation of A_v should therefore cause less than a 1 percent change (<0.1 dB) in A_{vc} within this frequency range.

20.4 THIRD-ORDER HIGH-PASS FILTER

ALTERNATE NAMES Unity-gain high-pass filter, active high-pass filter, active inductorless high-pass filter, ac-coupled voltage follower, active *RC* filter.

PRINCIPLES OF OPERATION The circuit shown in Fig. 20.7 provides zero response at dc and unity gain from f_{cp} up to the frequency where the op-amp gain crosses unity. By proper choice of the six passive components the transfer function can be made to exhibit the range of characteristics shown in Fig. 20.8. These curves were obtained using an actual circuit comprising a 741 op amp. The f_{cp} was chosen to be 100 Hz. It should be noted that this circuit is a high-pass filter only for frequencies between f_{cp} and f_u of the op amp. Experience tells us

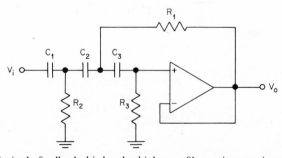

Fig. 20.7 A single-feedback third-order high-pass filter using a unity-gain amplifier.

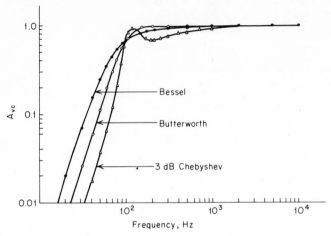

Fig. 20.8   Test data of three typical 3-pole high-pass filters using the circuit of Fig. 20.7.

that f_u varies with temperature. The filter will accordingly have more error drift in the region near f_u.

The transfer function of Fig. 20.7 is

$$A_{vc} = \frac{V_o}{V_i} = \frac{s^3}{s^3 + Ds^2 + Es + F}$$

where $D = \dfrac{1}{R_3}\left(\dfrac{1}{C_1} + \dfrac{1}{C_2} + \dfrac{1}{C_3}\right) + \dfrac{1}{R_2C_1}$

$E = \dfrac{1}{R_3}\left(\dfrac{1}{R_1C_2C_3} + \dfrac{1}{R_1C_1C_3} + \dfrac{1}{R_2C_1C_3} + \dfrac{1}{R_2C_1C_2}\right)$

$F = \dfrac{1}{R_1R_2R_3C_1C_2C_3}$

DESIGN PARAMETERS

Parameter	Description
A_v	Op-amp voltage gain as function of frequency
A_{vc}	Voltage gain of circuit as function of frequency
C	Initial capacitor size
C_1, C_2, C_3	Final capacitor sizes
D, E, F	Variables used in Eq. (1)
f_{cp}	Corner frequency of filter
K	Constant used in filter design
R_1, R_2, R_3	Final resistor sizes
R_1', R_2', R_3'	Initial unscaled resistor sizes
V_i	Input voltage to circuit
V_o	Output voltage from circuit

DESIGN EQUATIONS

Description	Equation
Transfer function (voltage gain) of circuit	$A_{vc} = \dfrac{V_o}{V_i} = \dfrac{s^3}{s^3 + Ds^2 + Es + F}$ (1)

$$\text{where } D = \frac{1}{R_3}\left(\frac{1}{C_1} + \frac{1}{C_2} + \frac{1}{C_3}\right) + \frac{1}{R_2 C_1}$$

$$E = \frac{1}{R_3}\left(\frac{1}{R_1 C_2 C_3} + \frac{1}{R_1 C_1 C_3} \right.$$
$$\left. + \frac{1}{R_2 C_1 C_3} + \frac{1}{R_2 C_1 C_2}\right)$$

$$F = \frac{1}{R_1 R_2 R_3 C_1 C_2 C_3}$$

| Relationship between initial and final resistor values | $R_1 = KR_1'$ $R_2 = KR_2'$ $R_3 = KR_3'$ (2) |

$$\text{where } K = \frac{1}{2\pi f_{cp} C_1} = \frac{1}{2\pi f_{cp} C_2} = \frac{1}{2\pi f_{cp} C_3}$$

| Required op-amp open-loop gain to assure accuracy | $A_v \geq 100$ at all frequencies where high-pass operation is required   (3) |

DESIGN PROCEDURE

The following design steps require only simple calculations even though the characteristic equation (1) of a third-order filter is quite cumbersome. A good feature of the procedure is that it gives equal-sized capacitors. This is usually convenient since it is easier to have a wide range of resistors available than capacitors.

Step 1.   Choose R_1', R_2', and R_3' from Table 20.4 according to the type of filter required.

Step 2.   Using the chosen corner frequency f_{cp}, perform the frequency scaling

$$C = \frac{1}{2\pi f_{cp}}$$

Step 3.   Choose a constant K which will provide practical capacitor sizes for C_1, C_2, and C_3 according to

$$C_1 = C_2 = C_3 = \frac{C}{K}$$

Step 4.   Calculate values for resistors with

$$R_1 = KR_1' \qquad R_2 = KR_2' \qquad R_3 = KR_3'$$

These last two steps are called *impedance scaling.*

Step 5.   Determine the range of frequencies where $A_v \geq 100$ using the op-amp data sheet.   This value of A_v must be maintained in order to keep the actual frequency response within 0.1 dB of the theoretical response.   This is even more important as temperature changes since A_v is always a function of temperature.

TABLE 20.4   Unscaled Resistor Values for Fig. 20.7

Type of 3-pole high-pass filter	R_1',Ω	R_2',Ω	R_3',Ω
Bessel	0.7027	1.012	3.940
Butterworth	0.2820	0.7184	4.941
Chebyshev, 0.1-dB peak	0.1503	0.5479	7.435
0.25-dB peak	0.1169	0.4955	9.017
0.5-dB peak	0.08905	0.4444	11.17
1-dB peak	0.06180	0.3896	15.56
2-dB peak	0.03595	0.3212	25.69
3-dB peak	0.02303	0.2756	39.48

20.5   MULTIPLE-FEEDBACK BANDPASS FILTER

ALTERNATE NAMES   Dual-feedback bandpass filter, active resonator, active filter, active bandpass amplifier, active RC filter.

PRINCIPLES OF OPERATION   This circuit, shown in Fig. 20.9, is useful for several reasons.
1. It requires only one op amp.
2. Adjustment of the resonant frequency f_0 can be performed with one resistor R_2.
3. If Q is less than 10, the sensitivity of Q and f_0 to component variations is not large.
4. If Q is less than 10, a large spread in calculated component values will not occur.
5. One resistor, R_3, can be used to adjust both Q and the midband

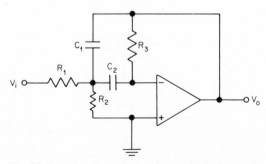

Fig. 20.9   A bandpass filter using multiple feedback.

gain H. Since this resistor also affects f_0, Q and H should always be adjusted before f_0.

This is an inverting circuit with a transfer function expressed as

$$A_{vc} = \frac{V_o}{V_i} = \frac{-A_s}{s^2 + Bs + C}$$

where $\quad A = \dfrac{1}{R_1 C_1} \qquad B = \dfrac{1/C_1 + 1/C_2}{R_3} \qquad C = \dfrac{1/R_1 + 1/R_2}{R_3 C_1 C_2}$

Because capacitors are more difficult to trim, the design of this circuit often begins by assuming $C = C_1 = C_2$, where C is some practical value. We can now state the effect of the three resistors on f_0, H and, Δf ($\Delta f = f_0/Q$).

$$R_1 = \frac{1}{2\pi \, \Delta f \, HC} \qquad R_2 = \frac{1}{2\pi \, C \left(\dfrac{2 f_0^2}{\Delta f} - \Delta f H \right)}$$

$$R_3 = \frac{1}{\pi \, \Delta f \, C}$$

We note from the above that
1. R_1 affects both Δf and H.
2. R_2 affects f_0, Δf, and H, but the effect on Δf and H is small.
3. R_3 affects only Δf.

We can also invert these three equations to get

$$f_0 = \frac{1}{2\pi} \left[\frac{1}{R_3 C_1 C_2} \left(\frac{1}{R_1} + \frac{1}{R_2} \right) \right]^{1/2}$$

$$Q = \frac{f_0}{\Delta f} = \frac{[R_3(1/R_1 + 1/R_2)]^{1/2}}{(C_2/C_1)^{1/2} + (C_1/C_2)^{1/2}}$$

$$H = \frac{R_3 C_2}{R_1(C_1 + C_2)}$$

DESIGN PARAMETERS

Parameter	Description
A_{v,f_0}	Open-loop voltage gain of op amp at frequency f_0
A_{vc}	Closed-loop voltage gain of circuit as a function of frequency
C, C_1, C_2	Common value for capacitors in circuit
$\Delta C/C$	Fractional change in capacitance of a capacitor C over a specified temperature range
ΔF	Frequency difference between -3-dB points on A_{vc} response curve (bandwidth)
f_0	Resonant frequency of circuit
H	Voltage gain of circuit at f_0

Parameter	Description
I_b	Input bias current of op amp
Q	Quality factor of circuit
R_1	Resistor which controls input resistance of circuit
R_2	Resistor which principally controls resonant frequency of circuit
R_3	Resistor which affects only Q of circuit
$\Delta R/R$	Fractional change in the resistance of a resistor R over a specified temperature range
S_A^B	Sensitivity of parameter B to changes in parameter A
V_i	Input voltage to circuit
V_o	Output voltage from circuit
V_{oo}	Circuit output offset voltage

DESIGN EQUATIONS

Description	Equation	
Voltage gain of circuit	$$A_{vc}=\frac{V_o}{V_i}=\frac{-As}{s^2+Bs+C}$$	(1)
	where $A=\dfrac{1}{R_1C_1}$ $B=\dfrac{1/C_1+1/C_2}{R_3}$ $$C=\frac{1/R_1+1/R_2}{R_3C_1C_2}\qquad s=2\pi f$$	
Voltage gain of circuit at f_0	$$H=\frac{R_3C_2}{R_1(C_1+C_2)}$$	(2)
Bandpass-filter center frequency	$$f_0=\frac{1}{2\pi}\left(\frac{1/R_1+1/R_2}{R_3C_1C_2}\right)^{1/2}$$	(3)
Q of circuit	$$Q=\frac{[R_3(1/R_1+1/R_2)]^{1/2}}{(C_2/C_1)^{1/2}+(C_1/C_2)^{1/2}}$$	(4)
Bandwidth (3 dB down from gain at f_0)	$$\Delta f=\frac{f_0}{Q}=\frac{1/C_1+1/C_2}{2R_3}$$	(5)
R_1 in terms of other parameters	$$R_1=\frac{1}{2\pi\,\Delta f\,HC}$$	(6)
R_2 in terms of other parameters	where $C=C_1=C_2$ $$R_2=\frac{1}{2\pi C\left(\dfrac{2f_0^2}{\Delta f}-\Delta f H\right)}$$	(7)
R_3 in terms of other parameters	$$R_3=\frac{1}{\pi\,\Delta f\,C}$$	(8)
Sensitivity of f_0 to component parameter changes	$$S_{R_3}^{f_0}=S_{C_1}^{f_0}=S_{C_2}^{f_0}=-\tfrac{1}{2}$$	(9)
	$$S_{R_1}^{f_0}=\frac{-1}{8\pi^2f_0^2R_1R_3C_1C_2}$$	(10)
	$$S_{R_2}^{f_0}=\frac{-1}{8\pi^2f_0^2R_2R_3C_1C_2}$$	(11)
Sensitivity of Q to component parameter changes	$$S_{R_1}^Q=\frac{R_1}{2(R_1+R_2)}-\frac{1}{2}$$	(12)

$$S_{R_3}^Q = \frac{1}{2} \tag{13}$$

$$S_{C_1}^Q = \frac{Q}{2\pi f_o R_3 C_1} - \frac{1}{2} \tag{14}$$

$$S_{C_2}^Q = \frac{Q}{2\pi f_o R_3 C_2} - \frac{1}{2} \tag{15}$$

DESIGN PROCEDURE

This list of design steps assumes that capacitor values are more difficult to trim than resistor values. It also assumes that both capacitor values are identical. The algebra is greatly simplified since several variables disappear from the equations.

Step 1. Choose values for f_o, H, and Q. As mentioned previously, this circuit design should not be attempted if $Q > 10$ is required. The chosen op amp places some restrictions on the choices for f_o and H. Using the op-amp open-loop frequency plot, make sure that $H < 0.01 A_v$ at f_o. This will guarantee that a 100 percent change in A_v will have much less than a 1 percent effect on f_o and H.

Step 2. Let $C = C_1 = C_2$ be some practical value. Compute

$$R_3 = \frac{2Q}{2\pi f_o C}$$

If R_3 is too large, then I_b, the op-amp input bias current, will cause a dc offset at V_o with a magnitude of

$$V_{oo} = I_b R_3$$

If this offset is larger than allowed for the application, choose a higher C. Recompute R_3 and V_{oo}.

Step 3. Find R_1 from

$$R_1 = \frac{Q}{2\pi f_o C H}$$

Step 4. Compute R_2 from

$$R_2 = \frac{Q}{(2\pi f_o C)(2Q^2 - H)}$$

Step 5. Determine the sensitivity of f_o to component parameter variations using Eqs. (9) to (11).

Step 6. Determine the sensitivity of Q to component parameter variations using Eqs. (12 to (15).

Step 7. Verify that H is correct using Eq. (2).

Step 8. Verify that f_o is correct using Eq. (3).

Step 9. Verify that Q is correct using Eq. (4).

Step 10. Verify that Δf is correct using Eq. (5).

EXAMPLE OF BANDPASS-FILTER DESIGN   A bandpass filter with a center frequency of 1000 Hz will be designed using the 10 design steps. Fig. 20.10 shows the measured frequency response of the resulting circuit.

Design Requirements

$$f_o = 1000 \text{ Hz} \qquad H = 10 \qquad Q = 5$$

$$\Delta f = \frac{f_o}{Q} = 200 \text{ Hz} \qquad V_{oo} \text{ (max)} = \pm 1 \text{ V}$$

Device Data

$$A_v \text{ (1000 Hz)} = 1000 \qquad I_b = 10^{-8} \text{ A}$$

$$\frac{\Delta R_1}{R_1} = \frac{\Delta R_2}{R_2} = \frac{\Delta R_3}{R_3} = +0.02 \qquad -55 \text{ to } +125°C$$

$$\frac{\Delta C_1}{C_1} = \frac{\Delta C_2}{C_2} = -0.03 \qquad -55 \text{ to } +125°C$$

Step 1.   The required parameters of the circuit are $f_o = 1000$ Hz, $H = 10$, and $Q = 5$. We next verify that

$$H \leq 0.01 A_v(f_o)$$
$$10 \leq 0.01(1000)$$
$$10 \leq 10$$

This is obviously satisfied.

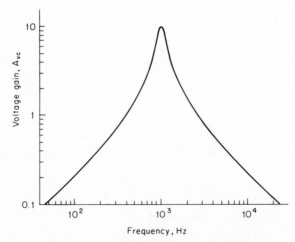

Fig. 20.10   Test data from a typical bandpass filter using the circuit shown in Fig. 20.9.

Step 2. Our initial choice for C will be 0.01 μF. We next compute

$$R_3 = \frac{2Q}{2\pi f_o C} = \frac{2(5)}{(2\pi)(1000 \times 10^{-8})} = 159 \text{ k}\Omega$$

The output offset is now checked

$$V_{oo} = I_b R_3 = 10^{-8} (1.6 \times 10^5) = 1.6 \text{ mV}$$

Step 3. R_1 is computed:

$$R_1 = \frac{Q}{2\pi f_o C H} = \frac{5}{2\pi(1000 \times 10^{-8})(10)} = 7960 \ \Omega$$

Step 4. R_2 is determined:

$$R_2 = \frac{Q}{(2\pi f_o C)(2Q^2 - H)} = \frac{5}{2\pi(1000 \times 10^{-8})(2 \times 5^2 - 10)} = 1990 \ \Omega$$

Step 5. The sensitivity of f_o to various parameter variations is computed:

R_3: $\qquad\qquad\qquad\qquad S_{R_3}^{f_o} = -\frac{1}{2}$

Since $\Delta R_3/R_3$ increases by 0.02 as the temperature increases from -55 to $+125°$C, f_o will change by $(-\frac{1}{2})(0.02)(1000 \text{ Hz}) = -10$ Hz.

C_1 and C_2: $\qquad\qquad\qquad S_{C_1}^{f_o} = S_{C_2}^{f_o} = -\frac{1}{2}$

We have indicated that $\Delta C_1/C_1$ and $\Delta C_2/C_2$ both change by -0.03 as the temperature changes from -55 to $+125°$C. This will make f_o increase by $(-\frac{1}{2})(-0.03)(1000) = +15$ Hz for each capacitor.

R_1: $\quad S_{R_1}^{f_o} = \dfrac{-1}{8\pi^2 f_o^2 R_1 R_3 C_1 C_2}$

$$= \frac{-1}{(8\pi^2 \times 10^6)(7960)(159,000 \times 10^{-8}) \times 10^{-8}} = -0.1$$

If $\Delta R_1/R_1 = +0.02$, as the temperature varies from -55 to $+125°$C, f_o will change $(-0.1)(0.02)(1000) = -10$ Hz.

R_2: $\quad S_{R_2}^{f_o} = \dfrac{-1}{8\pi^2 f_o^2 R_2 R_3 C_1 C_2}$

$$= \frac{-1}{(8\pi^2 \times 10^6)(1990)(159,000 \times 10^{-8}) \times 10^{-8}} = -0.4$$

This will cause a -8-Hz shift in f_o.

The total shift in frequency caused by variations in R_1, R_2, R_3, C_1, and C_2 is $-10 + 15 + 15 - 10 - 8 = 2$ Hz.

Step 6. The sensitivity of Q to various parameter variations is computed:

R_1: $\qquad S_{R_1}^Q = \dfrac{R_1}{2(R_1 + R_2)} - \dfrac{1}{2} = \dfrac{7960}{2(7960 + 1990)} - \dfrac{1}{2} = -0.1$

The $\Delta R_1/R_1 = +0.02$ change will cause Q to change by $(-0.1)(0.02)(5)$ $= -0.01$ as the temperature increases from -55 to $+125°C$:

R_3: $\qquad\qquad\qquad\qquad S_{R_3}^Q = \dfrac{1}{2}$

This will cause Q to change by $(\frac{1}{2})(0.02)(5) = +0.05$ as the temperature increases from -55 to $+125°C$:

C_1 or C_2: $\qquad S_{C_1}^Q = \dfrac{Q}{2\pi f_o R_3 C_1} - \dfrac{1}{2}$ $\qquad$ identical
$\qquad\qquad\qquad\qquad\qquad\qquad\qquad\qquad\qquad$ since
$\qquad\qquad\qquad S_{C_2}^Q = \dfrac{Q}{2\pi f_o R_3 C_2} - \dfrac{1}{2}$ $\qquad$ $C_1 = C_2$

$$= \frac{5}{2\pi(1000)(159{,}000 \times 10^{-8})} - \frac{1}{2}$$

$$= 4.86 \times 10^{-4}$$

If C_1 and C_2 each decrease by -0.03 as the temperature changes from -55 to $+125°C$, Q will change by $(4.86 \times 10^{-4})(-0.03)(5) = -7.2 \times 10^{-6}$.

The total shift in Q caused by variations in R_1, R_3, C_1, and C_2 is $-0.01 + 0.05 - 2(7.2 \times 10^{-6}) = +0.04$.

Step 7. Equation (2) is used to verify H:

$$H = \frac{R_3 C_2}{R_1(C_1 + C_2)} = \frac{159{,}000 \times 10^{-8}}{7960(10^{-8} + 10^{-8})} = 987$$

Step 8. Equation (3) is used to verify f_o:

$$f_o = \frac{1}{2\pi}\left(\frac{1/R_1 + 1/R_2}{R_3 C_1 C_2}\right)^{1/2}$$

$$= \frac{1}{2\pi}\left[\frac{1/7960 + 1/1990}{(159{,}000 \times 10^{-8}) \times 10^{-8}}\right]^{1/2} = 1000.29 \text{ Hz}$$

Step 9. Equation (4) is used to verify Q:

$$Q = \frac{[R_3(1/R_1 + 1/R_2)]^{1/2}}{(C_2/C_1)^{1/2} + (C_1/C_2)^{1/2}}$$

$$= \frac{[159{,}000(1/7960 + 1/1990)]^{1/2}}{(10^{-8}/10^{-8})^{1/2} + (10^{-8}/10^{-8})^{1/2}} = 4.997$$

Step 10. Equation (5) is used to verify Δf:

$$\Delta f = \frac{1/C_1 + 1/C_2}{2\pi R_3} = \frac{1/10^{-8} + 1/10^{-8}}{2\pi(159,000)} = 200.19$$

20.6 STATE-VARIABLE BANDPASS FILTER

ALTERNATE NAMES Biquadratic filter, biquad filter, active resonator, active filter, active *RC* filter, active bandpass amplifier.

PRINCIPLES OF OPERATION One of the most common monolithic filters is the state-variable configuration. A state-variable bandpass filter is shown in Fig. 20.11. This circuit requires three op amps, which preferably should be in one package for thermal tracking. The state-variable filter has several advantages over single op-amp bandpass filters: (1) if components are properly selected, the passband center frequency f_0 can be made independent of circuit Q; (2) the sensitivity of f_0 and Q

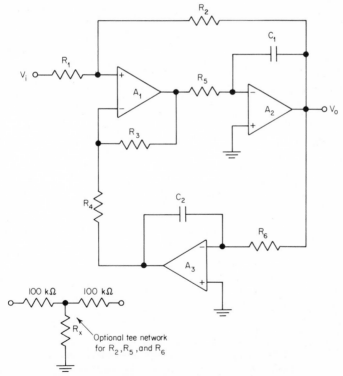

Fig. 20.11 An inverting state-variable bandpass filter using three op amps in a common package.

to parameter variations is very low; and (3) high circuit Q's are possible ($Q \gg 5$).

The filter is made up of two integrators (A_2 and A_3) and a summing amplifier. The passband center frequency is

$$f_o = \frac{1}{2\pi} \left(\frac{R_3}{R_4 R_5 R_6 C_1 C_2} \right)^{1/2}$$

The circuit Q is

$$Q = \frac{1 + R_2/R_1}{1 + R_3/R_4} \left(\frac{R_3 R_5 C_1}{R_4 R_6 C_2} \right)^{1/2}$$

If we initially set $R_3 = R_4$ (fixed resistors) and $C = C_1 = C_2$ (fixed capacitors), these equations reduce to

$$f_o = \frac{1}{2\pi (R_5 R_6 C^2)^{1/2}} \quad \text{and} \quad Q = \frac{(1 + R_2/R_1)(R_5/R_6)^{1/2}}{2}$$

Suppose we let R_5 and R_6 be ganged potentiometers with identical resistances. In this case R_5/R_6 always equals unity and Q depends only on R_1 and R_2. If the common value of R_5 and R_6 is R, the equations reduce to

$$f_o = \frac{1}{2\pi RC} \quad Q = \frac{R_1 + R_2}{2R_1}$$

The ganged potentiometers R_5 and R_6 are used to set f_o while R_2 is used for Q adjustment.

The transfer function for the circuit shown in Fig. 20.11 is

$$A_{vc} = \frac{V_o}{V_i} = \frac{-sA}{s^2 + sB + C}$$

where $A = \dfrac{1}{R_5 C_1} \dfrac{1 + R_3/R_4}{1 + R_1/R_2}$

$\qquad B = \dfrac{1}{R_5 C_1} \dfrac{1 + R_3/R_4}{1 + R_2/R_1}$

$\qquad C = \dfrac{R_3}{R_4} \dfrac{1}{R_5 R_6 C_1 C_2}$

DESIGN PARAMETERS

Parameter	Description
A_{vc}	Closed-loop voltage gain of circuit as a function of frequency
C_1, C_2	Capacitors which help determine bandpass center frequency
f_o	Bandpass center frequency

H	Voltage gain of circuit at resonance
I_b	Op-amp input bias current
Q	Quality factor of circuit
R_1, R_2	Resistors which determine Q of circuit
R_3, R_4	Resistors which set gain of A_1
R_5, R_6	Resistors which control f_0 of circuit
R_x	Resistor used in tee networks to make large simulated resistances in place of R_5 and R_6
S_A^B	Sensitivity of parameter B to variations in parameter A
ω_0	Radian frequency of passband ($\omega_0 = 2\pi f_0$)

DESIGN EQUATIONS

Description	Equation	
Passband center frequency	$$f_0 = \frac{1}{2\pi}\left(\frac{R_3}{R_4 R_5 R_6 C_1 C_2}\right)^{1/2}$$	(1)
Circuit Q	$$Q = \frac{1 + R_2/R_1}{1 + R_3/R_4}\left(\frac{R_3 R_5 C_1}{R_4 R_6 C_2}\right)^{1/2}$$	(2)
Circuit gain at f_0	$$H = \frac{R_2}{R_1}$$	(3)
Voltage gain of circuit (in classical form)	$$A_{vc} = \frac{V_o}{V_i} = \frac{sH\omega_0/Q}{s^2 + s\omega_0/Q + \omega_0^2}$$	(4)
	where $\omega_0 = 2\pi f_0 \qquad s = j2\pi f$	
Sensitivity of f_0 to component parameter variations	$$S_{R_3}^{f_0} = \tfrac{1}{2}$$	(5a)
	$$S_{R_4}^{f_0} = S_{R_5}^{f_0} = S_{R_6}^{f_0} = -\tfrac{1}{2}$$	(5b)
	$$S_{C_1}^{f_0} = S_{C_2}^{f_0} = -\tfrac{1}{2}$$	(5c)
Sensitivity of Q to component parameter variations	$$S_{R_1}^{Q} = S_{R_2}^{Q} = \frac{R_2}{R_1 + R_2}$$	(6a)
	$$S_{R_3}^{Q} = S_{R_4}^{Q} = \frac{1}{2} - \frac{R_3}{R_3 + R_4}$$	(6b)
	$$S_{R_5}^{Q} = S_{R_6}^{Q} = \tfrac{1}{2}$$	(6c)
	$$S_{C_1}^{Q} = S_{C_2}^{Q} = \tfrac{1}{2}$$	(6d)
Voltage gain of circuit	$$A_{vc} = \frac{V_o}{V_i} = \frac{-sA}{s^2 + sB + C}$$	(7)
	where $A = \dfrac{1}{R_5 C_1}\dfrac{1 + R_3/R_4}{1 + R_1/R_2}$	
	$B = \dfrac{1}{R_5 C_1}\dfrac{1 + R_3/R_4}{1 + R_2/R_1}$	
	$C = \dfrac{R_3}{R_4}\dfrac{1}{R_5 R_6 C_1 C_2}$	

DESIGN PROCEDURE

This circuit has two design requirements, the passband center frequency f_o and the circuit Q. The gain H at f_o is fixed once Q is chosen. The design steps to follow will show an optimum way of calculating circuit component values using the required f_o and Q. The sensitivity of f_o and Q to component parameter variations depends only on the quality of passive components used.

Step 1. Compute nominal values for R_3 and R_4 using

$$R_3 = R_4 = \frac{10^8}{f_o}$$

These values are not critical and may be selected from one-half to twice the computed R_3. However, $R_3 = R_4$ must be maintained. If A_1 is a conventional bipolar monolithic op amp, do not allow $R_3 = R_4$ to go above 1 MΩ. If A_1 has a low I_b, R_3 and R_4 may be 10 MΩ or more.

Step 2. Select a common value for C_1 and C_2 in the vicinity of

$$C_1 = C_2 = \frac{10^{-7}}{f_o}$$

Again, as in step 1, a value for $C_1 = C_2$ from one-half to twice the computed value may be used.

Step 3. Compute the required common values for R_5 and R_6 from

$$R_5 = R_6 = \frac{1}{2\pi f_o C_1}$$

As in step 1, these resistors may cause offset problems if they become much larger than 1 MΩ (or 10 MΩ if A_2 and A_3 have low input bias currents). If this is a problem, C_1 and C_2 can be adjusted upward and $R_5 = R_6$ recalculated. If C_1 and C_2 are already too large, R_5 and R_6 can each be replaced with the tee network shown in Fig. 20.11. The value for R_x in each tee network is

$$R_x = \frac{10^{10}}{R_5 - 2 \times 10^5}$$

Offsets at the output of A_2 can be further reduced by returning the noninverting input of A_2 to ground through a resistor with the same resistance as R_5. Likewise, A_3 offsets can be reduced by returning the noninverting input of A_3 to ground through a resistor equal to R_6.

Step 4. Set $R_1 = R_3$. Compute R_2 from

$$R_2 = R_1(2Q - 1)$$

Step 5. Use Eq. (3) to compute the circuit gain H at resonance.

Step 6. Compute the sensitivity of f_o to component parameter variations using Eqs. (5).

Step 7. Compute the sensitivity of Q to component parameter variations using Eqs. (6).

EXAMPLE OF BANDPASS-FILTER DESIGN A filter with a center frequency of 100 Hz will be designed using the seven steps. This example will illustrate the ease with which this circuit can be designed even though the transfer function is quite complex.

Design Requirements

$$f_o = 100 \text{ Hz} \qquad Q = 50$$

Device Data

$$\Delta T = -55 \text{ to } +125°\text{C}$$

$$\frac{\Delta R}{R} = +0.018 \qquad \text{all resistors}$$

$$\frac{\Delta C}{C} = -0.027 \qquad \text{all capacitors}$$

$$\text{Op-amp type} = \text{quad 741} \qquad I_b = 0.5 \text{ } \mu\text{A (maximum)}$$

DESIGN STEPS

Step 1. Nominal values for R_3 and R_4 are $10^8/f_o = 10^8/100 = 1$ MΩ.

Step 2. Nominal values for C_1 and C_2 are $10^{-7}/f_o = 10^{-7}/100 = 1000$ pF.

Step 3. Required values for R_5 and R_6 are

$$R_5 = R_6 = \frac{1}{2\pi f_o C_1} = \frac{1}{2\pi(100 \times 10^{-9})} = 1.59 \text{ MΩ}$$

These resistors will cause an output offset voltage in A_2 and A_3 of $I_b R_5$ = $(0.5 \times 10^{-6})(1.59 \times 10^6) = 0.8$ V. To prevent this offset at the output of A_2 and A_3 we will use two of the tee circuits shown in Fig. 20.11:

$$R_x = \frac{10^{10}}{R_5 - 2 \times 10^5} = \frac{10^{10}}{1.59 \times 10^6 - 2 \times 10^5} = 7194\Omega$$

As an additional guard against offsets we can return the noninverting inputs of A_2 and A_3 to ground through 100 kΩ + 7194 = 107-kΩ resistors.

Step 4.   We set $R_1 = R_3 = 1$ MΩ. R_2 is found from

$$R_2 = R_1(2Q - 1) = 10^6(2 \times 50 - 1) = 99 \text{ MΩ}$$

This value of R_2 appears to be impractical.   We can use the same tee recommended in step 3 to solve this problem.   The required value of R_x is

$$R_x = \frac{10^{10}}{R_2 - 2 \times 10^5} = \frac{10^{10}}{9.9 \times 10^7 - 2 \times 10^5} = 101Ω$$

Step 5.   The circuit gain H at resonance is

$$H = \frac{R_2}{R_1} = \frac{9.9 \times 10^7}{10^6} = 99$$

Step 6.   Sensitivity function computations for f_0 variations are as follows:

Resistors: $S_{R3}^{f_0} = \frac{1}{2}$ $S_{R4}^{f_0} = S_{R5}^{f_0} = S_{R6}^{f_0} = -\frac{1}{2}$

The $\Delta R/R = +0.018$ (-55 to $+125°C$) specified for all resistors will cause f_0 to increase by $\frac{1}{2}(0.018)(100 \text{ Hz}) = 0.9 \text{ Hz}$ owing to R_3.   Resistors R4, R5, and R6 will each cause f_0 to decrease by the same amount.

Capacitors: $S_{C1}^{f_0} = S_{C2}^{f_0} = -\frac{1}{2}$

The $\Delta C/C = -0.027$ specified for each capacitor will cause f_0 to increase by $(-\frac{1}{2})(-0.027)(100 \text{ Hz}) = 1.35 \text{ Hz}$.

The total shift in f_0 due to parameter variations of the above six components will be $0.9 - 0.9 - 0.9 - 0.9 + 1.35 + 1.35 = 0.9 \text{ Hz}$.   This 0.9 percent positive frequency shift will occur as the temperature increases from -55 to $+125°C$.

Step 7.   Sensitivity-function computations for Q variations are as follows:

Resistors R$_1$ and R$_2$: $S_{R1}^{Q} = S_{R2}^{Q} = \dfrac{R_2}{R_1 + R_2} = \dfrac{9.9 \times 10^7}{10^6 + 9.9 \times 10^7} = 0.99$

This will cause Q to vary by $0.99(0.018)(50) = 0.891$ as the temperature varies from -55 to $+125°C$.

Resistors R$_3$ and R$_4$: $S_{R3}^{Q} = S_{R4}^{Q} = \dfrac{1}{2} - \dfrac{R_3}{R_3 + R_4} = \dfrac{1}{2} - \dfrac{10^6}{10^6 + 10^6} = 0$

Resistors R$_5$ and R$_6$: $S_{R5}^{Q} = S_{R6}^{Q} = \frac{1}{2}$

The change in Q due to this sensitivity function is $\frac{1}{2}(0.018)(50) = 0.45$.

Capacitors: $S_{C1}^{Q} = S_{C2}^{Q} = \frac{1}{2}$

The changes in Q due to changes in each capacitance is $\frac{1}{2}(-0.027)(50)$ $= -0.675$.

The total change in Q as the temperature varies from -55 to $+125°C$ will be

$$0.891 + 0.891 + 0.45 + 0.45 - 0.675 - 0.675 = 1.332$$

This is a 2.7 percent change in Q.

20.7 ACTIVE-INDUCTOR BANDSTOP FILTER

ALTERNATE NAMES Notch filter, active bandstop filter, active RC notch filter, parasitic suppressor, hum-reduction circuit.

PRINCIPLES OF OPERATION The circuit shown in Fig. 20.12 provides unity gain for all frequencies from dc to f_u except at f_o, the notch frequency. The voltage gain of the circuit at f_o may be 50 or 60 dB below unity with a careful selection of components. The notch frequency is tuned using C_1 or C_2. These components actually affect both notch frequency and notch depth. The final notch depth (sharpness) is con-

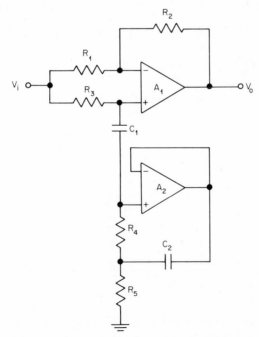

Fig. 20.12 A bandstop filter which uses an active inductor.

trolled using R_1. C_1 or C_2 may become fairly large with a low notch frequency. In this case C_1 or C_2 may use a large fixed capacitor in parallel with a trimmer capacitor.

The transfer function of the circuit is

$$A_{vc} = \frac{V_o}{V_i} = \frac{j2\pi C_2 R^2(f^2 - f_o^2)}{f(R + R_3) + j2\pi C_2 R^2(f^2 - f_o^2)}$$

where $R = R_4 = R_5$ and f_o, the notch frequency, is

$$f_o = \frac{1}{2\pi R(C_1 C_2)^{1/2}}$$

When $f = f_o$, the numerator goes to zero and the gain A_{vc} ideally should also equal zero. In practice, the gain of this circuit will be 50 or 60 dB ($\frac{1}{316}$ to $\frac{1}{1000}$) below unity if

$$\frac{R_1}{R_2} = \frac{R_3}{R_4 + R_5} = \frac{R_3}{2R}$$

is exactly satisfied. The notch-sharpness adjustment R_1 is used to satisfy this equation.

The network made up of C_1, C_2, R_4, R_5, and A_2 simulates a series RLC network. At its resonant frequency f_o it becomes a pure resistance $R_4 + R_5$. Zero filter output occurs only at the resonant frequency, where $R_1/R_2 = R_3/(R_4 + R_5)$. At all other frequencies the differential amplifier is out of balance and the circuit gain is $+1$. Figure 20.13 shows a typical response for this circuit.

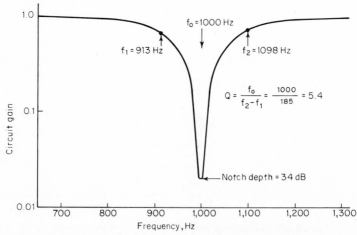

Fig. 20.13   Frequency-response curve of an active-inductor bandstop filter.

DESIGN PARAMETERS

Parameter	Description
A_1	Op amp for differential amplifier
A_2	Op amp for simulated inductance
A_{vc}	Voltage gain of circuit
C_1	Capacitor which is C portion of effective series RLC circuit; this part affects f_o only
C_2	Part of simulated inductance in RLC circuit; this part affects both f_o and Q
f_1, f_2	Frequencies where circuit response is 3 dB below the frequencies far removed from f_o
f_o	Notch center frequency
f_u	Unity-gain crossover frequency of A_1
Q	Quality factor of circuit, $Q = f_o/(f_2 - f_1)$
R	Common value for R_4 and R_5 ($R = R_4 = R_5$)
R_1, R_2, R_3	Gain-determining resistors
R_4, R_5	Resistive portion of effective RLC circuit
R_{in}	Input resistance of circuit
V_i	Input voltage to circuit
V_o	Output voltage from circuit

DESIGN EQUATIONS

Description	Equation	
Voltage gain of circuit	$$A_{vc} = \frac{V_o}{V_i} = \frac{j2\pi C_2 R^2(f^2 - f_o^2)}{f(R + R_3) + j2\pi C_2 R^2(f^2 - f_o^2)}$$ where $R = R_4 = R_5$	(1)
Notch frequency	$$f_o = \frac{1}{2\pi R(C_1 C_2)^{1/2}}$$	(2)
Resistor ratios required for proper operation	$$\frac{R_1}{R_2} = \frac{R_3}{R_4 + R_5} = \frac{R_3}{2R}$$	(3)
Q of circuit	$$Q = \frac{\pi f_o C_2 R}{2}$$	(4)
Input resistance of circuit at f_o	$$R_{in} = \frac{R_1(R_3 + 2R)}{R_1 + R_3 + 2R}$$	(5)
Input resistance of circuit at all frequencies not near f_o	$R_{in} = R_1$	(6)

DESIGN PROCEDURE

In order to simplify the calculations for this bandstop filter we begin by fixing the nominal value of all resistors. The two capacitors then depend only on the resistor choices, the notch frequency, and the notch sharpness.

Step 1. Set $R_1 = R_2 = R_3 = 2R_{in}$, where R_{in} is equal to or above the minimum required input resistance.

Step 2.   Set $R = R_4 = R_5 = R_1/2$.
Step 3.   Compute

$$C_2 = \frac{2Q}{\pi f_o R}$$

Step 4.   Compute

$$C_1 = \frac{1}{(2\pi f_o R)^2 C_2}$$

The smaller of C_1 and C_2 can be used to tune f_o.

Step 5.   Compute Eq. (2) to verify that the correct notch frequency has been implemented.

EXAMPLE OF BANDSTOP-FILTER DESIGN   We will design a medium-Q bandstop filter for 1000 Hz to illustrate the five design steps.

Design Requirements

$$f_o = 1000 \text{ Hz} \qquad Q = 5 \qquad R_{\text{in, min}} = 10,000 \ \Omega$$

Step 1.

$$R_1 = R_2 = R_3 = 2R_{\text{in}} = 2(10,000 \ \Omega) = 20,000 \ \Omega$$

Step 2.

$$R = R_4 = R_5 = \frac{R_1}{2} = \frac{20,000}{2} = 10,000 \ \Omega$$

Step 3.

$$C_2 = \frac{2Q}{\pi f_o R} = \frac{10}{\pi(10,000 \times 10^4)} = 0.318 \ \mu\text{F}$$

Step 4.

$$C_1 = \frac{1}{(2\pi f_o R)^2 C_2} = \frac{1}{(2\pi \times 10^3 \times 10^4)^2(3.18 \times 10^{-7})} = 795 \text{ pF}$$

Step 5.

$$f_o = \frac{1}{2\pi R(C_1 C_2)^{1/2}} = \frac{1}{(2\pi \times 10^4)(0.159 \times 10^{-6})(1.59 \times 10^{-9})^{1/2}}$$

$$= 1000 \text{ Hz}$$

20.8   TWIN-TEE BANDSTOP FILTER

ALTERNATE NAMES   Notch filter, active bandpass filter, active RC notch filter, parasitic suppressor, hum-reduction circuit.

PRINCIPLES OF OPERATION   The circuit shown in Fig. 20.14 provides a means of adjusting circuit Q without affecting notch frequency. The

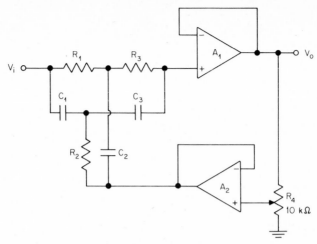

Fig. 20.14 A twin-tee bandstop filter with adjustable Q.

circuit Q is adjustable from approximately 0.3 to 50 using R_4. The minimum Q is obtained when the R_4 wiper is at ground potential. Notch depth and frequency are controlled with the six components in the twin tee. The basic six-component twin tee (C_1, C_2, C_3, R_1, R_2, and R_3) typically provides a maximum Q of approximately 0.3. A_1 and A_2 provide bootstrapping back to the twin-tee ground point, thus making a maximum Q of 50 possible. Figure 20.15 shows the range of adjustment R_4 provides.

The transfer function of the circuit is

$$A_{vc} = \frac{V_o}{V_i} = \frac{s^3 + As^2 + Bs + C}{s^3 + Ds^2 + Es + C}$$

where $A = \dfrac{R_2(R_1 + R_3)C_1C_3}{\Delta}$

$B = \dfrac{R_2(C_1 + C_3)}{\Delta}$

$C = \dfrac{1}{\Delta}$

$D = \dfrac{R_2(R_1 + R_3)C_1C_3 + R_1R_3C_2C_3 + R_1R_2C_2(C_1 + C_3)}{\Delta}$

$E = \dfrac{R_2(C_1 + C_3) + R_1C_2 + (R_1 + R_3)C_3}{\Delta}$

$\Delta = R_1R_2R_3C_1C_2C_3$

The notch frequency is

$$f_0 = \frac{1}{2\pi}\left(\frac{C_1 + C_3}{C_1C_2C_3R_1R_3}\right)^{1/2}$$

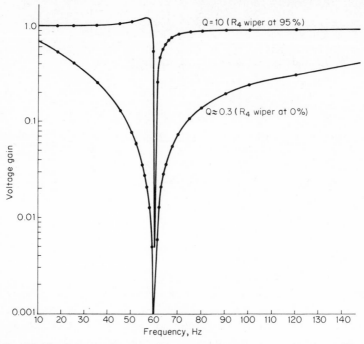

Fig. 20.15 Gain as a function of frequency for the twin-tee bandstop filter.

The design of this circuit is simplified if the following relations are used

$$R_1 = R_3 = 2R_2 \qquad C_1 = C_3 = \frac{C_2}{2}$$

$$R_1 C_1 = R_2 C_2 = R_3 C_3$$

DESIGN PARAMETERS

Parameter	Description
A_1	Buffer amplifier with high input resistance which does not load twin-tee network
A_2	Low-output-resistance buffer which bootstraps ground return of twin tee to circuit output voltage
A_{vc}	Voltage gain of circuit
C_1, C_2, C_3	Capacitors which determine notch frequency
f_0	Notch frequency
Q	Quality factor of circuit
R_1, R_2, R_3	Resistors which determine notch frequency
R_4	Potentiometer used to adjust the Q
R_g	Generator output resistance
V_i	Circuit input voltage
V_o	Circuit output voltage

DESIGN EQUATIONS

Description	Equation

Voltage gain of circuit

$$A_{vc} = \frac{V_o}{V_i} = \frac{s^3 + As^2 + Bs + C}{S^3 + Ds^2 + Es + C} \tag{1}$$

where $A = \dfrac{R_2(R_1 + R_3)C_1C_3}{\Delta}$ $\quad B = \dfrac{R_2(C_1 + C_3)}{\Delta}$ $\quad C = \dfrac{1}{\Delta}$

$$D = \frac{R_2(R_1 + R_3)C_1C_3 + R_1R_3C_2C_3 + R_1R_2C_2(C_1 + C_3)}{\Delta}$$

$$E = \frac{R_2(C_1 + C_3) + R_1C_2 + (R_1 + R_3)C_3}{\Delta}$$

$$\Delta = R_1R_2R_3C_1C_2C_3$$

Notch frequency

$$f_0 = \frac{1}{2\pi}\left(\frac{C_1 + C_3}{C_1C_2C_3R_1R_3}\right)^{1/2} \tag{2}$$

Recommended relationship between resistors

$$R_1 = R_3 = 2R_2 \geqq 100R_g \tag{3}$$

Recommended relationship between capacitors

$$C_1 = C_3 = \frac{C_2}{2} \tag{4}$$

Recommended relationships between resistors and capacitors

$$R_1C_1 = R_2C_2 = R_3C_3 \tag{5}$$

DESIGN PROCEDURE

Since the equations describing this circuit are so complex, a simplified approach is necessary. The steps can be used for a "first cut" design and refinements made afterward.

Step 1. Choose $R_1 = R_3$ equal to a practical value greater than $100R_g$. Set $R_2 = R_1/2$.

Step 2. C_1 and C_3 are found by combining Eqs. (2) to (5):

$$C_1 = C_3 = \frac{1}{4\pi f_0 R_2}$$

Step 3. Using Eq. (4), we now determine

$$C_2 = 2C_1 \qquad C_3 = C_1$$

Step 4. Verify that f_0 is correct using Eq (2):

$$f_0 = \frac{1}{2\pi}\left(\frac{C_1 + C_3}{C_1C_2C_3R_1R_3}\right)^{1/2}$$

EXAMPLE OF BANDSTOP-FILTER DESIGN   A 60-Hz bandstop filter using the four design steps will be designed.

Design Requirements

$$f_o = 60 \text{ Hz} \qquad \text{largest resistor} = 2\text{M}\Omega$$

Design Parameters

$$R_g = 600\Omega$$

DESIGN STEPS

Step 1. We first compute $100 R_g = 100(600) = 60$ kΩ. Set $R_1 = R_3 = 2$ MΩ to satisfy Eq (3). Then set $R_2 = R_1/2 = 1$ MΩ.

Step 2. We determine C_1 and C_3 as follows:

$$C_1 = C_3 = \frac{1}{4\pi f_o R_2} = \frac{1}{4\pi(60 \times 10^6)} = 1320 \text{ pF}$$

Step 3. C_2 is simply $C_2 = 2C_1 = 2(1320) = 2640$ pF.

Step 4. The resonant frequency is double-checked:

$$f_o = \frac{1}{2\pi}\left(\frac{C_1 + C_3}{C_1 C_2 C_3 R_1 R_3}\right)^{1/2}$$

$$= \frac{1}{2\pi}\left[\frac{1.32 \times 10^{-9} + 1.32 \times 10^{-9}}{(1.32 \times 10^{-9})(2.64 \times 10^{-9})(1.32 \times 10^{-9})(2 \times 10^6)(2 \times 10^6)}\right]^{1/2}$$

$$= 60.2 \text{ Hz}$$

REFERENCES

1. Shepard, R. R.: Active Filters, 12: Short Cuts to Network Design, *Electronics,* Aug. 18, 1969, p. 82.
2. Al-Nasser, F.: Tables Shorten Design Time for Active Filters, *Electronics,* Oct. 23, 1972, p. 113.
3. Graeme, J. G., G. E. Tobey, and L. P. Huelsman, "Operational Amplifiers: Design and Applications," McGraw-Hill, New York, 1971, pp. 291–307.
4. Geffe, P. R.: Designer's Guide to Active Bandpass Filters, pts. 1–5, *EDN,* Feb. 5, 1974, p. 68; Mar. 5, 1974, p. 40; Apr. 5, 1974, p. 46; May 5, 1974, p. 63; June 5, 1974, p. 65.
5. Kerwin, W. J., L. P. Huelsman, and R. W. Newcomb: State Variable Synthesis for Insensitive Integrated Circuit Transfer Functions, *IEEE J. Solid State Circuits,* vol. SC-2, September 1967, pp. 87–92.
6. Harris, R. J.: The Design of an Operational Amplifier Notch Filter, *Proc. IEEE,* October 1968, p. 1722.
7. Dobkin, B.: High Q Notch Filter, National Semiconductor Corp., *Linear Brief* LB-5, 1969.
8. Ramey, R. L., and E. J. White: "Matrices and Computers in Electronic Circuit Analysis," McGraw-Hill, New York, 1961, p. 36.

Chapter 21

Regulators

21.1 INTRODUCTION TO MONOLITHIC REGULATORS

The monolithic regulator is an extension of monolithic op-amp technology. "Monolithic" means that all elements of the regulator are fabricated on a single integrated-circuit (IC) silicon chip. Most regulators, however, require one or two capacitors, as a minimum, in addition to the IC regulator device. If additional current capability is needed, or if switching-mode operation is to be implemented, 10 to 20 additional parts are required.

As shown in Fig. 21.1, the main element in an IC regulator is the op amp, which compares a fraction V_2 of the output voltage V_L with a fraction V_1 of the reference voltage V_R. The loop formed by A1 and Q1 forces V_L to that voltage which makes $V_1 = V_2$. The output voltage can therefore be only as stable as V_R. Consequently, zener diode Z1 must be a high-quality temperature-compensated device. In most regulator chips this diode is not a single part but a complex circuit made up of 10 or more parts.

All devices shown in Fig. 21.1 are included in most fixed-voltage monolithic regulators. Many regulators permit external selection of R_3 and R_4 so that the output voltage can be adjusted over a wide range.

Transistor Q1 in Fig. 21.1 is physically the largest device on the chip. In power regulators this device can handle many amperes. In small general-purpose regulators it is typically capable of several tens of milliamperes. This transistor is designed to withstand 40 or 50 V $(V_{CC} - V_L)$ in most regulators.

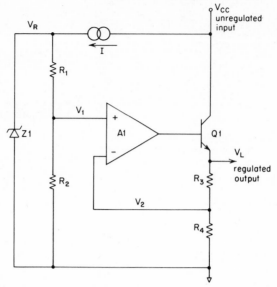

Fig. 21.1 Basic elements of a monolithic voltage regulator. Fixed-voltage regulators contain R_3 and R_4 on the chip, while variable regulators require external resistors.

21.2 PARAMETERS OF MONOLITHIC REGULATORS

Data sheets for monolithic regulators provide information on at least 10 important parameters. In this section we define and discuss each parameter.

LOAD REGULATION

Definition

$$\Delta V_L(\%) = \% \text{ change in load voltage as load}$$
$$\text{resistance varies from minimum to maximum}$$

$$= \frac{\Delta V_L \times 100\%}{V_{L,\text{av}}} \tag{21.1}$$

where ΔV_L = change in output voltage as load
resistance varies from minimum to maximum
$V_{L,\text{av}}$ = average load voltage
= ½ $(V_{L,\text{max}} + V_{L,\text{min}})$

The load regulation is related to the regulator output resistance R_o by

$$\Delta V_L(\%) = \frac{\Delta I_L R_o}{V_{L,\text{av}}} 100\% \tag{21.2}$$

where ΔI_L = change in load current

Some data sheets provide load regulation in millivolts instead of percentages. Equation (21.1) is used to convert from one system of units to the other. A good load-regulation test is one where the load current varies from zero to maximum. The current-limiting circuit in the regulator is usually disabled for this test.

Load regulation is a dc test and does not tell us much about the transient performance of a regulator. The output resistance as a function of frequency provides additional information useful for transient-performance calculations. This will be discussed on the following pages. Equation (21.2) is used to convert from output resistance to load regulation.

LINE REGULATION (INPUT REGULATION, RIPPLE REJECTION)

Definition

$$\Delta V_L(\%/V) = \text{percent change in load voltage for}$$
$$\text{1-V change in input voltage}$$

$$= \frac{\Delta V_L}{\Delta V_{CC} \; V_L} \; 100\% \qquad (21.3)$$

where ΔV_L = peak-to-peak change in load voltage due
        to change in V_{CC}
ΔV_{CC} = peak-to-peak change in input voltage
  V_L = nominal load voltage

This parameter is a measure of the isolation properties of the regulator. The output voltages of most regulators have better than 60-dB (1000:1) isolation from changes in the input voltage. As the frequency of the input ripple increases, the rejection properties of the regulator are degraded. At 1 kHz the ripple rejection may be only 50 dB (316:1). Figure 21.2 shows how this parameter is measured using a transformer to modulate the dc input.

OUTPUT RESISTANCE

Definition

$$R_o = \text{effective resistance between regulator output}$$
$$\text{terminal and its internal ideal voltage source}$$

$$= \frac{V_{L2} - V_{L1}}{I_{L1} - I_{L2}} = \left| \frac{\Delta V_L}{\Delta I_L} \right| \qquad (21.4)$$

where I_{L1} = nominal load current
  I_{L2} = load current 5 or 10% lower than I_{L1}
  V_{L1} = load voltage corresponding to load current I_{L1}
  V_{L2} = load voltage corresponding to load current I_{L2}

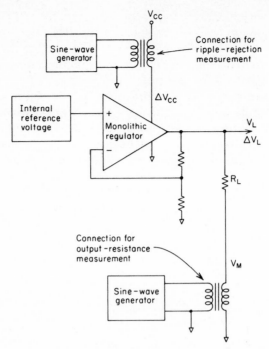

Fig. 21.2 Test circuit for measuring the output resistance of a monolithic regulator as a function of frequency.

The output resistance can be measured using two dc levels [as implied in Eq. (21.4)] or by using ac measurements. In the ac case the load current must be modulated by placing a transformer in series with the load resistor. As shown in Fig. 21.2, the primary of the transformer is driven with an ac generator. Since the ac rms load current is V_M/R_L, we have

$$R_o = \frac{V_{L,\text{rms}}}{V_{M,\text{rms}}/R_L} \tag{21.5}$$

OUTPUT VOLTAGE This is the primary independent parameter of a monolithic regulator. For the circuit of Fig. 21.1 it is equal to

$$V_L = V_R \frac{R_2}{R_1 + R_2} \frac{R_3 + R_4}{R_4} \tag{21.6}$$

The maximum possible value of V_L in adjustable regulators is within 2 to 4 V of the maximum V_{CC}. The lower limit of V_L is usually within 1 or 2 V of ground.

The minimum allowable input-output voltage difference is called the *dropout voltage.* It is usually 1 or 2 V. If the input-output voltage differential is less than the dropout voltage, regulation ceases and the output voltage tracks the input voltage.

OUTPUT-VOLTAGE THERMAL STABILITY

Definition

$$TC(\%/°C) = \text{average temperature coefficient of } V_L$$
$$\text{over specified temperature range}$$

$$= \pm \frac{(V_{L,\text{max}} - V_{L,\text{min}})\, 100\%}{V_{L,\text{av}}\, (T_{\text{max}} - T_{\text{min}})} \qquad (21.7)$$

where T_{max}, T_{min} = two temperatures at which load voltage
                     is measured

$V_{L,\text{max}}$, $V_{L,\text{min}}$ = load voltages measured at
                     above two temperatures

$V_{L,\text{av}}$ = reference load voltage, usually at +25°C

Some regulator data sheets also provide a curve of load voltage as a function of ambient temperature. For some devices a curve of this type indicates a temperature coefficient near zero over a given temperature range. For other less desirable devices, this curve may warn us that the *TC* slope is too steep at temperature extremes.

OUTPUT-VOLTAGE LONG-TERM STABILITY   This parameter is expressed in percentage, percentage per 1000 h, millivolts, or millivolts per 1000 h. Sometimes it is not provided on a data sheet. Occasionally the manufacturer will volunteer the test conditions for this test, but usually the design engineer is left guessing. If the long-term stability is −1 percent per 1000 h one would expect the load voltage to drop to zero in 100,000 h (11.4 years). However, this will not happen because the long-term stability improves with age. Most electronic devices age rapidly for the first 1000 h then level off to a much better stability for many years thereafter. The designer generally does not need to consider aging effects beyond the first 1000 h.

INPUT-VOLTAGE RANGE   The silicon transistors on the monolithic regulator chip all have finite collector-emitter breakdown voltages. This generally limits the input voltage of regulators to 40 or 50 V. The maximum allowed input-output voltage differential is usually about 10 V less than the maximum allowed input voltage.

High-voltage regulator circuits using a monolithic regulator are possible using a circuit as shown in Fig. 21.3.

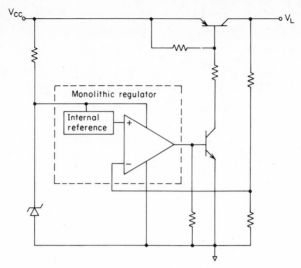

Fig. 21.3 A typical method for extending the voltage rating of a monolithic regulator [1].

SHORT-CIRCUIT OUTPUT CURRENT   Figure 21.4 conveniently defines this parameter. Some regulators allow the designer to choose the magnitude of the short-circuit current. Other regulators have a built-in current limit designed so that the device cannot burn out from excessive dissipation. The designer must exercise caution with adjustable regulators

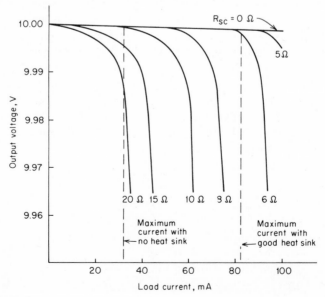

Fig. 21.4 A set of typical output *V-I* curves (RCA CA3085) for a monolithic regulator which has an adjustable short-circuit limit [1].

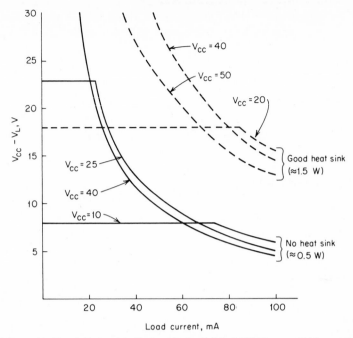

Fig. 21.5 A family of dissipation limits for the RCA CA3085 monolithic regulator [1].

to make sure that the dissipation limit of the device will not be exceeded when the load resistance is zero.   This is done using a family of dissipation curves, as shown in Fig. 21.5.   For example, if $V_{CC} - V_L$ is 20 V and V_{CC} is 40 V, Fig. 21.5 shows that the CA3085 device can provide 73 mA.   Referring back to Fig. 21.4, we note that R_{SC} must not be less than 9 Ω if the short-circuit output current is to stay below 73 mA.

MAXIMUM DISSIPATION

Definition

$$P_D = \text{total device dissipation, W}$$
$$= V_{CC}I_Q + (V_{CC} - V_L)I_L \qquad (21.8)$$

where I_L = load current
I_Q = regulator quiescent current
V_L = load voltage
V_{CC} = input voltage

The typical maximum dissipation curves shown in Fig. 21.5 do not include the $V_{CC}I_Q$ term.   When this term is added, we obtain the 1.6-W maximum published for the regulator.   It is obvious from Fig. 21.5

that a heat sink more than doubles the capability of this regulator. However, if the monolithic regulator is used only as the control element for a large series pass transistor, heat sinking may not be required.

QUIESCENT CURRENT   The current from V_{CC} to ground is defined as the quiescent current.   It typically adds several tenths of a watt to the device dissipation.   In applications where the regulator is used as a low-output-current bias source the dissipation caused by quiescent current may predominate.

OUTPUT NOISE VOLTAGE   This parameter is specified in several ways. Some data sheets specify output noise in percentage (peak to peak) or millivolts (peak to peak).   These are the least desirable methods of presentation because the designer does not know which frequency components of the noise are dominant.   Some manufacturers provide the rms noise over some broad range of frequencies.   This is a little more useful. The best data sheets provide the output spot-noise voltage as a function of frequency.   A typical presentation of this type of data is shown in Fig. 21.6.   A discussion of spot noise is found in Sec. 19.4.

Most of the noise in a monolithic regulator is caused by the reference diode (Z1 in Fig. 21.1).   Consequently, some regulators provide a terminal, called *reference bypass,* which connects directly to the internal reference diode or to the junction of R_1 and R_2 (see Fig. 21.1).   A capacitor is connected from this terminal to ground if the designer must reduce the noise contributed by the reference diode.   This capacitor also significantly improves the ripple rejection of the regulator.

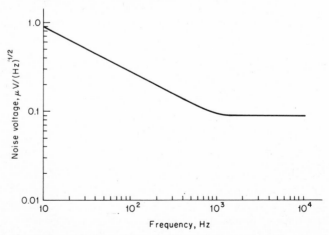

Fig. 21.6 A typical curve of output spot-noise voltage vs. frequency for the LM309 5-V regulator (National Semiconductor Corp.).

21.3 MONOLITHIC CURRENT LIMITERS

Several types of current limiters are incorporated in various monolithic regulators in order to make these devices essentially burnoutproof. The current-limiter circuit causes the regulator output characteristics to behave as shown in Fig. 21.4. Some monolithic regulators monitor both output current I_L and the input-output voltage differential $V_{CC} - V_L$. The $I_L(V_{CC} - V_L)$ product is maintained within the safe operating area of the device by limiting the peak I_L. This technique is used in the LM340 by National Semiconductor. The LM340 also senses chip temperature and adjusts the current limit accordingly. Figure 21.7 graphically displays the interrelationship of peak output current, input-output voltage differential, device dissipation, and chip temperature for the LM340 device.

Figure 21.8 illustrates two current-sensing techniques utilized in monolithic regulators. The first circuit, shown in Fig. 21.8a, is widely used by many manufacturers. The current-sense resistor R_6 is often an integral part of the chip. Adjustable regulators usually allow the designer a choice for R_6. When current flows through R_6, a voltage is developed across the emitter-base junction of Q3. When this voltage reaches 0.6 V, Q3 begins to conduct and thereby steals drive current from Q1. With a short-circuited load ($R_L = 0$) we have Q3 fully ON

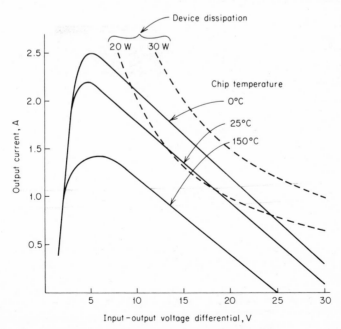

Fig. 21.7 Curves showing how peak load current is controlled by input-output voltage differential and chip temperature (National Semiconductor LM340).

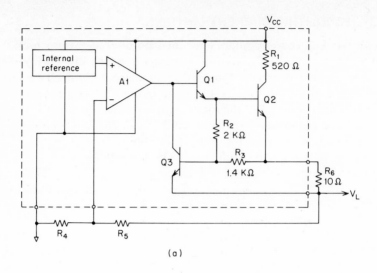

(a)

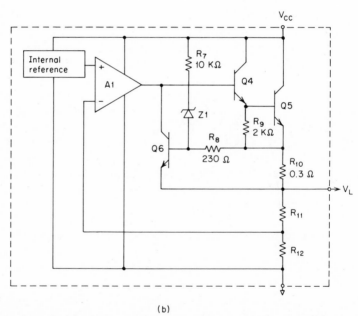

(b)

Fig. 21.8 Current-limiter circuits used by two different monolithic regulators: *(a)* LM100 connected for a 31-mA limit; *(b)* LM340 set up for a 2-A limit.

and Q1 completely OFF. The load current is reduced to zero in this limiting case.

The point at which Q3 begins to steal current from Q1 is determined as follows: Q3 is already partially forward-biased from R_2 and R_3. Assuming Q2 is fully ON, we have approximately 0.7 V across R_2 and

R_3. The voltage across R_3 (and the base-emitter junction of Q3) is then approximately 0.29 V. Assume that 0.7 V across the base-emitter junction of Q3 is sufficient to initiate the removal of base current from Q1. The voltage across R_6 will linearly add to the 0.29-V forward bias already established by R_3. Thus, $0.6-0.29 = 0.31$ V is needed across R_6 to initiate the current limiting. The load current required to do this is

$$I_{L,\max} = \frac{0.31 \text{ V}}{10 \text{ }\Omega} = 31 \text{ mA}$$

The current limiter shown in Fig. 21.8b is a slight modification of the basic current limiter in Fig. 21.8a. Zener diode Z1 and R_7 have been added to make the current limiter sensitive to the input-output voltage differential. As $V_{CC} - V_L$ increases, the current through R_7 likewise increases, causing Q6 to pull more base current away from Q4. As Fig. 21.7 indicates, the peak load current allowed is therefore reduced as $V_{CC} - V_L$ increases. This keeps the dissipation in the safe operating area for all values of $V_{CC} - V_L$.

21.4 CURRENT-BOOSTER TRANSISTORS FOR AN IC REGULATOR

ALTERNATE NAMES External-pass transistor regulator, linear regulator with booster transistors, high-current voltage regulator with foldback current limiting.

PRINCIPLES OF OPERATION The output-current capability of any regulator can be boosted by 100 or more using external-pass transistors. Figure 21.9 shows one possible configuration for a two-transistor booster circuit using the 105 regulator. The 105 can supply approximately 20 mA worst case over temperature. We need booster transistors with a worst-case composite current gain of 100 to provide an output current of 2 A. A third transistor connected to Q1 in the Darlington configuration could boost the output current to 10 A.

The basic principles of the circuitry inside the 105 were presented in Secs. 21.1 and 21.3. The addition of external-pass transistors will increase the current capability of the 105 up to several amperes. However, operation at 2 A or above requires very large heat sinks if the regulator is designed to handle a short-circuited load. Because designers are reluctant to use large heat sinks, two alternative circuits have been developed: (1) switching regulators, which we will discuss in Secs. 21.5 and 21.6, and (2) foldback current limiting, which we discuss in the following paragraphs.

By using a foldback current limiter it is assumed that under short-circuit load conditions the designer does not care if the regulator output voltage drops to zero. The output V-I characteristics of a foldback type

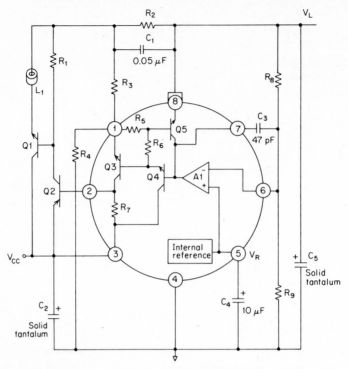

Fig. 21.9 A double-boost transistor circuit for the 105 IC regulator. The output-current capability is 2 A. Foldback current limiting is incorporated using R_3 and R_4. L_1 = ferroxcube K5-001-00/38 [3].

of regulator are shown in Fig. 21.10, giving a comparison with a standard current-limited regulator. We note that if the output terminal is shorted to ground, the Q1 dissipation in the foldback circuit (Fig. 21.9) is less than one-third the dissipation of the standard current-limited circuit.

Resistors R_3 and R_4 convert the standard regulator into a foldback regulator. These resistors supply a voltage to the base of Q5 which acts in a direction opposite to the current-limit voltage developed across R_2. When the output terminal is shorted to ground, the effects of R_3 and R_4 disappear, and R_2 controls the ultimate current limit of the regulator.

Transistors Q1 and Q2 provide current gains of β_1 and β_2, respectively. Except for the currents lost in R_1 and R_7, most of the $\beta_1\beta_2$ current gain is used to increase the output capability of the 105. Inductor L_1 (a ferrite bead) and capacitor C_1 are installed to prevent oscillations caused by certain Q1-Q2-load combinations. If Q1 is a device with a low f_T, such as the 2N3740 ($f_T = 4$ MHz), L_1 and C_1 are not required.

Capacitors C_2 and C_5 are required principally to stabilize the regulator feedback loop. They should be low-inductance, i.e., solid tantalum,

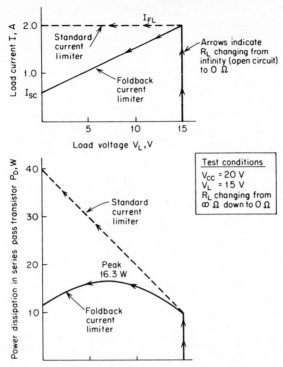

Fig. 21.10 Comparison of the output *V-I* characteristics and Q1 power dissipation for a standard current regulator and the foldback current regulator shown in Fig. 21.9.

capacitors. Electrolytic capacitors should not be used in these locations. Capacitor C_5 is also useful as an output filter since it improves the regulator high-frequency transient performance. Its positive terminal should be close to pin 8 of the 105. Capacitor C_3 should remain at 47 pF for all circuit configurations. This capacitor makes the op amp act as an integrator, thus improving high-frequency stability. Capacitor C_4 performs two functions: (1) it filters out the noise generated by the internal reference diode, and (2) it improves the ripple rejection of the regulator by a factor of 4.

Resistor R_1 is required to minimize the effect of a large collector-base leakage current from Q1. It also provides a minimum collector current for Q2, thus stabilizing the regulator under light load conditions. Resistor R_2 senses current passing out to the load. It establishes the final foldback current after the output voltage has collapsed to zero. Resistors R_3 and R_4 (along with R_2) establish the peak current available at the regulated output voltage. Resistors R_8 and R_9 are used to set the regulated output voltage. For the 105 regulator the parallel combination of R_8 and R_9 should equal 2 kΩ ($\pm$ 30 percent).

CURRENT-BOOSTED-REGULATOR DESIGN PARAMETERS

Parameter	Description
A1	Op amp inside 105 regulator
C_1	Compensation capacitor needed to stabilize loop if Q2 is high-frequency device ($f_T > 50$ MHz)
C_2	Input bypass capacitor required to isolate Q1 from low-level circuitry inside the 105
C_3	Compensation capacitor which transforms op amp into integrator
C_4	Filter capacitor to reduce noise caused by internal reference diode and to improve ripple rejection
C_5	Bypass capacitor to suppress oscillations in the minor feedback loop of Q1, Q2, and Q3; this capacitor also provides high-frequency output filtering for transient load currents
f_T	Frequency of unity current gain for a transistor
I_B	Base current for a transistor
I_{CBO}	Collector-to-base leakage current of a transistor
I_{FL}	Full-load output current (see Fig. 21.10)
I_L	Load current
I_S	Standby current from V_{CC} into the 105
I_{SC}	Design value of short-circuit current
P_D	Total power dissipation in Q1, Q2, or the 105
Q1	Series pass transistor which handles most of the load current
Q2	Buffer pass transistor required for phase inversion and increase in output current
R_1	Minimizes effect of I_{CBO} from Q1
R_2	Current-sense resistor
R_3, R_4	Establishes level of maximum load current I_{FL}
R_5–R_7	Internal 105 resistors
R_8, R_9	Resistors which establish magnitude of V_L
T_A	Ambient temperature, °C
T_J	Junction temperature of Q1, Q2, or 105
V_{CC}	Input voltage
V_{CL}	Voltage required across R_2 to initiate current limiting
V_L	Load voltage
V_R	Regulator internal reference voltage (1.8 V)
V_S	Maximum voltage across R_2 for which 0.1% load regulation is possible
θ_{JA}	Thermal resistance from junction to ambient (Q1, Q2, or 105)

CURRENT-BOOSTED-REGULATOR DESIGN EQUATIONS

Description	Equation	
Output voltage of circuit	$$V_L = \frac{1.8(R_8 + R_9)}{R_9}$$	(1)
Power dissipation in Q1	$P_{D(Q1)} = P_D$ in Fig. 21.10	(2)
Power dissipation in Q2	$$P_{D(Q2)} \approx (V_{CC} - V_L - 0.7)\left(\frac{I_{FL}}{\beta_1} + \frac{0.7}{R_1}\right)$$	(3)
Power dissipation in 105	$P_{D,105} \approx V_{CC}I_S + (V_{CC} - V_L - 0.7)I_{B2}$	(4)
Junction temperature of each device (105, Q1, or Q2)	$T_J = T_A + P_D\theta_{JA}$	(5)

Maximum voltage allowed across current-sense resistor R_2 if load regulation is is to remain within 0.1%

$$V_S = 0.262 - (1.08 \times 10^{-3})\, T_{J,105} \qquad (6)$$

Voltage across current-sense resistor R_2 at which current limiting begins

$$V_{CL} = 0.425 - 10^{-3}\, T_{J,105} \qquad (7)$$

Resistor values:

R_1

$$R_1 << \frac{1}{I_{CBO,\text{max, }Q1}} \qquad (8)$$

R_2

$$R_2 = \frac{V_{CL}}{I_{SC}} \qquad (9)$$

R_3

$$R_3 = \frac{(I_{FL}R_2 - V_S)R_4}{V_L} \qquad (10)$$

R_4

$$R_4 = \frac{V_L + V_S}{2 \times 10^{-3} + I_{B2,\text{max}}} \qquad (11)$$

R_5–R_7

$$R_5 \approx 1400\Omega \quad R_6 \approx 2000\Omega \quad R_7 \approx 620\Omega \quad \text{105 internal resistors} \qquad (12)$$

R_8

$$R_8 = 1129 V_L \qquad (13)$$

R_9

$$R_9 = \frac{2031 V_L}{V_L - 1.8} \qquad (14)$$

Capacitor values:

C_1

$$C_1 = \begin{cases} 0.05\ \mu\text{F} & \text{if } f_T \text{ of Q2} > 50\text{ MHz} \\ 0 & \text{otherwise} \end{cases} \qquad (15)$$

C_2

$$C_2 \geq I_F \quad \mu\text{F} \qquad (16)$$

C_3

$$C_3 = 47\text{ pF} \qquad (17)$$

C_4

$$C_4 = \begin{cases} 10\mu\text{F} & \text{if line regulation must be} < 0.02\ \%/\text{V} \\ 0 & \text{otherwise} \end{cases} \qquad (18)$$

C_5

$$C_5 \geq 1 + I_{FL}^{2.7}\ \mu\text{F} \qquad (19)$$

DESIGN PROCEDURE

We begin our design by considering the thermal problems of Q1, Q2, and the 105. Once these are solved, the resistor and capacitor values can be determined. Initial data must be available on T_A, $T_{J,\text{max}}$ for Q1, Q2, and 105, θ_{JA} for Q1, Q2, and 105, I_S, I_{FL}, I_{SC}, V_L, and V_{CC}.

DESIGN STEPS

Step 1. Use Eq. (5) to find $P_{D,\text{max}}$ of the 105.

Step 2. Determine the maximum base current the 105 can deliver to Q2 by use of Eq. (4).

Step 3. Make a graphical plot of power dissipation in Q1 similar to that shown in Fig. 21.10. The only data required are I_{FL}, I_{SC}, V_{CC}, and V_L.

Step 4. Choose a device for Q1 which satisfies the $P_{D,\max}$ found in step 3 with a good margin. Check to see that the collector-to-emitter breakdown voltage is at least 20 percent larger than V_{CC}. Use Eq. (5) to make sure T_J is safely within the manufacturer's limits.

Step 5. Calculate the current gain required of Q2:

$$\beta_2 = \frac{I_{FL}}{\beta_1 I_{B2,\max}}$$

Base current I_{B2} was found in step 2.

Step 6. Find a value for R_1 from Eq. (8). Substitute the result into Eq. (3), which provides the dissipation in Q2. Choose a device for Q2 which can easily handle this P_D and also V_{CC}. Equation (5) can be used to verify that the T_J of Q2 is within manufacturer's limits.

Step 7. Values for V_{CL} and V_S are found using Eqs. (6) and (7).

Step 8. Calculate resistor values in the following order:

R_2 using Eq. (9)

R_4 using Eq. (11)

R_3 using Eq. (10)

R_8 using Eq. (13)

R_9 using Eq. (14)

Step 9. Capacitor values are determined by using Eqs. (15) to (19).

EXAMPLE OF A 2-A-REGULATOR DESIGN   Although power monolithic regulators exist which can supply 2 A, it is instructional to follow the design steps through a numerical example. The ideas presented here can be used to increase the current capability of any regulator, low-power or high-power.

Design Requirements and Initial Device Data

$$I_{FL} = 2 \text{ A} \qquad I_{SC} = 0.6 \text{ A}$$

$$I_S = 2 \text{ mA (max)} \qquad T_A = 60°\text{C (max)}$$

$$T_{J,\max,\,105} = 110°\text{C} \qquad \theta_{JA,105} = 150°\text{C/W} \qquad \text{(TO-5 in air)}$$

$$V_L = 15 \text{ V} \qquad V_{CC} = 20 \text{ V}$$

Step 1. Equation (5) is rearranged to find $P_{D,\max}$ of the 105.

$$P_{D,\max,105} = \frac{T_{J,\max} - T_{A,\max}}{\theta_{JA}}$$

$$= \frac{110 - 60}{150} = 333 \text{ mW}$$

Step 2. Equation (4) can be modified to find the maximum base current available from the 105.

$$I_{B2,\max} \approx \frac{P_{D,\max,105} - V_{CC}I_S}{V_{CC} - V_L - 0.7}$$

$$\approx \frac{0.333 - 20(2 \times 10^{-3})}{20 - 15 - 0.7} \approx 68 \text{ mA}$$

To be on the conservative side we will limit the maximum 105 output current to 20 mA.

Step 3. A plot of power dissipation in Q1 is shown in Fig. 21.10.

Step 4. The maximum power dissipation in Q1 is 16.3 W. As shown in Fig. 21.10, this occurs at $V_{CC} - V_L = 13$ V. We choose the 2N3773 device for Q1 since it is readily available from many manufacturers and reasonably priced. The β_1 of this device is 40 (min) at $-55°$ C. It can dissipate 150 W at 25°C case temperature. Derating linearly to zero P_D at 200°C, we find that it can safely dissipate the 16.3 W found in Fig. 21.10 at a case temperature of 181°C. The specified collector-to-emitter breakdown voltage of 140 V is sufficiently above V_{CC} in this example.

The thermal resistance of the 2N3773 from junction to case is $\theta_{JC} = 1.17°$ C/W (max). Assume the mica insulator and heat sink add another 1.5°C/W. This is substituted into Eq. (5) with the following result

$$T_{JQ1} = T_A + P_D\theta_{JA}$$

$$= 60°C + 16.3 \text{ W } (2.67°C/W) = 104°C$$

This is a safe temperature for the 2N3773 since its data sheet claims it will operate with a junction temperature of 200°C.

Step 5. The current gain required of Q2 is found from

$$\beta_2 = \frac{I_{FL}}{\beta_1 I_{B2,\,\max}} = \frac{2}{40(20 \times 10^{-3})} = 2.5$$

We should have no problem meeting this requirement.

Step 6. R$_1$ is found from Eq. (8):

$$R_1 \ll \frac{1}{I_{CBO,\max,Q1}} = \frac{1}{2 \text{ ma}}$$

$$\ll 500\Omega \qquad \text{use } 68 \text{ }\Omega$$

We obtained I_{CBO} from the 2N3773 data sheet. Equation 3 is now used to provide the dissipation in Q2:

$$P_{D,Q2} \approx (V_{CC} - V_L - 0.7)\left(\frac{I_{FL}}{\beta_1} + \frac{0.7}{R_1}\right)$$

$$\approx (20 - 15 - 0.7)\left(\frac{2}{40} + \frac{0.7}{68}\right) \approx 0.25 \text{ W}$$

We might consider using the 2N2905 for this application. It has a dissipation capability of 0.60 W at an ambient temperature of 25°C. No thermal resistance is given for this device. However, a derating factor of 3.43 mW/°C is provided for ambient temperatures above 25°C. Thus,

$$P_{D,\text{capability},Q2} = 0.6 \text{ W} - (60 - 25°\text{C})(3.43 \text{ mW/°C}) = 0.48 \text{ W}$$

The 0.26-W maximum dissipation is comfortably below this 0.48-W capability. This transistor also has a collector-to-emitter breakdown voltage of 40 V. This is 200 percent of the required 20 V.

Step 7. We next calculate V_{CL} and V_S using Eqs. (6) and (7):

$$V_S = 0.262 - 1.08 \times 10^{-3} T_{J,105}$$

$$= 0.262 - 1.08 \times 10^{-3}(110) = 0.143 \text{ V}$$

$$V_{CL} = 0.425 - 10^{-3} T_{J,105}$$

$$= 0.425 - 10^{-3}(110) = 0.315 \text{ V}$$

Step 8. Resistor values are

$$R_2 = \frac{V_{CL}}{I_{SC}} = \frac{0.315}{0.6} = 0.525 \ \Omega \qquad \text{use } 0.5 \ \Omega$$

$$R_4 = \frac{V_L + V_S}{2 \times 10^{-3} + I_{B2, \text{ max}}} = \frac{15 + 0.143}{2 \times 10^{-3} + 0.02}$$

$$= 626 \ \Omega \qquad \text{use } 620 \ \Omega$$

$$R_3 = (I_{FL}R_2 - V_S)\frac{R_4}{V_L} = [2(0.5) - 0.143] \ (^{620}\!/_{15})$$

$$= 35.4 \ \Omega \qquad \text{use } 36 \ \Omega$$

$$R_8 = 1129 V_L = 1129(15) = 16{,}935 \ \Omega$$

$$R_9 = \frac{2031 V_L}{V_L - 1.8} = \frac{2031(15)}{15 - 1.8} = 2308 \ \Omega$$

Step 9. The f_T of a 2N2905 is 200 MHz. In this case we let

$$C_1 = 0.05 \ \mu\text{F} \qquad C_2 \geqq I_{FL} \ \mu\text{F} = 2 \ \mu\text{F} \qquad C_3 = 47 \text{ pF}$$

If good line regulation is required,

$$C_4 = 10 \ \mu\text{F}$$
$$C_5 \geqq 1 + I_{\text{FL}}^{2.7} \ \mu\text{F} = 1 + 2^{2.7} \ \mu\text{F} = 7.5 \ \mu\text{F}$$

21.5 MONOLITHIC SWITCHING REGULATORS

Monolithic switching regulators are being considered in many new systems since they provide a low-cost, compact means of designing switching-mode circuits. There are two prime motivations behind the development of switching regulators:

1. The pass transistors dissipate substantially less heat than linear regulators for a wide range of input-output voltages and currents.

2. Size and weight are much less than linear regulators since dissipation is reduced and operating frequency is much higher.

The pass transistor in a linear regulator must dissipate

$$P_{D, \text{ linear}} = (V_{CC} - V_L)I_L \qquad (21.9)$$

where V_{CC} = input voltage

$\quad V_L$ = load voltage

$\quad I_L$ = load current

If we have a 10-A, 5-V logic supply which must be connected directly to a 28-V bus, the dissipation in the pass transistor is

$$P_{D, \text{ linear}} = (28 - 5)(10) = 230 \text{ W}$$

The power delivered to the load is $P_L = V_L I_L = 5(10) = 50$ W. The efficiency of this regulator is only

$$\eta = \frac{P_L}{P_{CC}} = \frac{V_L I_L}{V_{CC} I_L} = \frac{5(10)}{28(10)} = 18\%$$

In a switching regulator the series pass transistor is always totally ON or totally OFF. In the OFF state its dissipation is near zero. In the ON state the transistor is saturated. At the 10-A load level the saturation voltage (collector to emitter) is typically 1.5 V. The dissipation of the pass transistor during the ON time is therefore 1.5 V × 10 A = 15 W. While the transistor is OFF, the catch diode (D1 in Fig. 21.14) is fully ON. It also typically dissipates 15 W at the 10-A level.

The pass transistor has finite rise and fall times (t_r and t_f) which must be considered. During these transitions the dissipation is approximately one-half the dissipation of a pass transistor in a linear regulator since the current falls as the voltage rises (and vice versa). Assume that the switching frequency is 40 kHz (25-μs period) and $t_r = t_f = 1$ μs. The regulator dissipation from this source will be (230 W)/2 = 115 W for 2 μs out of each 25 μs.

The final dissipation to be considered in a switching regulator is the resistive losses in L_1 and L_2 (see Fig. 21.14). If these inductors have a total resistance R_T of 0.05 Ω, the dissipation at the 10-A load level is $I_L^2 R_T = 10^2 \times 0.05 = 5$ W. The total dissipation for the 10-A regulator is

$$P_{D,sw} = f_o V_{CE} I_L (t_r + t_f) + I_L \left[\frac{V_{\text{sat}} V_L}{V_{CC}} + \frac{V_D(V_{CC} - V_L)}{V_{CC}} \right] + I_L^2 R_T \qquad (21.10)$$

$$= \tfrac{1}{2}(40,000)(23)(10)(2 \times 10^{-6}) + 10 \left[\frac{1.5(5)}{28} + \frac{1.5(23)}{28} \right] + 10^2 \times 0.05$$

$$= 9.2 + 15 + 5 = 29.2 \text{ W}$$

where f_o = switching-regulator operating frequency
V_{sat} = saturation voltage of pass transistor
V_D = forward voltage drop of catch diode
V_{CE} = average voltage across pass transistor $\approx (V_{CC} - V_L)/2$
The efficiency of this regulator is

$$\eta = \frac{P_L}{P_{in}} = \frac{P_L}{P_L + P_D} = \frac{50}{50 + 29.2} = 63\%$$

It is worth emphasizing that the switching-regulator heat sink must be designed to handle only 29 W whereas the linear-regulator heat sink must dissipate 230 W. This is a 7:1 improvement.

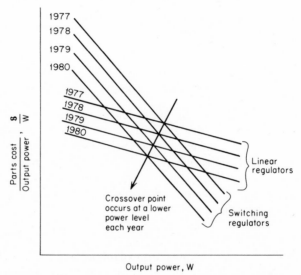

Fig. 21.11 The relationship of cost per watt to output power for both linear and switching regulators. At the crossover point their costs are equivalent for a given year.

Until the past few years switching regulators had not been widely utilized in commercial equipment because of cost and complexity. With the advent of monolithic regulators and low-cost high-performance semiconductors the cost trade-offs are changing. As shown in Fig. 21.11 the parts cost per output power ($/W) is lowering each year. Also, the cost per output power is lower for higher-power regulators. We note that the slope of each linear-regulator curve is shallow compared with the switching-regulator curves. This is due primarily to the heat-sink problems in high-power linear regulators. Thus, for a particular year, an output power level can be found wherein the costs per output power for the two types of regulators are equivalent.

Figure 21.12 shows a typical plot of regulator power dissipation as a function of input voltage. This plot assumes that $I_L = 10$ A, $V_L = 5$ V, $f_o = 40$ kHz, and $t_r = t_f = 1$ μs. The curves stop at 7 V since we need at least 2 V across the pass transistors for proper operation. We note that for input voltages of 20 V or more the dissipation differences of linear and switching regulators become quite significant.

If we plot converter dissipation as a function of load current, a typical result is as shown in Fig. 21.13. For this plot it is assumed that $V_{CC} = 28$V, $V_L = 5$ V, $f_o = 40$ kHz, and $t_r = t_f = 1$ μs. It is obvious that the switching regulator has significant thermal advantages for all load currents above 2 A.

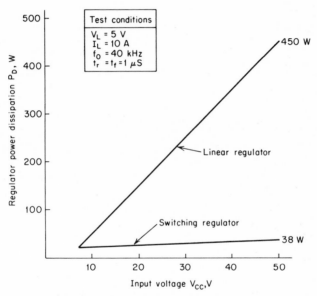

Fig. 21.12 Power dissipation of linear and switching regulators as a function of input voltage [plots of Eqs. (21.9) and (21.10)].

Circuit diagrams of the three most common switching regulators are shown in Figs. 21.14 to 21.16. Of these three, the most widely used circuit is the buck converter (or down converter) in Fig. 21.14. The output voltage of this circuit is $V_L = V_R(1 + R_1/R_2)$. We discuss the buck converter in detail in Sec. 21.6.

The boost regulator shown in Fig. 21.15 provides an output voltage $V_L = V_R(1 + R_1/R_2)$. This is the same equation as that used for the buck regulator. In the boost regulator, however, R_1 is much larger than R_2, so that a large voltage gain above the reference voltage V_R is achieved.

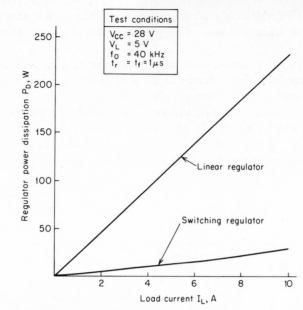

Fig. 21.13 Power dissipation of linear and switching regulators as a function of load current [plots of Eqs. (21.9) and (21.10)].

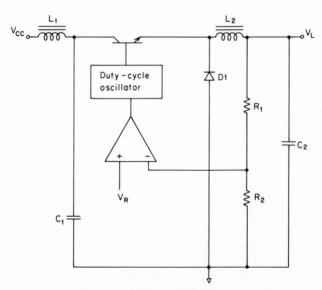

Fig. 21.14 The buck regulator, often called the down regulator or down converter because the output voltage is less than the input voltage.

The circuit shown in Fig. 21.16 is called the buck-boost regulator. With the *npn* transistor shown it converts a positive input voltage to a negative output voltage. If a *pnp* transistor is used (and the diode is reversed), it converts a negative input voltage to a positive output voltage. The magnitude of the output voltage is $V_L = -V_R(1 + R_1/R_2)$.

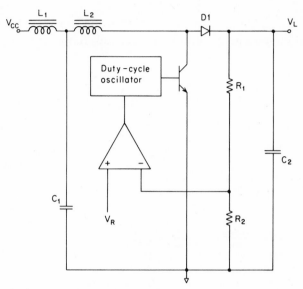

Fig. 21.15   The boost regulator, which provides an output voltage larger than the input voltage.

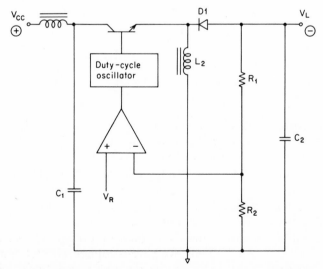

Fig. 21.16   The buck-boost regulator, which provides an output voltage having a polarity opposite to that of the input voltage.

21.6 A 5-A SWITCHING REGULATOR

ALTERNATE NAMES   Buck regulator, buck converter, down regulator, down converter, step-down regulator.

PRINCIPLES OF OPERATION   A buck regulator using the Texas Instruments TL497 monolithic switching regulator is shown in Fig. 21.17. This device is capable of supplying 500 mA using its internal pass transistor (Q2).   If higher-current operation is required, we must add an exter-

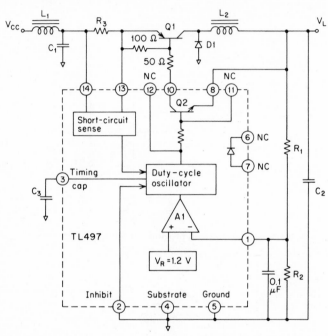

Fig. 21.17   A 5-A switching regulator utilizing the Texas Instruments TL497 monolithic switching voltage regulator.

nal *pnp* transistor Q1, also shown in Fig. 21.17.   The TL497 has an internal high-speed catch diode capable of 500 mA.   For higher-current operation an external high-power high-speed diode is required.

Transistors Q1 and Q2 are always fully ON or OFF except for the short time during transition from one state to the other.   Figure 21.18 shows the voltage waveform V_{C1} at the emitter of Q1.   While Q1 is ON, it impresses V_{CC} on the left side of L_2.   (For simplicity we neglect V_{sat} in the following discussion.)   A constant voltage V_L is stored on C_2 at the right side of L_2.   Thus, we have a voltage $V_{CC} - V_L$ impressed across L_2 while Q1 is ON.   The current through L_2 will linearly increase during this period by a factor of

$$\Delta i_{L2} = \frac{1}{L_2} \int_{t_1}^{t_2} (V_{CC} - V_L) \, dt = \frac{(V_{CC} - V_L)(t_2 - t_1)}{L_2}$$

This generates a positive ramp, as shown in Fig. 21.18.

Transistor Q1 is OFF during the period t_2 to t_3. However, since the inductor resists any abrupt change in current, diode D1 becomes forward-biased and delivers current into L_2. During this period the left side of L_2 will be at a potential of one diode drop below ground.

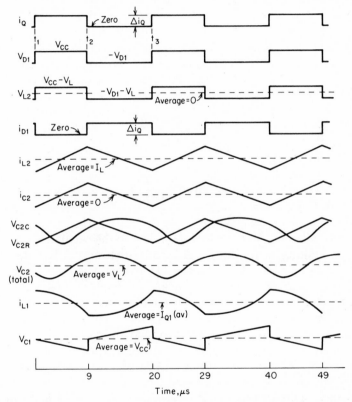

Fig. 21.18 Voltage and current waveforms at various locations in the switching regulator.

Since the right side of L_2 remains at V_L, the voltage across L_2 is $-V_{D1} - V_L$. This negative potential causes i_{L2} to slew downward by the quantity

$$\Delta i_{L2} = \frac{1}{L_2} \int_{t_2}^{t_3} -(V_{D1} - V_L) \, dt = \frac{(-V_{D1} - V_L)(t_3 - t_2)}{L_2}$$

The average value of the i_Q pulses is equal to the load current I_L. Likewise, the average value of i_{L2} during the positive and negative ramps is also equal to the load current. The current through C_2 is identically equal to the ac portion of i_{L2}. Figure 21.18 therefore shows that these current waveforms are identical except that i_{C2} has an average value of zero.

The ON and OFF times of Q1 are determined by the frequency of oscillation and by the size of V_{CC} relative to V_L. Load current has no bearing on the ON time in the TL497. In other regulators the ON time, OFF time, and/or f_o may change according to the load requirements. In general, the ON and OFF times of Q1 are

$$T_{on} = \frac{f_o V_L}{V_{CC}} \qquad T_{off} = \frac{1 - V_L/V_{CC}}{f_o}$$

Regulation is achieved by comparing a fraction $R_2/(R_1 + R_2)$ of V_L with the reference voltage V_R. This reference voltage is generated inside the chip in regulators such as the TL497. The output voltage relative to V_R is

$$V_L = V_R\left(1 + \frac{R_1}{R_2}\right)$$

For the TL497 we have

$$V_L = 1.2\left(1 + \frac{R_1}{R_2}\right)$$

The output of op amp A1 controls the duty cycle of the switching-regulator oscillator. An external inhibit line to the oscillator is provided if current limiting is required. A circuit such as R_2, R_3, and Q3 in Fig. 21.8 can be utilized, where the Q3 collector is connected to the inhibit line. If current limiting on the input side of the TL497 is desired, R_3 and the internal short-circuit sense circuit are used. Input-current limiting has the advantage of protecting the V_{CC} supply from shorts in the monolithic regulator in addition to load shorts.

The magnitude of load ripple voltage v_{C2} depends entirely on the quality and size of the $L_2 C_2$ product. A large $L_2 C_2$ product gives a low ripple if the effective series resistance of these two devices (R_{L2} and R_{C2}) is low. This means using large-gauge wire in L_2 and choosing one or more capacitors for C_2 which have published data indicating low ESR (effective series resistance). Banks of solid-tantalum capacitors are highly recommended for this application.

In addition to the large $L_2 C_2$ product a small L_2/C_2 ratio is also desirable because (1) for a given amount of stored energy, capacitors have fewer losses, are less expensive, smaller, and lighter and (2) a small L_2 and a large C_2 provide the best transient behavior for step changes in load current. The inductor ripple current i_{L2} increases in amplitude as L_2 decreases. The final choice for L_2 will therefore depend on the maximum allowable peak-to-peak excursions of i_{L2}.

The input filter (L_1 and C_1) is designed in two steps:

1. Capacitor C_1 is selected considering its ripple voltage rating, ripple current rating, and effective series resistance R_{C1}. We assume that the entire current pulse into Q1 flows out of C_1. This current causes i^2R heating of the capacitor.

2. Inductor L_1 is selected to satisfy the switching-regulator input-ripple-current requirement. We assume that the entire C_1 ripple voltage appears across L_1 for this calculation.

SWITCHING-REGULATOR DESIGN PARAMETERS

Parameter	Description
A1	Op amp inside monolithic regulator used as comparator
C_1	Part of input filter
C_2	Part of output filter
C_3	Timing capacitor
D1	Catch diode used to maintain current through L_2 when Q1 is in OFF state
ESR	Effective series resistance of capacitor
f_o	Oscillation frequency of switching regulator
Δi_{C2}	Peak-to-peak ripple current through C_2
Δi_{CC}	Maximum allowed peak-to-peak ripple current into regulator
i_{D1}	Current through D1
I_L	Load current
i_{L1}	Current through L_1
i_{L2}	Current through L_2
Δi_{L2}	Peak-to-peak ripple current through L_2
i_Q	Current through pass transistors
Δi_Q	Peak-to-peak current through pass transistors
L_1	Part of input filter
L_2	Part of output filter
P_{DD}	Power dissipation in the catch diode
P_{DL}	Power dissipation in inductors and current sense resistor
P_{DQ}	Power dissipation in series pass transistor
P_L	Power delivered to load
Q1	High-power external pass transistor
Q2	Internal pass transistor
R_1, R_2	Resistors used to adjust output voltage
R_{C1}, R_{C2}	Equivalent series resistance (ESR) of C_1 and C_2, respectively
R_{L1}, R_{L2}	DC resistance of L_1 and L_2, respectively
R_{Q1}	ON resistance of Q1
R_S	Sum of all source resistances driving the output filter (source resistance + R_{L1} + R_{L2} + R_{Q1} + wiring resistance)
t_r, t_f	Rise and fall times of current through Q1
$t_R(\uparrow)$, $t_R(\downarrow)$	Recovery time of load voltage from decrease or increase in load current
T_{on}	ON time of Q1
T_{off}	OFF time of Q1
V_{C1}, V_{C2}	AC voltage across C_1 or C_2
ΔV_{C1}	Peak-to-peak ac voltage across C_1
V_{CC}	Source voltage for switching regulator
V_{C2C}, V_{C2R}	Capacitive and resistive components of ac voltage across C_2
V_{CE}	Average voltage across pass transistor $\approx (V_{CC} - V_L)/2$

SWITCHING-REGULATOR DESIGN PARAMETERS *(Continued)*

Parameter	Description
V_{D1}	AC voltage across D1
V_D	Forward ON voltage of D1
V_L	Output voltage of switching regulator
ΔV_L	Peak-to-peak output ripple of circuit
V_{L2}	AC voltage across L_2
$\Delta V_L(\uparrow)$, $\Delta V_L(\downarrow)$	Positive or negative overshoot output voltages caused by step decrease or increase in load current
V_R	Reference voltage for comparator (internally generated in most monolithic switching regulators)
V_{sat}	ON saturation voltage of Q1
η	Efficiency of regulator

SWITCHING-REGULATOR DESIGN EQUATIONS

Description	Equation	
Output voltage	$V_L = V_{CC}T_{on}f_o$	(1a)
	or	
	$V_L = V_R\left(1 + \dfrac{R_1}{R_2}\right)$	(1b)
ON time of Q1	$T_{on} = \dfrac{V_L}{V_{CC}f_o}$	(2)
	(T_{on} also equals $9 \times 10^4 C_3$ in TL497)	
OFF time of Q1	$T_{off} = \dfrac{1 - V_L/V_{CC}}{f_o}$	(3)
Frequency of oscillation	$f_o = \dfrac{1}{T_{on} + T_{off}}$	(4)
Peak-to-peak ripple voltage across C_1	$\Delta V_{C1} = \Delta i_{Q1}\dfrac{R_{C1} + T_{on}T_{off}f_o}{C_1}$	(5)
Peak-to-peak ripple current through L_2	$\Delta i_{L2} = \dfrac{(V_{CC} - V_L)T_{on}}{L_2}$	(6a)
	or	
	$\Delta i_{L2} = \dfrac{(V_L + V_{D1})T_{off}}{L_2}$	(6b)
Positive transient overshoot of load voltage ΔV_L for a step decrease in load current Δi_L	$\Delta V_L(\uparrow) = \dfrac{L_2(\Delta i_L)^2}{C_2 V_L}$	(7)
Negative transient overshoot of load voltage ΔV_L for a step increase in load current Δi_L	$\Delta V_L(\downarrow) = \dfrac{L_2(\Delta i_L)^2}{C_2(V_{CC} - V_L)}$	(8)

Recovery time of load voltage from a step decrease in load current Δi_L	$t_R(\uparrow) = \dfrac{2L_2\,\Delta i_L}{V_L}$	(9)
Recovery time of load voltage from a step increase in load current Δi_L	$t_R(\downarrow) = \dfrac{2L_2\,\Delta i_L}{V_{CC} - V_L}$	(10)

Resistor values:

R_1	$R_1 = R_2\!\left(\dfrac{V_L}{V_R} - 1\right)$	(11)
R_2	$R_2 \approx \dfrac{500\,R_{L,\min}V_R}{V_L}$	(12)
R_3	See monolithic regulator data sheet	
	$R_3 \approx \dfrac{0.6}{I_{L,\max}}\qquad$ for TL497	(13)

Capacitor values:

C_1	$C_1 > \dfrac{T_{on}\,T_{off}\,f_o\,\Delta i_Q}{\Delta V_{C1} - \Delta i_Q R_{C1}}$	(14)
C_2	$C_2 = \dfrac{V_L}{L_2\,\Delta V_L}\left(\dfrac{V_{CC} - V_L}{f_o - V_{CC}}\right)^2$	(15)
C_3	See monolithic regulator data sheet	
	$C_3 = \dfrac{T_{on}}{9 \times 10^4}\qquad$ for TL497	(16)

Inductor values:

L_1	$L_1 > \dfrac{T_{on}R_{C1}\,\Delta i_Q}{\Delta i_{CC,\max}}$	(17)
L_2	$L_2 = \dfrac{(V_{CC} - V_L)T_{on}}{\Delta i_{L2}}$	(18)

Inequality required to prevent oscillation when output filter is underdamped	$\dfrac{R_S^{\,2}}{4L_2} > \dfrac{1}{C_2}\qquad$ no load	(19a)
	$\dfrac{1}{4R_L^{\,2}C_2} > \dfrac{1}{L_2}\qquad$ with load	(19b)
Approximate power dissipation in pass transistor Q1	$P_{DQ} = f_o[\,V_{CE}I_L(t_r + t_f)\,] + \dfrac{I_L V_L V_{sat}}{V_{CC}}$	(20)
Approximate power dissipation in catch diode D1	$P_{DD} = \dfrac{I_L V_D(V_{CC} - V_L)}{V_{CC}}$	(21)
Power dissipation in inductors and current-sense resistor	$P_{DL} = I_L^{\,2}(R_{L1} + R_{L2} + R_3)$	(22)
Efficiency of regulator	$\eta = \dfrac{P_L}{P_L + P_{DQ} + P_{DD} + P_{DL}}$	(23)

DESIGN PROCEDURE

Although the foregoing discussion appears to be centered around a specific monolithic switching regulator (TL497), the list of equations [except for Eqs. (13) and (16)] applies to any buck regulator of the class shown in Figs. 21.14 and 21.17.

DESIGN STEPS

Step 1. Choose a capacitor C_3 according to the IC regulator data sheet which provides the required f_o.

Step 2. Calculate values for R_1 and R_2 using Eqs. (11) and (12). These values provide less than 1 percent loading of the regulator. Solve Eq. (12) first.

Step 3. Choose a current sense resistor R_3 from the IC data sheet.

Step 4. Calculate T_{on} using Eq. (2) and T_{off} using Eq. (3).

Step 5. The value for the filter capacitor C_1 is difficult to calculate. As indicated in Eq. (14), its required capacitance depends on its own maximum allowable ripple voltage ΔV_{C1} and its own equivalent series resistance R_{C1}. An iterative design approach is recommended. Examine some data sheets to glean an idea of ΔV_{C1} and R_{C1} for a capacitor in the 10- to 100-μF range. Calculate a C_1 value and then refer back to the spec sheet to see if that capacitance can be purchased with the assumed ΔV_{C1} and R_{C1}. If not, use new values for ΔV_{C1} and R_{C1} and recompute C_1.

Step 6. Use Eq. (17) to compute a value for L_1. The core size and saturation flux density of L_1 must be large enough to ensure that saturation does not occur with $i_{Q,max}$.

Step 7. Select a peak-to-peak ripple current Δi_{L2} for L_2. This is usually in the range of $0.1 I_{L,max}$ up to $0.5 I_{L,max}$. Calculate a value for L_2 using Eq. (18). The core size and saturation flux density of this inductor must be sufficiently large to prevent saturation.

Step 8. Use Eq. (15) to calculate a value for C_2.

Step 9. With Eqs. (7) and (8) calculate the overshoot and undershoot of the load voltage for step changes in load current.

Step 10. Using Eqs. (9) and (10), compute the recovery time of the load voltage for step changes in load current.

Step 11. Determine the amount of preloading required, i.e., the maximum allowed R_L, using both parts of Eq. (19).

Step 12. Perform the calculations shown in Eqs. (20) to (23) to determine the regulator efficiency.

EXAMPLE OF AN IC SWITCHING-REGULATOR DESIGN   A 5-A switching regulator circuit will be designed using the 12 design steps.   The TL497

IC is assumed to be the primary control device. Other monolithic micro-circuits which use the classical buck-regulator configuration (Fig. 21.14) can also be used.

Design Requirements and Initial Device Data

$$V_{CC} = 11.4 \text{ V}$$

(i.e., the full-wave-rectified 12.6-V rms output from a filament trans-former)

$$V_L = 5.0 \text{ V} \qquad I_L = 5 \text{ A (max), that is } R_{L,\text{min}} = 1 \text{ } \Omega$$

$$\Delta V_L = 200 \text{ mV} \qquad \Delta i_{CC,\text{max}} = 0.01 \text{ A}$$

$$R_S = 1.5 \text{ } \Omega$$

$$V_D = 1 \text{ V} \qquad \text{at } 5 \text{ A}$$

$$2\text{N}6127: \ t_r = t_f = 0.5 \text{ } \mu s \qquad \beta_{\text{min}} = 30 \text{ at } 5 \text{ A} \qquad V_{\text{sat}} = 0.4 \text{ V}$$

$$\text{TL497: } V_R = 1.2 \text{ V}$$

Step 1. We initially choose a maximum operating frequency of 50 kHz. The TL497 spec sheet recommends $C_3 \approx 200 \text{ pF}$ for an operating frequency in this range.

Step 2. Equations (11) and (12) provide values for R_1 and R_2:

$$R_2 \approx 500 \ R_{L,\text{min}} \frac{V_R}{V_L} = 500(1)\frac{1.2}{5} = 120 \text{ } \Omega$$

$$R_1 = R_2 \left(\frac{V_L}{V_R} - 1 \right) = 120 \left(\frac{5}{1.2} - 1 \right) = 380 \text{ } \Omega$$

Step 3. The TL497 data sheet says R_3 will cause the regulator to begin current limiting when it develops a voltage of approximately 0.6 V. Thus,

$$R_3 \approx \frac{0.6}{I_{L,\text{max}}} \approx \frac{0.6}{5} \approx 0.12$$

Step 4. The nominal Q1 ON and OFF times are

$$T_{\text{on}} = \frac{V_L}{V_{CC} f_o} = \frac{5}{(11.4)(5 \times 10^4)} = 8.8 \text{ } \mu s$$

$$T_{\text{off}} = \frac{1 - V_L/V_{CC}}{f_o} = \frac{1 - 5/11.4}{5 \times 10^4} = 11.2 \text{ } \mu s$$

Step 5. Suppose we have a class of capacitors with a minimum ESR of 0.2 Ω in the 1- to 100-μF range. These capacitors can also withstand 4 V (peak to peak) ac ripple in the 50-kHz frequency range. Applying Eq. (14), we get

$$C_1 > \frac{T_{on} T_{off} f_0 \Delta i_Q}{\Delta V_{C1} - \Delta i_Q R_{C1}}$$

$$> \frac{(8.8 \times 10^{-6})(11.2 \times 10^{-6})(5 \times 10^4)(5)}{4 - 5(0.2)} > 8.2 \; \mu F$$

Step 6. Equation (17) provides a minimum value for L_1:

$$L_1 > \frac{T_{on} R_{C1} \Delta i_Q}{\Delta i_{CC,max}} > \frac{(8.8 \times 10^{-6})(0.2)(5)}{0.01} > 880 \; \mu H$$

Step 7. Let $\Delta i_{L2} = 0.3 \, I_{L,max} = 0.3(5) = 1.5$ A. We obtain L_2 using Eq. (18):

$$L_2 = \frac{(V_{CC} - V_L) T_{on}}{\Delta i_{L2}} = \frac{(11.4 - 5)(8.8 \times 10^{-6})}{1.5} = 37.5 \; \mu H$$

Step 8. The computed value of C_2 is

$$C_2 = \frac{V_L}{L_2 \, \Delta V_L} \left(\frac{V_{CC} - V_L}{f_0 V_{CC}} \right)^2$$

$$= \frac{5}{(37.5 \times 10^{-6})(0.2)} \left[\frac{11.4 - 5}{(5 \times 10^4)(11.4)} \right]^2 = 84 \; \mu F$$

Step 9. Assume we have step changes of 2 A in load current. The overshoot and undershoot of the load voltage is

$$\Delta V_L (\uparrow) = \frac{L_2 (\Delta i_L)^2}{C_2 V_L} = \frac{(37.5 \times 10^{-6})(2)^2}{(84 \times 10^{-6})(5)} = 0.36 \; V$$

$$\Delta V_L (\downarrow) = \frac{L_2 (\Delta i_L)^2}{C_2 (V_{CC} - V_L)} = \frac{(37.5 \times 10^{-6})(2)^2}{(84 \times 10^{-6})(11.4 - 5)} = 0.28 \; V$$

Step 10. The recovery times from the 2-A step load changes in step 9 are

$$t_R (\uparrow) = \frac{2 L_2 \, \Delta i_L}{V_L} = \frac{2(37.5 \times 10^{-6})(2)}{5} = 30 \; \mu s$$

$$t_R (\downarrow) = \frac{2 L_2 \, \Delta i_L}{V_{CC} - V_L} = \frac{2(37.5 \times 10^{-6})(2)}{11.4 - 5} = 23.4 \; \mu s$$

Step 11. We first check to see if the output filter is sufficiently damped with no load:

$$\frac{R_S^2}{4L_2} \overset{?}{>} \frac{1}{C_2}$$

$$\frac{(1.5)^2}{4(37.5 \times 10^{-6})} \overset{?}{>} \frac{1}{84 \times 10^{-6}}$$

$$15,000 > 11,905$$

This inequality is satisfied so no preloading is required.

Step 12. Power dissipation in Q1 is

$$P_{DQ} = f_0 V_{CE} I_L (t_r + t_f) + \frac{I_L V_L V_{\text{sat}}}{V_{CC}}$$

$$= (5 \times 10^4)(3.2)(5 \times 10^{-6}) + \frac{5(5)(0.4)}{11.4} = 1.68 \text{ W}$$

Power dissipation in the catch diode is

$$P_{DD} = \frac{I_L V_D (V_{CC} - V_L)}{V_{CC}} = \frac{5(1)(11.4 - 5)}{11.4} = 2.81 \text{ W}$$

If we assume $R_{L1} = R_{L2} = 0.1 \ \Omega$, we get

$$P_{DL} = I_L^2 (R_{L1} + R_{L2} + R_3) = 5^2(0.1 + 0.1 + 0.12) = 8 \text{ W}$$

The regulator efficiency is

$$\eta = \frac{P_L}{P_L + P_{DQ} + P_{DD} + P_{DL}} = \frac{5 \times 5}{5 \times 5 + 1.68 + 2.81 + 8} = 0.67 = \text{ or } 67\%$$

REFERENCES

1. Sheng, A. C. N., and L. R. Avery: RCA Linear Integrated Circuits, *Appl. Note* ICAN-6157, December 1973.
2. Kesner, D.: Monolithic Voltage Regulators, *IEEE Spectrum,* April 1970, p. 24.
3. Widlar, R. J.: The LM105: An Improved Positive Regulator, National Semiconductor Corp., *Appl. Note* AN-23, January 1969.
4. Silber, D.: Simplifying the Switching Regulator Input Filter, *Solid-State Power Convers.,* May–June 1975, p. 23.
5. Dixon, L., and R. Patel: Designer's Guide to Switching Regulators, pts. 1 and 2, *EDN,* Oct. 20, 1974, p. 53; Nov. 5, 1974, p. 37.

Chapter 22

Nonlinear Analog Microcircuits

DEFINITION OF A NONLINEAR CIRCUIT

Any nondigital circuit with one or more outputs which are a nonlinear function of one or more inputs could be called a nonlinear analog circuit. Since hundreds of specialized microcircuits fall into this category, we will limit this chapter to the following common nonlinear analog microcircuits:

Peak detector
Amplitude limiter
Precision rectifier
Logarithmic and antilogarithmic amplifier
Analog multiplier and divider

All these circuits are available as packaged monolithic circuits, but data sheets for most of these devices tell the user little about how the function is implemented. If a schematic is provided, it is usually shown at a detailed level showing dozens of transistors. Without extensive analysis these schematics are difficult to interpret.

In this chapter we will present these nonlinear circuits using op amps and discrete parts. This approach makes it easier for the reader to visualize the nature of each nonlinear process. Many of the common nonlinear microcircuits use circuits very similar to those shown in the following pages. In most cases the only external components are the larger capacitors.

22.1 POSITIVE-PEAK DETECTOR

ALTERNATE NAMES Peak holder, peak-signal tracker.

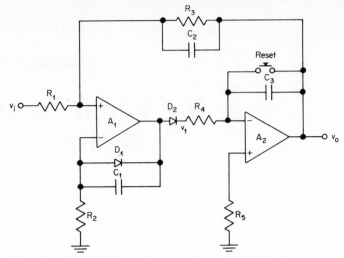

Fig. 22.1 Positive-peak detector with negative output.

PRINCIPLES OF OPERATION The circuitry of a peak detector can be arranged for positive- or negative-peak detection. For each of these cases the output can be made positive or negative. The circuit we have chosen to discuss, in Fig. 22.1, selects positive peaks and produces a negative output.

Peak detectors track the input signal and hold the output at the highest peak found since operation of the reset switch. They continuously compare the input waveform with the stored peak value to determine if the stored value must be updated. This is graphically illustrated in Fig. 22.2. A peak detector may be thought of as a type of sample-hold circuit. It samples and holds the peak value of the largest peak

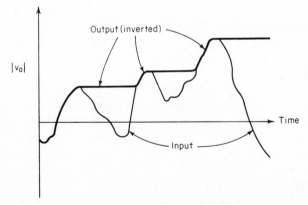

Fig. 22.2 Input and output waveforms of positive-peak detector.

in a given measurement interval. This is extremely useful in applications where widely spaced transients in a system must be measured.

This peak detector is actually a combination of two circuits described elsewhere in this handbook. The circuit of A_1 is similar to the precision rectifier discussed in Sec. 22.3, but the feedback resistor R_f and capacitor C_f of Fig. 22.7 have been replaced by an active-feedback network, namely, R_3, R_4, R_5, C_2, C_3, and A_2. The A_2 circuit is merely a fast integrator.

The circuit gain is defined as the ratio of peak output voltage to peak input voltage. In terms of circuit components the gain is

$$A_{vc} = \frac{V_{o,\text{peak}}}{V_{i,\text{peak}}} = -\frac{R_3}{R_1}$$

After a peak is stored on C_3, diode D_2 is reverse-biased for all succeeding lower amplitudes. This actually opens the feedback loop. The A_1 output will then try to saturate with negative v_i. Diode D_1 prevents this by holding the A_1 output near -0.7 V if v_i becomes negative.

This circuit may be unstable with some types of op amps because of the large phase shift around the loop (see Chap. 19). The gain A_{vc} must be critically damped or overdamped to prevent overshoot. Since an overshoot may be interpreted as a maximum peak, caution in the feedback design is recommended. C_1 and C_2 are two possible compensation capacitors. Since the size of these capacitors is critically dependent on the types of op amp, an experimental approach is recommended. Start with $C_1 = C_2 = 5$ pF and work up or down from that value while observing the overshoot in v_1 with v_i a step function.

If the peak must be stored for long periods of time, A_2 should be an FET-input op amp. C_3 should also be a low-leakage capacitor. The bias current of A_2 and the leakage current of C_3 will produce a peak-hold error of

$$\Delta v_o = \frac{I \times \text{hold time}}{C_3}$$

where I is the sum of A_2 input bias current and C_3 leakage current.

DESIGN PARAMETERS

Parameter	Description
A_{vc}	Voltage gain of entire circuit until first peak is reached; then ratio of output voltage to maximum input peak
C_1, C_2	Compensation capacitors for feedback stability
C_3	Integrating capacitor which holds peak output voltage
D1	Diode to prevent A_1 negative saturation during negative input voltages
I_b	Input bias current of A_2
I_c	Leakage current of C_3
$I_{o2,\text{max}}$	Maximum output current of A_2

Parameter	Description
R_1	Determines gain of circuit along with R_3
R_2	Used to cancel most of the offset caused by input bias current of A1
R_3	Determines gain of circuit along with R_1
R_4	Determines speed of response of circuit
R_5	Used to cancel most of the integrator error caused by the input bias current of A_2
S_2	Slew rate of A_2
t_r	Approximate speed of response of circuit
ΔT	Sampling time of circuit, i.e., from reset to reset
v_i	Input voltage to circuit
v_1	Output of rectifier circuit
v_0	Value of peak voltage determined during ΔT
Δv_0	Error in v_0 due to I_b and I_c

DESIGN EQUATIONS

Description	Equation	
Voltage gain of circuit, i.e., peak output to peak input	$A_{vc} = \dfrac{V_{o,\text{peak}}}{v_{i,\text{peak}}} = -\dfrac{R_3}{R_1}$	(1)
Approximate rise time of integrator (circuit cannot respond accurately to peaks having rise times faster than this)	$t_r \approx R_4 C_3$ This assumes the slew-rate limit S_2 of A2 is not exceeded: $S_2 = I_{o2.\text{max}}/C_3$	(2)
Optimum value for R_2	$R_2 = \dfrac{R_1 R_3}{R_1 + R_3}$	(3)
Optimum value for R_5	$R_5 = R_4$	(4)
Error in stored peak value of v_0 due to I_b and I_c	$\Delta v_0 = \dfrac{(I_b + I_c)\,\Delta T}{C_3}$	(5)

22.2 AMPLITUDE LIMITER

ALTERNATE NAMES   Limited amplifier, volume compressor, amplitude leveler, feedback limiter, precision limiter, limiting amplifier.

PRINCIPLES OF OPERATION   Amplitude limiters are required in many systems where the amplitude of a signal cannot be allowed to exceed given positive or negative limits. This function is often done utilizing resistor-zener networks. A zener in the feedback network of an op amp will also accomplish the same function without excessive loading of the input signals. Extensive coverage of these circuits is given in the literature. These circuits all suffer from a common disadvantage; i.e., any time or temperature variation of the zener breakdown voltage creates a circuit error of corresponding size. This may be acceptable for applications where amplitude limiting is performed only for protection or noise reduction, but in circuits where the limiting voltage is an important system parameter a precision limiter, as described below, is required.

In this section we will cover, in detail, the design of the precision bridge-type amplitude limiter shown in Fig. 22.3. The diode bridge is placed in the forward loop of the feedback network. This means that the effects of all nonlinear, forward-resistance, and temperature characteristics of the diodes will be divided by a factor of $1/\beta A_v$ in the closed-loop circuit. The transfer characteristics of the circuit, shown in Fig. 22.4, therefore depend only on resistor values and the two refer-

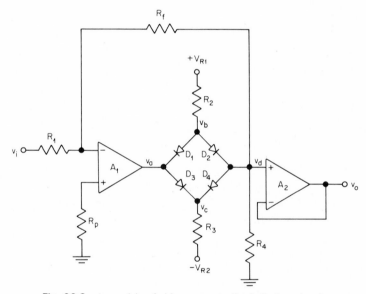

Fig. 22.3 A precision bridge-type amplitude-limiter circuit.

ence voltages V_{R1} and V_{R2}. The slope of the linear region is identical to that of an inverting amplifier, namely $-R_f/R_1$. The positive limiting output voltage is given by

$$V_{0,\,\text{sat,pos}} = \frac{V_{R1}R_fR_4}{R_fR_4 + R_fR_2 + R_4R_2}$$

The negative limiting output voltage is

$$V_{0,\,\text{sat,neg}} = \frac{-V_{R2}R_f\,R_4}{R_fR_4 + R_fR_3 + R_4R_3}$$

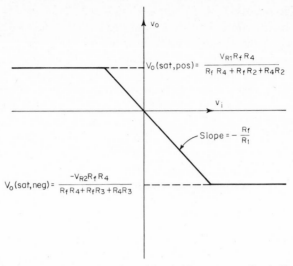

Fig. 22.4 Transfer function of precision bridge-type amplitude limiter.

In the linear region of circuit operation the four diodes are all forward-biased. The voltage v_a closely follows v_d, and the circuit gain is

$$A_{vc} = -\frac{R_f}{R_1}$$

Note that no power can flow directly from v_a to v_d. All power for v_d comes through R_2 or R_3. However v_b and v_c are controlled by v_a, since v_a is a stiff source and draws current through D_1 and D_3. If D_1 is conducting, v_b is one diode drop above v_a, and likewise v_d is one diode drop below v_b. If $D3$ is conducting, v_c is one diode drop below v_a and v_d is one diode drop above v_c. In either case, $v_a \approx v_d$. Utilizing back-to-back diodes in series with the signal flow allows cancellation of most of the temperature-induced changes of diode characteristics. Any residual error caused by a mismatch of diodes will be reduced by a factor of $1/\beta A_v$, since the diodes are within the forward loop of the feedback circuit.

If v_a is not present, the maximum positive v_d is set by the voltage divider composed of V_{R1}, R_2, R_4, and R_f. Likewise the most negative v_d is set by the voltage divider composed of V_{R2}, R_3, R_4, and R_f. v_a cannot swing v_d beyond these limits. If v_a swings positive beyond the upper limit, D_1 becomes reverse-biased and v_d remains at the upper limit shown in Fig. 22.4. If v_a swings negative, it reverse-biases D_3 at the limit shown in Fig. 22.4. Thus D_1 and D_3 are switches which cause the abrupt change in circuit characteristics when the limiting voltages are reached.

DESIGN PARAMETERS

Parameter	Description
A_v	Op-amp open-loop gain as a function of frequency
A_{vc}	Closed-loop gain of circuit in linear region of operation
D_1 to D_4	Diode bridge used to switch circuit from linear to limited
R_1	Input resistor which establishes input resistance of circuit
R_2 to R_3	Resistors which provide output current for diode bridge
R_4	Resistor used to refer output of diode bridge to ground
R_f	Feedback resistor
R_p	Resistor used to minimize effects of op-amp input bias current
v_a to v_d	Diode bridge voltages
v_i	Input voltage to circuit
v_o	Output voltage of circuit
$V_{o,\,\text{sat.pos}}$	Positive limited output voltage
$V_{o,\,\text{sat.neg}}$	Negative limited output voltage
V_{R1}	Magnitude of positive reference voltage
V_{R2}	Magnitude of negative reference voltage
β	Feedback ratio due to R_f and R_1

DESIGN EQUATIONS

Description	Equation	
Voltage gain of circuit in linear region assuming ideal op-amp and diode parameters	$$A_{vc} = \frac{v_o}{v_i} = -\frac{R_f}{R_1}$$	(1)
Voltage gain of circuit in linear region assuming finite op-amp gain	$$A_{vc} = \frac{-R_f/R_1}{1 + 1/\beta A_v}$$	(2)
Positive limiting output voltage	$$V_{o,\,\text{sat.pos}} = \frac{V_{R1} R_f R_4}{R_f R_4 + R_f R_2 + R_2 R_4}$$	(3)
Negative limiting output voltage	$$V_{o,\,\text{sat.neg}} = \frac{-V_{R2} R_f R_4}{T_f R_4 + R_f R_3 + R_3 R_4}$$	(4)

22.3 PRECISION HALF-WAVE RECTIFIER

ALTERNATIVE NAMES Polarity selector, ideal half-wave rectifier, ideal diode, zero-bound circuit, precision AM detector, precision ac-dc converter, low-level ac-dc converter.

PRINCIPLES OF OPERATION This circuit comes in four basic configurations. Figure 22.5 shows pictorially the input-output relationship of these four basic circuits along with a plot of each transfer function. A precision half-wave rectifier performs very closely to the expected response of an ideal diode. Figure 22.6b shows the response expected if one could produce an ideal diode. The ideal diode possesses two advantages over the silicon diode: (1) the ideal diode can rectify signals

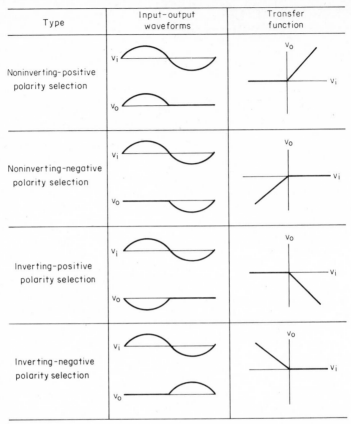

Type	Input-output waveforms	Transfer function
Noninverting-positive polarity selection		
Noninverting-negative polarity selection		
Inverting-positive polarity selection		
Inverting-negative polarity selection		

Fig. 22.5 Input-output waveforms and transfer functions of four basic precision half-wave rectifiers.

down to 0 V amplitude, and (2) the forward-conduction region of an ideal diode is linear.

In the inverting half-wave precision-rectifier circuit shown in Fig. 22.7 the two ideal properties discussed above can be approached with nearly zero error. The circuit will rectify low-level signals with peak voltages of only $0.7/A_v$. If $A_v = 1000$, precision linear rectification of a 0.7-mV signal is possible.

The circuit operates by providing two gains. For one polarity of input, D_1 is reverse-biased and D_2 is forward-biased. Under these conditions the gain of the circuit is $\pm R_f/R_1$ (+ for Fig. 22.8 and − for Fig. 22.7). If the opposite-polarity input is applied, D_1 is forward-biased and D_2 is reverse-biased. The gain of the circuit then becomes zero. The slope of the linear gain and the breakpoint are insensitive to temperature thanks to the $1/A_v$ factor.

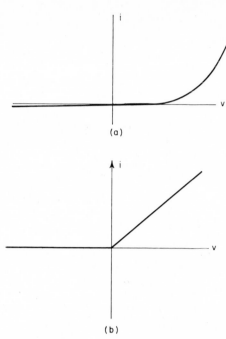

Fig. 22.6 Typical *V-I* curve of *(a)* a silicon diode and *(b)* an ideal diode.

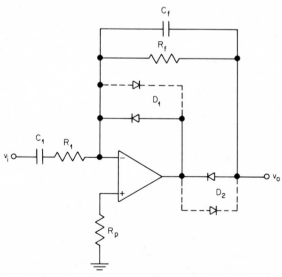

Fig. 22.7 Inverting-precision half-wave rectifier; solid lines are for positive-polarity selection and dashed lines for negative-polarity selection.

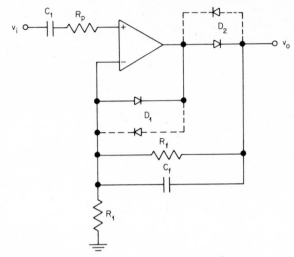

Fig. 22.8 Noninverting precision half-wave rectifier; solid lines are for positive-polarity selection and dashed lines for negative-polarity selection.

Figure 22.8 shows the circuit configuration for the noninverting precision half-wave rectifier.

Several sources of error are possible in the circuits of Figs. 22.7 and 22.8. If the op-amp output offset voltage approaches 0.7 V, D_1 or D_2 may begin to conduct. This will add a dc component to v_o which may be falsely interpreted as a rectified ac signal. The portion of this error voltage due to the op-amp input offset voltage can be eliminated by adding a coupling capacitor in series with R_1. Even though midband ac signals will be amplified by R_f/R_1, the input offset voltage will be multiplied by 1. If R_p is made equal to R_f, a further reduction in offset is possible by cancellation of the effects of each input bias current.

Capacitor C_f is added to the circuit if a dc output voltage proportional to the peak input voltage is required. The magnitude of this dc voltage is given in Eqs. (5).

DESIGN PARAMETERS

Parameter	Description
A_v	Op-amp open-loop gain (varies with frequency)
A_{vo}	Op-amp dc open-loop gain
A_{vc}	Closed-loop gain of circuit during linear-gain portion of cycle
B	Variable used in computations
C_1	Input isolation capacitor to reduce errors due to input offset voltage
C_f	Feedback capacitor used to provide dc output instead of half-wave-rectified ac

D_1	Provides an effective feedback resistance of 0 Ω during portion of waveform when no output voltage is required
D_2	Passes op-amp output on to v_0 terminal during portion of waveform when undistorted output is required
f_m	Modulating frequency of input carrier frequency f_c
f_c	Carrier frequency of input waveform
f_{max}	Maximum frequency at which high-accuracy performance can be achieved
f_u	Unity-gain frequency of the op amp (gain crossover frequency)
I_b	Input bias current of op amp
I_{io}	Input offset current of op amp
R_1	Controls gain and input resistance of circuit
R_f	Controls gain of circuit and degree of filtering provided by C
R_{id}	Op-amp differential input resistance
R_o	Op-amp output resistance
R_p	Resistor used to minimize offset due to input bias current
S	Slew rate of op amp
t_{rr}	Reverse recovery time of diodes
$\dfrac{\Delta v_i}{\Delta t}$	Fastest slew rate of input waveform
v_i	Input voltage
V_{io}	Input offset voltage of op amp
v_o	Output voltage

DESIGN EQUATIONS

Description	Equation	
Voltage gain of circuit if solid diode connections are used ($C_f = 0$)	Inverting: $$A_{vc} = \begin{cases} \dfrac{v_o}{v_i} = -\dfrac{R_f}{R_1} & \text{if } v_i > 0 \\ 0 & \text{if } v_i < 0 \end{cases}$$	(1a)
	Noninverting: $$A_{vc} = \begin{cases} \dfrac{v_o}{v_i} = 1 + \dfrac{R_f}{R_1} & \text{if } v_i > 0 \\ 0 & \text{if } v_i < 0 \end{cases}$$	(1b)
Voltage gain of circuit if dashed diode connections are used ($C_f = 0$)	Inverting: $$A_{vc} = \begin{cases} \dfrac{v_o}{v_i} = -\dfrac{R_f}{R_1} & \text{if } v_i < 0 \\ 0 & \text{if } v_i > 0 \end{cases}$$	(2a)
	Noninverting: $$A_{vc} = \begin{cases} \dfrac{v_o}{v_i} = 1 + \dfrac{R_f}{R_1} & \text{if } v_i < 0 \\ 0 & \text{if } v_i > 0 \end{cases}$$	(2b)
Input resistance if $f_{in} \gg 1/2\pi R_1 C_1$	Inverting: $R_{in} = R_1$	(3a)
	Noninverting: $R_{in} = A_v R_{id}$	(3b)
Size of filter capacitor C_f	$$\dfrac{1}{2\pi f_c R_f} \ll C_f \ll \dfrac{1}{2\pi f_m R_f}$$	(4)

Description	Equation
Magnitude of dc output voltage if C_f utilized	Inverting: $$V_{o, dc} = \pm \frac{0.45 v_{i, rms} R_f}{R_1} \qquad (5a)$$ Noninverting: $$V_{o, dc} = \pm 0.45 v_{i, rms}\left(1 + \frac{R_f}{R_1}\right) \qquad (5b)$$
Maximum high-accuracy frequency of circuit (error $< 1\%$)	$$f_{max} = \frac{f_u}{100|A_{vc}|} \qquad (6)$$
Required slew rate of op amp for a nonsinusoidal input waveform	$$S > |A_{vc}|\frac{\Delta v_i}{\Delta t} \qquad (7)$$
Required slew rate of op amp for sinusoidal input	$$S > 2\pi f_c |A_{vc}| v_{i, peak} \qquad (8)$$
Maximum dc offset voltage at op-amp output	$$V_{o, off, max} = V_{io}\left(1 + \frac{R_f}{R_1}\right) + \frac{2I_b + I_{io}}{2R_f} \qquad (9)$$
Voltage gain of circuit if op-amp gain is considered	In Eqs. (1) and (2) replace $\frac{R_f}{R_1}$ with $\frac{R_f A_v}{R_1 A_v + R_f}$ $\qquad (10)$
Optimum value of R_p	$$\begin{cases} R_p = R_f & \text{if } C_1 \text{ used} \\ R_p = \dfrac{R_1 R_f}{R_1 + R_f} & \text{if } C_1 \text{ not used} \end{cases} \qquad (11)$$
Maximum diode reverse recovery time so that output waveform will not be distorted	$$t_{rr} < \frac{0.01}{f_c} \qquad (12)$$
Effective forward-voltage drop of precision rectifier (increases with frequency as A_v drops)	$$V_f = \frac{0.7}{A_v} \qquad (13)$$ (silicon diodes assumed)

DESIGN PROCEDURE

Precision rectifiers are often utilized to extract a modulation frequency f_m from a carrier frequency f_c. This is called AM demodulation. In the following design procedure we will assume such an application. Other applications, such as AGC detectors, require a consideration of the topics to be discussed below. The inverting circuit (Fig. 22.7) will be assumed.

DESIGN STEPS

Step 1. Choose R_1 so that it equals the required input resistance of the circuit.

Step 2. Compute the maximum frequency f_{max} at which high-accuracy performance can be expected:

$$f_{max} = \frac{f_u}{100\,|A_{vc}|}$$

Step 3. Compute the required slew rate of the op amp

$$S > \begin{cases} 2\pi f_c\,|A_{vc}|\,v_{i,peak} & v_i \text{ sinusoid} \\ |A_{vc}|\dfrac{\Delta v_i}{\Delta t} & v_i \text{ nonsinusoid} \end{cases}$$

Step 4. Compute $R_f = A_{vc} R_1$.

Step 5. If filtering of the rectified output is not required, skip steps 5 to 7 and do not install C_f. Compute

$$B = \left(\frac{f_c}{f_m}\right)^{1/2}$$

Step 6. Compute

$$C_f = \frac{B}{2\pi f_c R_f}$$

In this equation, B sets the ripple level of f_c in v_o.

Step 7. Verify

$$C_f = \frac{1}{2\pi f_m R_f B}$$

In this equation, B sets the degree of attenuation of f_m in v_o. *Note:* C_f was found by a compromise above. It must be large enough to keep the f_c ripple low but low enough to ensure that the f_m modulation does not vanish. The above computation results in a geometric-mean value for C_f.

Step 8. Compute the size of R_p required to minimize the effects of bias-current offset

$$R_p = \frac{R_1 R_f}{R_1 + R_f}$$

Step 9. Compute the dc output level for the median $v_{i,rms}$ input.

$$v_{o,dc} = \frac{\pm 0.45\,v_{i,rms} R_f}{R_1}$$

Polarity depends on the direction of D_1 and D_2.

Step 10. If a negative v_o (filtered or unfiltered) is required, the solid diode connections shown in Fig. 22.7 are required. A positive output requires the dashed diode connections. The reverse recovery time t_{rr} of the diodes must be less than $0.01/f_c$ or the output waveform will be distorted.

Step 11. Compute the dc output error caused by V_{io}, I_b, and I_{io}. This will vary with temperature.

$$V_{o,\text{offset, max}} = (A_{vc} + 1)V_{io,\text{max}} + \frac{2I_b + I_{io}}{2R_f}$$

Step 12. If voltage-gain accuracy over temperature and power-supply variations is important, determine the magnitude of closed-loop gain for different values of open-loop gain using

$$A_{vc} = \frac{-R_f A_v}{R_1 A_v + R_f}$$

DESIGN EXAMPLE Assume a demodulator design is required which will efficiently extract a low-frequency modulation from a 7-kHz carrier frequency. The upper frequency component of the modulation is 2 Hz. A positive output waveform is required, but the negative portion of the input waveform contains the required information. Therefore, an inverting circuit is used.

Tentative Circuit-Performance Requirements

$A_{vc} = -3$
$R_{\text{in}} = 2000\ \Omega$
$f_m = 2$-Hz sine wave
$f_c = 7$-kHz sine wave
$v_{i,\text{peak-to-peak}} = 1$ V
$v_{o,\text{peak-to-peak}} = 3$ V
Positive output with filtering
Maximum offset without external offset adjustment $= 0.5$ V

Op-Amp Parameters (741)

$f_u = 800$ kHz
Maximum 3-V (peak-to-peak) frequency $= 100$ kHz
$S_{\text{min}} = 0.5$ V/μs
$V_{io,\text{max}} = 6$ mV
$I_{io,\text{max}} = 0.5\ \mu$A
$I_{b,\text{max}} = 1.5\ \mu$A
$\phi_{m,\text{open-loop}} = 80°$
$A_v = 25,000$ if $V^{(\pm)} = \pm2$ V to $250,000$ if $V^{(\pm)} = \pm20$ V
Closed-loop bandwidth (normalized) $= \begin{cases} 1 & \text{at } 25°\text{C} \\ 1.12 & \text{at } -55°\text{C} \\ 0.8 & \text{at } 125°\text{C} \end{cases}$

DESIGN STEPS

Step 1

$$R_1 = R_{in} = 2000 \Omega$$

Step 2

$$f_{max} = \frac{f_u}{100 \, |A_{vc}|} = \frac{8 \times 10^5}{100 \times 3} = 2667 \text{ Hz}$$

Step 3

$$S > 2\pi f_c \, |A_{vc}| \, v_{i,peak} = 6.28(7 \times 10^3)(3) \, (0.5) = 0.066 \text{ V}/\mu s$$

Step 4

$$R_2 = |A_{vc}| \, R_1 = 3(2000) = 6000 \Omega$$

Step 5

$$B = \left(\frac{f_c}{f_m}\right)^{1/2} = \left(\frac{7000}{2}\right)^{1/2} = 59.16$$

Step 6

$$C_1 = \frac{B}{2\pi f_c R_2} = \frac{59.16}{6.28(7000)(6000)} = 0.22 \ \mu F$$

Step 7

$$C_f = \frac{1}{2\pi f_m R_f B} = \frac{1}{6.28(2)(6000)(59.16)} = 0.22 \ \mu F$$

Step 8

$$R_p = \frac{R_1 R_f}{R_1 + R_f} = \frac{2000(6000)}{2000 + 6000} = 1500 \ \Omega$$

Step 9

$$V_{o,dc} = \frac{+0.45 \, v_{i,rms} R_f}{R_1} = \frac{0.45(0.35)(6000)}{2000} = +0.473 \text{ V}$$

Step 10. Appropriate D_1 and D_2 direction chosen for a positive output. The diode chosen is the IN191, which has a t_{rr} of 0.5 μs. This satisfies the requirement $t_{rr} < 0.01/7000 = 1.4 \ \mu s$. Laboratory tests showed no distortion.

Step 11. The maximum output offset is

$$\Delta V_o = (A_{vc} + 1) \, V_{io,max} + \frac{2 \, I_b + I_{io}}{2 \, R_f}$$

$$= 4(6 \times 10^{-3}) + \frac{2(1.5 \times 10^{-6}) + 0.5 \times 10^{-6}}{2(6000)} = 24 \text{ mV}$$

Step 12. The changes of dc closed-loop gain resulting from changes of dc open-loop gain are computed as follows:

$$A_{vc}(A_{vo} = 250 \text{ kV,dc}) = \frac{-R_f A_v}{R_1 A_v + R_f} = \frac{-6000(250,000)}{2000(250,000) + 6000} = 2.99996$$

$$A_{vc}(A_{vo} = 25 \text{ kV,dc}) = \frac{-6000(25,000)}{2000(25,000) + 6000} = 2.9996$$

The closed-loop gain changes only 0.01 percent from a 90 percent reduction of open-loop gain.

22.4 DIFFERENTIAL LOGARITHMIC AMPLIFIER

ALTERNATE NAMES Log ratio circuit, log amplifier, log converter, data compressor, log subtracting circuit.

PRINCIPLES OF OPERATION Logarithmic and antilogarithmic amplifiers are basic building blocks for many nonlinear circuits. They are also intrinsically useful by themselves as analog compressors and expanders. Log and antilog amplifiers are mathematical inverses of each other. Their principal usefulness is in such applications as multipliers, dividers, and square-root circuits.

The differential log amplifier shown in Fig. 22.9 provides several useful types of transfer functions. Since it has differential inputs and adjustable gain, transfer functions of the following forms are possible:

$$v_0 = K \log_a v_1$$

$$v_0 = K \log_a \frac{v_1}{v_2} = K \log_a v_1 - K \log_a v_2$$

$$v_0 = K \log_a \frac{1}{v_2} = -K \log_a v_2$$

where K is the gain of the circuit and a the base of logarithm, which can be set at e, 10, or any other useful number. There is really only one constant in the above equations, since logarithms of different bases are related to each other by a constant. For example, the relationship between logarithms to the bases 10 and e is

$$\frac{\log x}{\ln x} = 0.4343$$

To simplify the following discussion, we will therefore work exclusively with natural logarithms (base $e = 2.71828$).

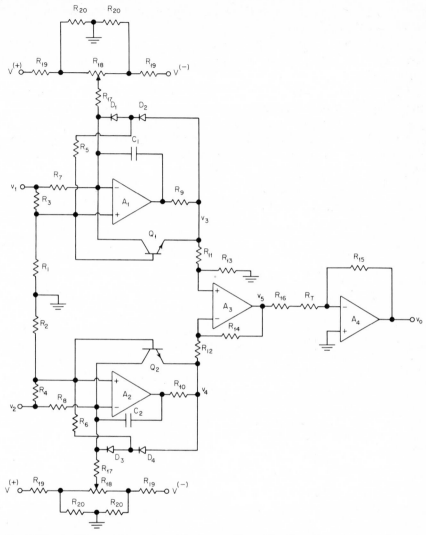

Fig. 22.9 A differential-input logarithmic amplifier with adjustable gain and adjustable logarithm base.

In Fig. 22.9 if we assume $R_1 = R_2$, $R_3 = R_4$, $R_5 = R_6$, $R_9 = R_{10}$, $R_{11} = R_{12}$, and $R_{13} = R_{14}$, the transfer function of the circuit is

$$v_0 = \frac{kTR_{15}R_{13}}{q(R_{16} + R_T)R_{11}} \ln \frac{v_1/R_7}{v_2/R_8}$$

Resistor R_T is a device having a positive linear temperature coefficient (a silicon resistor). It is used to cancel the T in the transfer function.

Otherwise v_0 would vary linearly with temperature. T is the temperature in kelvins (273 K = 0°C).

Several other design tricks will keep this circuit from drifting with temperature. Transistors Q_1 and Q_2 should be a matched pair of devices on one chip. Ideally they should be gain-regulated, e.g., the μA726 temperature-controlled differential pair (Fairchild). This device has active temperature-regulating circuitry on the same chip as the matched pair so that external temperature sources have no effect on transistor parameters. A_1 and A_2 should also be a matched pair of op amps. Perhaps A_1 to A_4 could be a high-quality quad set of op amps on one chip.

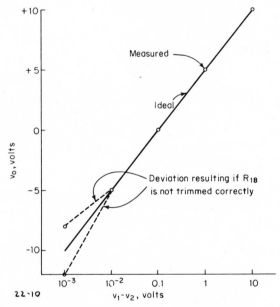

Fig. 22.10 Transfer function of the differential logarithmic amplifier.

This circuit is designed only for positive input voltages. Diodes D_1 to D_4 clamp the outputs of A_1 and A_2 to zero if negative input voltages are accidentally applied. The output v_0, however, can swing positive or negative, as shown in the plotted transfer function (Fig. 22.10).

Feedback stability of A_1 and A_2 is controlled with R_9, R_{10}, C_1, and C_2. Selection of these parts is quite difficult, since stability depends on the feedback factor and the feedback factor depends on the input voltages v_1 and v_2. Experience has shown that the following equations provide reasonable values for these four components:

$$R_9 = R_{10} = \frac{v_{3,\max} - 0.7}{v_{1,\max}/R_7 + v_{3,\max}/R_{11}}$$

$$C_1 = C_2 = \frac{1}{\pi f_u R_9}$$

A tolerance of ± 20 percent is sufficient for these parts.

DESIGN PARAMETERS

Parameter	Description
A	Apparent resistance of R at 0 K ($-273°C$)
A_1, A_2	Differential-input log amplifiers
A_3	Summing differential amplifier
A_4	Temperature-compensating circuit
B	Slope of R_r as a function of temperature
C_1 to C_2	Feedback-stabilizing capacitors
D_1 to D_4	Diodes used to clamp A_1 and A_2 if a negative input voltage is accidentally applied to v_1 or v_2
E	Fractional error in transfer function
f_u	Frequency of open-loop unity gain (A_1 and A_2)
I_{b1} to I_{b2}	Input bias current of A_1 and A_2
ΔI_{b1} to ΔI_{b2}	Change of input bias current of A_1 and A_2
k	Boltzmann's constant $= 1.380 \times 10^{-23}$ J/K
q	Electronic charge $= 1.60 \times 10^{-19}$ C
Q_1 to Q_2	Transistors which provide logarithmic characteristics to circuit
R_1 to R_4	Resistors to compensate for bulk-resistance effects of Q_1 and Q_2 at high levels
R_5 to R_6	Part of diode clamp circuits
R_7 to R_8	Gain-determining resistors
R_9 to R_{10}	Feedback-compensating resistors
R_{11} to R_{14}	Gain-determining resistors of differential summing amplifier A_3
R_{15} to R_{16}	Determine gain of temperature-compensating circuit
R_{17} to R_{20}	A_1 and A_2 input offset adjustment to trim circuit at $v_0 = 0$
R_{Q1} to R_{Q2}	Effective collector-emitter resistance of Q_1 and Q_2
R_T	Positive-temperature-coefficient resistor
T	Temperature in kelvins (K)
$V^{(\pm)}$	Power-supply voltages
v_0	Circuit output voltage
v_1 to v_2	Circuit input voltages
Δv_1 to Δv_2	Measurement errors in v_1 or v_2
V_{io1} to V_{io2}	Input offset voltages of A_1 and A_2
ΔV_{io1} to ΔV_{io2}	Change in input offset voltages of A_1 and A_2
β	Current gain of Q_1 and Q_2

DESIGN EQUATIONS

Description	Equation	
Output-circuit voltage assuming $R_1 = R_2$, $R_3 = R_4$, $R_5 = R_6$, $R_9 = R_{10}$, $R_{11} = R_{12}$, and $R_{13} = R_{14}$	$v_0 = \dfrac{kTR_{15}R_{13}}{q(R_{16} + R_T)R_{11}} \ln\left(\dfrac{v_1/R_7}{v_2/R_8}\right)$	(1)

Description	Equation
Error in measurement of v_1 due to A_1 input errors (important at low v_1 levels)	$\Delta v_1 = \pm V_{io1} + R_7 I_{b1}$ (2)
Error in measurement of v_2 due to A_2 input errors (important at low v_2 levels); these two errors can usually be canceled out at one temperature with the R_{18} potentiometers; if A_1 and A_2 have identical parameter drifts with temperature, Δv_1 and Δv_2 will cancel over temperature	$\Delta v_2 = \pm V_{io2} + R_8 I_{b2}$ (3)
Required relationship between R_1, R_3, and R_7 to cancel the effects of bulk resistance in Q_1 (important at high v_1 levels)	$R_{Q1} = \dfrac{R_1 R_7}{R_1 + R_3}$ (4)
Required relationship between R_2, R_4, and R_8 to cancel effects of bulk resistance in Q_2 (important at high v_2 levels)	$R_{Q2} = \dfrac{R_2 R_8}{R_2 + R_4}$ (5)
Approximate dynamic range possible for v_1 input (assuming no cancellation of errors in A_3)	$\dfrac{v_{1.max}}{v_{1.min}} \approx \dfrac{kTE^2 R_7/q R_{Q1.min}}{\Delta V_{io1.max} + R_7\, \Delta I_{b1.max}}$ (6)
Approximate dynamic range possible for v_2 input (assuming no cancellation of errors in A_3)	$\dfrac{v_{2.max}}{v_{2.min}} \approx \dfrac{kTE^2 R_8/q R_{Q2.min}}{\Delta V_{io2.max} + R_8\, \Delta I_{b2.max}}$ (7)

Resistor values:

R_1, R_2	$R_1 = R_2 = 10\,\Omega$	(8)
R_3	$R_3 = \dfrac{R_1 R_7}{R_{Q1.min}} - R_1$	(9)
R_4	$R_4 = \dfrac{R_2 R_8}{R_{Q2.min}} - R_2$	(10)
R_5, R_6	$R_5 = R_6 = 10\text{ k}\Omega$	(11)
R_7, R_8	$R_7 = R_8 \approx \dfrac{\Delta V_{io1}/\Delta T}{\Delta I_{b1}/\Delta T}$	(12)

R_9, R_{10} $\qquad R_9 = R_{10} = \dfrac{(kT/q) \ln (v_1/v_2)_{max}}{(v_{1,max}/R_7) - [kT \ln (v_1/v_2)_{max}]/qR_{11}} \qquad$ (13)

$\qquad$ (assume $T = 300$ K)

R_{11}, R_{12} $\qquad R_{11} = R_{12} = 10$ kΩ $\qquad$ (14)

R_{13}, R_{14} $\qquad R_{13} = R_{14} = R_{11} \left[\dfrac{qv_{o,max}}{kT \ln (v_1/v_2)_{max}} \right]^{1/2} \qquad$ (15)

$\qquad$ (assume $T = 300$ K)

R_{15} $\qquad R_{15} = BT \left[\dfrac{qv_{o,max}}{kT \ln (v_1/v_2)_{max}} \right]^{1/2} \qquad$ (16)

$\qquad$ (assume $T = 300$ K)

R_{16} $\qquad R_{16} = -A \qquad$ (17)

R_{17} $\qquad R_{17} \geqq 10R_7 \qquad$ (18)

R_{18} $\qquad R_{18} \geqq \dfrac{R_{17}}{10} \qquad$ (19)

R_{19} $\qquad R_{19} \approx \dfrac{V^{(+)} R_{20}}{100 V_{io1,max}} \qquad$ (20)

R_{20} $\qquad R_{20} \leqq \dfrac{R_{18}}{100} \qquad$ (21)

Resistance of positive-
temperature-
coefficient device $\qquad R_T \approx A + BT \qquad$ (22)

Capacitor values C_1, C_2 $\qquad C_1 = C_2 \approx \dfrac{1}{\pi f_u R_9} \qquad$ (23)

Bulk resistance of Q_1 $\qquad R_{Q1} = \dfrac{kTR_7}{qv_1 \beta} \qquad$ (24)

Bulk resistance of Q_2 $\qquad R_{Q2} = \dfrac{kTR_8}{qv_2 \beta} \qquad$ (25)

DESIGN PROCEDURE

Many complicating factors must be considered to design a high-quality log amplifier. We will assume that linearity is the most important parameter and let the other parameters be controlled by physical limitations imposed by the design equations.

DESIGN STEPS

Step 1. If A_1 and A_2 are identical op amps, R_7 and R_8 are determined from Eq. (12).

Step 2. Calculate the minimum expected bulk resistances of Q_1 and Q_2 using Eqs. (24) and (25).

Step 3. Compute values for R_3 and R_4 using Eqs. (8) to (10) and the results of steps 1 and 2.

Step 4.   Let $R_{11} = R_{12} = 10$ kΩ. Determine values for R_9 and R_{10} using Eq. (13).

Step 5.   Compute values for R_{13} and R_{14} from Eq. (15).

Step 6.   The resistance of most positive-coefficient temperature-sensitive resistors (silicon resistors) can be described by the form shown in Eq. (22). If the temperature could be extended down to 0 K, its resistance would theoretically be A. The slope of resistance as a function of temperature is B. Use these numbers to compute R_{15} and R_{16} using Eqs. (16) and (17).

Step 7.   Calculate values for C_1 and C_2 using Eq. (23).

Step 8.   Calculate approximate values for R_{17} to R_{20} using Eqs. (18) to (21).

EXAMPLE OF LOG-AMPLIFIER DESIGN   Suppose we wish to precondition analog data before they are applied to an analog-to-digital converter. Our goal is to keep the A/D error small by compressing the input voltage (which encompasses four orders of magnitude) into ± 10 V (slightly more than one order of magnitude). The A/D converter accepts voltages from -10 to $+10$ V. The input voltage to the log amplifier ranges from 0.001 to 10 V. The output voltage is to be zero when the input voltage is 0.1 V. The required transfer function, along with test data from a circuit designed with the following steps, is shown in Fig. 22.10.

Design Requirements

$$v_{o,\,\text{max}} = \pm 10 \text{ V}$$
$$v_{1,\,\text{min}} = 10^{-3} \text{ V}$$
$$v_{1,\,\text{max}} = 10 \text{ V}$$
$$v_2 = 0.1 \text{ V (fixed)}$$
$$V^{(\pm)} = \pm 15 \text{ V}$$

Device Data

$$\Delta V_{io1} = \Delta V_{io2} = \pm 0.3 \text{ mV (0 to } + 75°\text{C)}$$
$$V_{io1,\,\text{max}} = 1 \text{ mV}$$
$$\Delta I_{b1} = \Delta I_{b2} = 2 \text{ nA} \qquad (0 \text{ to } + 75°\text{C})$$
$$f_u = 10^6 \text{ Hz}$$
$$R_T = 500 \ \Omega \qquad \text{at 300 K}$$
$$A = -730 \ \Omega$$
$$B = 4.1 \ \Omega/\text{K}$$
$$\beta = \begin{cases} 150 & \text{(max)} \\ 100 & \text{at } I_{c,\,\text{max}} \end{cases}$$

Step 1. R_7 and R_8 must be approximately

$$R_7 = R_8 \approx \frac{\Delta V_{io1}/\Delta T}{\Delta I_{b1}/\Delta T} \approx \frac{3 \times 10^{-4}}{2 \times 10^{-9}} = 150 \text{ k}\Omega$$

Step 2.   The minimum bulk resistances of Q_1 and Q_2 are

$$R_{Q1, \min} = R_{Q2, \min} = \frac{kTR_7}{qv_{1, \max}\, \beta \text{ (at } I_{c, \max})}$$

$$= \frac{(1.38 \times 10^{-23} \text{ J/K})(273 \text{ K})(1.5 \times 10^5 \text{ }\Omega)}{(1.6 \times 10^{-19} \text{ C})(10 \text{ V})(100)}$$

$$= 3.53 \text{ }\Omega$$

Step 3.   From Eq. (8) $R_1 = R_2 = 10 \text{ }\Omega$.   Equations (9) and (10) provide

$$R_3 = R_4 = \frac{R_1 R_7}{R_{Q1, \min}} - R_1 = \frac{10(1.5 \times 10^5)}{3.53} - 10 = 425 \text{ k}\Omega$$

Step 4.   We let $R_{11} = R_{12} = 10 \text{ k}\Omega$, as recommended. R_9 and R_{10} are found from

$$R_9 = R_{10} = \frac{(kT/q)\ln (v_1/v_2)_{\max}}{(v_{1, \max}/R_7) - [kT\ln (v_1/v_2)_{\max}]/qR_{11}}$$

$$= \frac{[(1.38 \times 10^{-23})(300)/(1.6 \times 10^{-19})]\ln (10/0.1)}{[10/(1.5 \times 10^5)] - [(1.38 \times 10^{-23})(300)\ln (10/0.1)]/(1.6 \times 10^{-19})(10^4)}$$

$$= 2176 \text{ }\Omega$$

Step 5. R_{13} and R_{14} are now found:

$$R_{13} = R_{14} = R_{11}\left[\frac{qv_{o, \max}}{kT\ln (v_1/v_2)_{\max}}\right]^{1/2}$$

$$= 10^4\left[\frac{(1.6 \times 10^{-19})(10)}{(1.38 \times 10^{-23})(300)\ln (10/0.1)}\right]^{1/2} = 91.6 \text{ k}\Omega$$

Step 6.   Equation (16) gives

$$R_{15} = BT\left[\frac{q\,v_{o, \max}}{kT\ln (_1/v_2)_{\max}}\right]^{1/2}$$

$$= (4.1)(300)\left[\frac{(1.6 \times 10^{-19})(10)}{(1.38 \times 10^{-23})(300)\ln (10/0.1)}\right]^{1/2}$$

$$= 11.3 \text{ k}\Omega$$

During the test this resistor is trimmed so that $v_o = 10$ when $v_1 = 10$. Equation (17) provides

$$R_{16} = -A = -(-730) = 730 \text{ }\Omega$$

Step 7.   Capacitor values are

$$C_1 = C_2 \approx \frac{1}{\pi f_u R_9} = \frac{1}{\pi(10^6 \times 2176)} = 146 \text{ pF}$$

Step 8. R_{17} to R_{20} are determined using Eqs. (18) to (21):

$$R_{17} \gtreqqless 10R_7 = 10(150 \text{ k}\Omega) = 1.5 \text{ M}\Omega$$

$$R_{18} \gtreqqless \frac{R_{17}}{10} = \frac{1.5 \times 10^6}{10} = 150 \text{ k}\Omega$$

$$R_{20} \lesseqqgtr \frac{R_{18}}{100} = \frac{150{,}000}{100} = 1500 \ \Omega$$

$$R_{19} \approx V^{(\pm)} \frac{R_{20}}{100 \ V_{io1,\max}} \approx \frac{15(1500)}{100 \times 10^{-3}} \approx 225 \text{ k}\Omega$$

22.5 ANTILOGARITHMIC AMPLIFIER

ALTERNATE NAMES Antilog amplifier, inverse-log amplifier, antilog converter, data expander, exponential amplifier.

PRINCIPLES OF OPERATION This circuit is merely a modification of the log circuit described in the last section. The antilog function is implemented in Fig. 22.11 by changing the connections to Q_1 and Q_2 in the log circuit of Fig. 22.9.

The relationship between input and output voltages in this circuit is

$$v_o = De^{-Ev_1}$$

where $\qquad D = \dfrac{R_5 V_R (R_7 + R_8)}{R_1 R_7} \qquad E = -\dfrac{q(R_T + R_3)}{kT(R_T + R_2 + R_3)}$

The D term is equal to the required output voltage when the input voltage is zero.

The R_T and R_2 terms in E establish the dynamic range of the antilog amplifier. For a given $\pm v_{1,\max}$ input range, the total number of decades excursion of v_o increases as E increases in magnitude. The maximum input voltage range $\pm v_{1,\max}$ is controlled mainly by the $R_T/(R_T + R_2)$ ratio. In some applications this ratio can be made quite large, so that $\pm v_{1,\max}$ may be many tens or hundreds of volts if required. The resistor R_T must have a positive temperature coefficient to cancel the T term in Eq. (1) (see Design Equations).

The emitter saturation currents I_{E1} and I_{E2} will have no effect on v_o if they track over temperature. If they are not equal, they merely affect D [see Eqs. (1) and (2)]; however, they still should be selected so that their ratio remains constant over temperature.

The dynamic range of v_o is limited by V_n, I_n, V_{io}, I_b, and I_{io} of both op amps. In practice, only three or four decades of dynamic range are possible without going to expensive op amps and complex temperature-compensating circuits. Measured data from an antilog amplifier built according to the design equations are shown in Fig. 22.12.

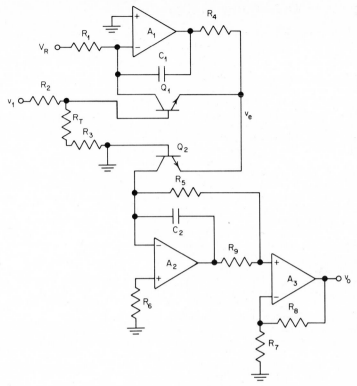

Fig. 22.11 Antilog amplifier formed by changing the input and feedback circuits of Fig. 22.9.

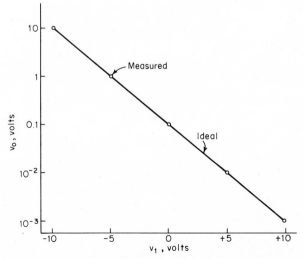

Fig. 22.12 Response of an antilog amplifier designed by using equations listed in this section.

DESIGN PARAMETERS

Parameter	Description
A	Apparent resistance of R_T at 0 K ($-273°C$)
A_1 to A_2	Nonlinear amplifiers
A_3	Buffer to extend output-voltage range
B	Slope of R_T as a function of temperature
C_1 to C_2	Feedback-stabilizing capacitors
D	Output voltage when $v_i = 0$
E	Determines dynamic range of circuit
I_b	Op-amp input bias current (A_1 to A_2)
I_{io}	Op-amp input offset current (A_1 to A_2)
I_n	Equivalent rms input noise current of an op amp (A_1 to A_2)
k	Boltzmann's constant = 1.380×10^{-23} J/K
N	Number of decades response of circuit on either side of $v_o = D$
q	Electronic charge = 1.60×10^{-19} C
Q_1 to Q_2	Transistors used to provide nonlinear transfer function
R_1	Sets reference current into A_1
R_2 to R_3	Part of temperature-compensation circuit
R_4, R_9	Provide feedback stability
R_5	Establishes gain of A_2 circuit
R_6	Cancels effect of I_b in A_2
R_7 to R_8	Establishes gain of A_3 buffer stage
R_T	Provides temperature compensation
T	Temperature in kelvins (K)
v_o	Circuit output voltage
v_1	Circuit input voltage
v_e	Voltage at emitters of Q_1 and Q_2
V_{io}	Input offset voltage of an op amp
V_n	Equivalent rms input voltage noise of an op amp
V_R	Reference voltage

DESIGN EQUATIONS

Description	Equation
Circuit output voltage	$v_o = De^{-Ev_1}$
	where $D = \dfrac{R_5 V_R (R_7 + R_8)}{R_1 R_7}$ $E = \dfrac{q(R_T + R_3)}{kT(R_T + R_2 + R_3)}$ (1)
Voltage at junction of transistor emitters	$v_e = \dfrac{v_1(R_T + R_3)}{R_T + R_2 + R_3} - \dfrac{kT}{q} \ln \dfrac{V_R I_{E2} R_5}{v_o I_{E1} R_1}$ (2)
R_T, R_2, and R_3 required to provide $\pm N$ decades response of v_o about D (nominal v_o)	$\dfrac{R_T + R_3}{R_T + R_2 + R_3} = \dfrac{kTN \ln 10}{q|v_{1,\text{max}}|}$ (3)
Resistor values:	
R_1	$R_1 = 10^6$ (4)
	chosen such that it does not load down V_R

R_2	$R_2 = (R_T + R_3)\left[\dfrac{q\lvert v_{1,\text{max}}\rvert}{kTN \ln 10} - 1\right]$	$T = 300$ K	(5)
R_3	$R_3 = -A$		(6)
R_4	$R_4 = 10 \text{ k}\Omega$		(7)
R_5, R_6	$R_5 = R_6 = \dfrac{DR_1}{10 V_R}$		(8)
Capacitor C_1	$C_1 = 100 \text{ pF}$		(9)
Resistance of R_T as a function of temperature	$R_T = A + BT$		(10)

22.6 FET-CONTROLLED MULTIPLIER

ALTERNATE NAMES Voltage-controlled amplifier, analog multiplier, linear multiplier.

PRINCIPLES OF OPERATION This section will provide design information on the voltage-controlled FET multiplier, probably the simplest and most easily implemented multiplier in the literature. Satisfactory performance can be achieved with this circuit only after the FET transfer characteristics are made linear. This is done with proper biasing and local feedback around the FET.

FETs make nearly ideal voltage-controlled resistors. Their range of operation, however, is limited by several constraints which must be understood before their usefulness can be exploited fully. FETs have a voltage-controlled drain-to-source resistance of $R_{DS} = V_P^2 / I_{DSS}\lvert v_c - 2V_P\rvert$. We will let R_{DS} be the drain-to-source resistance of both Q_1 and Q_2 since they are identical devices at the same temperature and use the same v_c.

The output of the A_1 stage in Fig. 22.13 is

$$v_o = \frac{v_1 R_2 R_{10}}{(R_1 + R_2)R_{DS}}$$

But R_{DS} depends on the control voltage v_c. This control voltage depends on v_2 and V_R. For proper circuit operation V_R must be positive and v_2 can range from zero to a specified negative limit. These polarities must be observed. The current through R_{DS} is identical to the current through R_9 since A_2 draws insignificant current. The control voltage v_c will force R_{DS} to the correct resistance so that these currents are equal. Since the inverting input of A_2 tries to remain at ground potential (because of feedback), we can draw the following conclusion:

$$I_{Q2} = \frac{v_4 - 0}{R_{DS}} = I_{R9} = \frac{0 - v_2}{R_9}$$

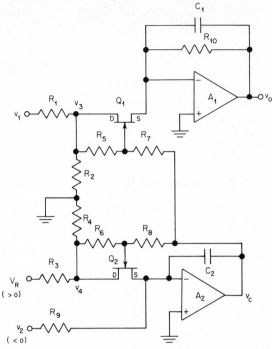

Fig. 22.13   A linear multiplier which utilizes the voltage-controlled resistance property of FETs.

The voltage divider R_3 to R_4 provides us with

$$v_4 = \frac{V_R R_4}{R_3 + R_4}$$

Combining these two equations gives

$$R_{DS} = \frac{R_4 R_9 V_R}{|v_2|(R_3 + R_4)}$$

Q_1 and Q_2 must be a matched pair on a single chip such as the 2N5196. The control voltage v_c drives the gates of both Q_1 and Q_2. This guarantees that $R_{DS1} = R_{DS2}$. The final result for v_0 is found by substituting the R_{DS} expression into the v_0 expression:

$$v_0 = -\frac{v_1|v_2| R_2 R_{10}(R_3 + R_4)}{V_R R_4 R_9(R_1 + R_2)}$$

If we allow $R_1 = R_3$, $R_2 = R_4$, and $R_9 = R_{10}$, the above equation reduces to

$$v_0 = -\frac{v_1|v_2|}{V_R}$$

This circuit can even be used as a divider by allowing V_R to be an input variable. However, the available range for V_R is only several volts, which makes this option inadvisable. Also, if one attempts to drive the R_3, R_9, R_{DS} circuit backward by letting V_R be negative and v_o positive, a lockup condition will occur. The A_2 circuit has positive feedback in this situation.

The circuit in Fig. 22.13 will exhibit less than 5 percent distortion only if the following are observed:

1. R_9 must be larger than $R_{DS,\text{min}}$.

2. The voltages at v_3 and v_4 must always be within the ± 1-V range.

3. The control voltage v_c must operate only in the range from zero down to $-2V_p$.

4. FETs with large V_p are used. Note, however, that A_2 must be able to drive v_c down to $-2V_p$.

5. $R_{DS,\text{min}}$ must be at least 100 times larger than R_2 or R_4 so that R_{DS} does not load the R_1/R_2 and R_3/R_4 dividers. This will cause an error which varies according to the magnitude of v_2.

DESIGN PARAMETERS

Parameter	Description
C_1, C_2	Capacitors required if A_1 or A_2 tend to be unstable (100 to 500 pF)
I_{DSS}	Drain-to-source current of an FET if $V_{GS} = 0$ and $V_{DS} = 5$ to 10 V
I_{Q1}, I_{Q2}	Drain-to-source current through Q_1 or Q_2
I_{R9}	Current through R_9
Q_1, Q_2	Field-effect transistors
R_1 to R_4	Input attenuators
R_5 to R_8	FET biasing and feedback resistors
R_9 to R_{10}	Determine overall gain of circuit
R_{DS}	Drain-to-source resistance of an FET at a specified gate-to-source voltage
R_{G1}, R_{GR}	Generator resistances for V_1 and V_R
v_0 to v_4, v_c	Voltages as noted in Fig. 22.13
V_p	FET pinch-off voltage (where R_{DS} approaches infinity)
V_R	Reference input voltage
$V^{(\pm)}$	Positive and negative supply voltages

DESIGN EQUATIONS

Description	Equation	
Output voltage of circuit	$$v_o = -\frac{v_1|v_2|R_2R_{10}(R_3 + R_4)}{V_R R_4 R_9(R_1 + R_2)}$$	(1)
Output voltage of circuit if $R_1 = R_3$, $R_2 = R_4$, $R_9 = R_{10}$	$$v_o = -\frac{v_1|v_2|}{V_R}$$	(2)

Description	Equation	
R_{DS} as a function of control voltage v_c	$$R_{DS} = \frac{V_p^2}{I_{DSS}|v_c - 2V_p|}$$	(3)
Minimum value for R_{DS} if nonlinearity is to be minimized	$$R_{DS,min} = \frac{|V_p|}{2I_{DSS}}$$	(4)
Maximum recommended value for R_{DS}	$$R_{DS,max} = 10R_{DS,min}$$	(5)
Minimum and maximum recommended $|v_2|$; use minimum and maximum R_{DS} from Eqs. (4) and (5)	$$v_2 = \frac{V_R R_4 R_9}{R_{DS}(R_3 + R_4)}$$	(6)
Optimum range for v_c	v_c range $= 2V_p$ to zero	(7)
Resistor values		
R_1, R_3	$R_1 = R_3 > 100R_{G1}$ or 100 R_{GR} whichever is larger	(8)
R_2, R_4	$$R_2 = R_4 < \frac{R_{DS,min}}{100}$$	(9)
R_5, R_6, R_7, R_8	$R_5 = R_6 = R_7 = R_8 > 1000R_2$	(10)
R_9, R_{10}	$$R_9 = R_{10} = \frac{|v_o|_{max}(R_1 + R_2)|V_p|}{2|v_1|_{max}R_2 I_{DSS}}$$	(11)

DESIGN PROCEDURE

The design of this circuit begins by choosing a good-quality matched FET pair with a high V_p. The op amp A_2 must be able to swing to a negative output voltage twice the value of V_p.

DESIGN STEPS

Step 1. Choose a single-chip set for Q_1 and Q_2 which track I_{DSS} and V_p over temperature. A high value for V_p is also desirable. Choose good-quality op amps for A_1 and A_2 which will drive v_c more negative than $2V_p$ with several volts' margin. Let the positive and negative power supplies be compatible with $\pm v_{o,max}$ and $2V_p$ with several volts' margin.

Step 2. Compute $R_{DS,min}$ from Eq. (4). Choose a common value for R_2 and R_4 which is less than 1 percent of $R_{DS,min}$. Choose a common value for R_1 and R_3 which is at least 100 times larger than R_{G1} or R_{GR}.

Step 3. Calculate a nominal value for R_5 to R_8 using Eq. (10).

Step 4. Compute values for R_9 and R_{10} using Eq. (11).

Step 5. Determine the allowable range of v_2 from Eqs. (4) to (6).

EXAMPLE OF MULTIPLIER DESIGN This circuit is a simple multiplier, and, as expected, it has several limitations which should be recognized. Both v_1 and v_2 have definite limits over which linearity of $v_o = -v_1|v_2|/V_R$ can be expected. We will assume for this example that Q_1 and

Q_2 are a 2N5196 dual-FET device and the op-amp outputs can drive ±15 V.

Design Requirements

$$v_o = \frac{-v_1|v_2|}{5}$$

$$\pm v_{o,\text{max}} = \pm5 \text{ V}$$
$$V^{(\pm)} = \pm15 \text{ V}$$
$$V_R = +5.00 \text{ V regulated}$$

Device Data

$$V_{p,Q_1,Q_2} = -3 \text{ V}$$
$$I_{DSS,Q_1,Q_2} = 0.85 \text{ mA}$$

both measured on a 2N5196.

Step 1. Dual-FET specification sheets do not usually specify the degree of matching and tracking over temperature of I_{DSS} and V_p. Fairly good results can be achieved if the dual device is on a single chip and has guaranteed tracking of 5 or 10 μV/°C for differential gate-source voltage. This is often the only parameter specified over temperature. For this example we choose the 2N5196 dual FET which has a specified pinch-off voltage of -0.7 to -4 V. The device tested in this example measures -3 V for both Q_1 and Q_2. The maximum required drive from A_2 is $2V_p = 2(-3) = -6$ V. We will use the 747 op amp for this application. The power supplies will be ±15 V so that the maximum $v_o = \pm5$ V can be achieved.

Step 2. The minimum R_{DS} is

$$R_{DS,\text{min}} = \frac{|V_p|}{2I_{DSS}} = \frac{3}{2(0.85 \times 10^{-3})} = 1765 \text{ }\Omega$$

The common value for R_2 and R_4 must be

$$R_2 = R_4 < \frac{1765}{100} = 17.65 \text{ }\Omega$$

We will use 17.4-Ω precision resistors.

Assume the source resistances for v_1 and v_2 are 50 Ω. We should make $R_1 = R_3 \geqq 100(50) = 5000 \text{ }\Omega$ at least. We will use 5110-Ω precision resistors.

Step 3. With Eq. (10) we get

$$R_5 = R_6 = R_7 = R_8 \geqq 1000R_2 = 1000(17.4) = 17.4 \text{ k}\Omega$$

We will use 100-kΩ resistors.

Step 4. R_9 and R_{10} are found with Eq. (11):

$$R_9 = R_{10} = \frac{|v_0|_{\max}(R_1 + R_2)|V_p|}{2|v_1|_{\max}R_2 I_{DDS}}$$

$$= \frac{5(5110 + 17.4)(3)}{2(5)(17.4)(8.5 \times 10^{-4})} = 520 \text{ k}\Omega$$

Step 5.   We first compute the minimum and maximum allowable R_{DS} from Eqs. (4) and (5):

$$R_{DS,\min} = \frac{V_p}{2 I_{DSS}} = \frac{3}{2(0.85 \times 10^{-3})} = 1765 \ \Omega$$

$$R_{DS,\max} = 10 R_{DS,\min} = 10(1765) = 17{,}650 \ \Omega$$

The range of allowable v_2 values is now determined with the help of Eq. (6):

$$v_{2,\min} = \frac{V_R R_4 R_9}{R_{DS,\max}(R_3 + R_4)} = \frac{5(17.4)(520{,}000)}{17{,}650(5110 + 17.4)} = 0.50 \text{ V}$$

$$v_{2,\max} = \frac{V_R R_4 R_9}{R_{DS,\min}(R_3 + R_4)} = \frac{5(17.4)(520{,}000)}{1765(5110 + 17.4)} = 5 \text{ V}$$

Figure 22.14 shows the response of a circuit built according to the above calculations.   The worst-case linearity error was 4.4 percent of full-scale output.

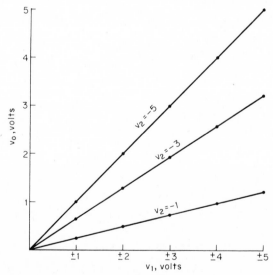

Fig. 22.14   Measured transfer function of multiplier shown in Fig. 22.13 built according to the design steps.

22.7   LOG-ANTILOG MULTIPLIER-DIVIDER

ALTERNATE NAMES   Analog multiplier, analog divider, analog multiplier-divider, one-quadrant multiplier-divider.

PRINCIPLES OF OPERATION   Although operation is restricted to the first quadrant (for v_1, v_2, v_3, and v_o), this circuit is extremely useful in a variety of applications.   Linearity errors of less than 1 percent can be achieved if input offsets of A_1, A_2, and A_3 are properly handled. This minimal error is possible over two to three decades of input voltages.

The circuits of A_1, A_2, and A_3 are single-ended log amplifiers.   Two of these circuits are utilized in the log ratio circuit shown in Fig. 22.9.

Note that Q_1 and Q_2 are effectively connected in series.   This results in addition of logarithms, i.e., multiplication of v_1 and v_2.   The output from A_1 and A_2 drives the A_4 antilog circuit and thereby produces the product of v_1 and v_2 at v_o.

The output from A_3 is treated differently.   This voltage subtracts current from Q_4.   Subtraction of logarithmic quantities results in division. The final result is

$$v_o = \frac{v_1 v_2}{v_3}$$

The foregoing remarks will now be treated in mathematical terminology.   Each output voltage from the three log amplifiers is equal to the emitter-base voltage of the transistor in its feedback loop.   The following can be stated:

$$v_4 = -\frac{kt}{q} \ln \frac{v_1}{R_1 I_{ES1}} = -v_{BE1}$$

$$v_5 = -\frac{kt}{q} \ln \frac{v_2}{R_2 I_{ES2}} = -v_{BE2}$$

$$v_6 = -\frac{kt}{q} \ln \frac{v_3}{R_3 I_{ES3}} = -v_{BE3}$$

The bases of Q_1 and Q_3 are at ground potential (through 10 Ω). Starting at the base of Q_1, we trace the voltage loop

$$v_{BE1} + v_{BE2} - v_{BE4} - v_{BE3} = 0$$

Transistor Q_4 has the following voltage-current relationship:

$$v_{BE4} = -\frac{kt}{q} \ln \frac{I_{C4}}{\alpha_4 I_{ES4}}$$

If this equation and the other equations for emitter-base voltages are substituted into the voltage-loop equation, we get

$$-\frac{kt}{q}\ln\frac{v_1}{R_1 I_{ES1}}-\frac{kt}{q}\ln\frac{v_2}{R_2 I_{ES2}}+\frac{kt}{q}\ln\frac{v_3}{R_3 I_{ES3}}+\frac{kt}{q}\ln\frac{I_{C4}}{\alpha_4 I_{ES4}}=0$$

If $I_{ES1}=I_{ES2}=I_{ES3}=I_{ES4}$ and $\alpha_4=1$, this reduces to

$$-\ln\frac{v_1}{R_1}-\ln\frac{v_2}{R_2}+\ln\frac{v_3}{R_3}=-\ln I_{C4}\qquad\text{or}\qquad\ln I_{C4}=\ln\frac{v_1 v_2 R_3}{R_1 R_2 v_3}$$

Taking the antilog of each side gives

$$I_{C4}=\frac{v_1 v_2}{v_3}\frac{R_3}{R_1 R_2}$$

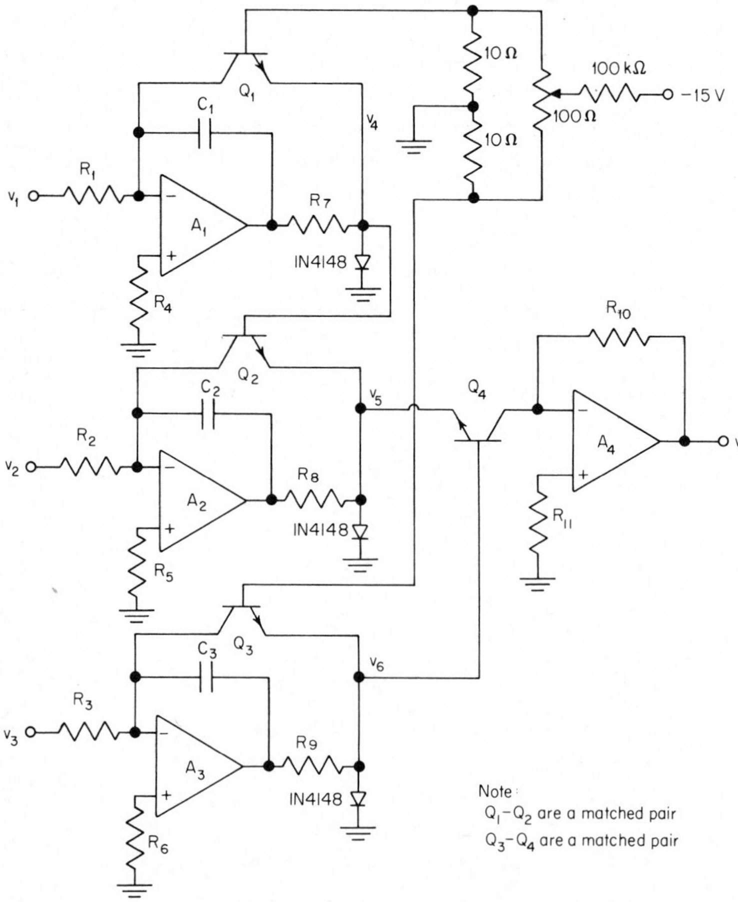

Fig. 22.15   A multiplier-divider which uses three logarithmic amplifiers and an antilog amplifier.

The output voltage v_o is related to I_{C4} by $v_o = I_{C4} R_{10}$. The output voltage now becomes

$$v_o = \frac{v_1 v_2}{v_3} \frac{R_3 R_{10}}{R_1 R_2}$$

If we let $R_1 = R_2 = R_3 = R_{10}$, we get

$$v_o = \frac{v_1 v_2}{v_3}$$

The circuit of Fig. 22.15 can easily be converted into a square-root circuit. If v_o is connected to v_3 and $v_2 = 1$ V,

$$v_o = \frac{v_1(1)}{v_o} \qquad \text{or} \qquad v_o = v_1^{1/2}$$

DESIGN PARAMETERS

Parameter	Description
A_1 to A_3	Log-amplifier-circuit op amps
A_4	Antilog-amplifier-circuit op amp
C_1 to C_4	Capacitors required for feedback stability
f_{u1} to f_{u3}	Unity-gain crossover frequency of A_1 to A_3
I_{C4}	Collector current of Q_4
I_{ES1} to I_{ES4}	Emitter saturation currents of Q_1 to Q_4
I_{io}	Input offset current of A_1, A_2, or A_3
k	Boltzmann's constant $= 1.38 \times 10^{-23}$ J/K
q	Electronic charge $= 1.6 \times 10^{-19}$ C
Q_1 to Q_3	Transistors used in log amplifiers
Q_4	Transistor used in antilog amplifier
R_1 to R_3	Resistors used to determine gain of input stages and to provide a high input resistance
R_4 to R_6, R_{11}	Compensate for input bias currents of A_1 to A_4
R_7 to R_9	Utilized to provide feedback stability in A_1 to A_3 circuits
R_{10}	Determines overall gain of circuit along with R_1 to R_3
R_{in}	Input resistance of any input
R_{S1} to R_{S3}	Source resistances of v_1 to v_3 sources
T	Temperature in kelvins (273 K $= 0°$C)
v_o to v_6	Voltages as noted in Fig. 22.15
V_{BE1} to V_{BE4}	Base-to-emitter voltages of Q_1 to Q_4
V_{io}	Input offset voltage of A_1, A_2, or A_3
α_4	Common-base current gain of Q_4
β_1 to β_4	Common-emitter current gains of Q_1 to Q_4

DESIGN EQUATIONS

Description	Equation
Output voltage of circuit as a function of the three input voltages	$v_o = \dfrac{v_1 v_2}{v_3} \dfrac{R_3 R_{10}}{R_1 R_2}$ (1)
Output voltages of three log amplifiers	$v_4 = -\dfrac{kt}{q} \ln \dfrac{v_1}{R_1 I_{ES1}}$ (2a)
	$v_5 = -\dfrac{kt}{q} \ln \dfrac{v_2}{R_2 I_{ES2}}$ (2b)
	$v_6 = -\dfrac{kt}{q} \ln \dfrac{v_3}{R_3 I_{ES3}}$ (2c)
Output voltage of antilog amplifier as a function of its input current	$v_o = I_{C4} R_{10}$
	where $I_{C4} = \dfrac{v_1 v_2}{v_3} \dfrac{R_3}{R_1 R_2}$ (3)
Resistor values:	
R_1, R_2, R_3 resistance must be at least 1000 times smaller than each input resistor	$R_1 = R_2 = R_3 = R_{\text{in}}$ required for each input (4)
R_4	$R_4 = R_1$ (5)
R_5	$R_5 = R_2$ (6)
R_6	$R_6 = R_3$ (7)
R_7	$R_7 \approx \dfrac{(kt/q) \ln v_{1,\text{max}}}{(v_{1,\text{max}}/R_1) - v_{2,\text{max}}/R_2 \beta_2}$ (8)
R_8	$R_8 \approx \dfrac{(kt/q) \ln v_{2,\text{max}}}{(v_{2,\text{max}}/R_2) + v_{o,\text{max}}/R_{10}}$ (9)
R_9	$R_9 \approx \dfrac{(kt/q) \ln v_{3,\text{max}}}{(v_{3,\text{max}}/R_3) - v_{o,\text{max}}/R_{10} \beta_4}$ (10)
R_{10}	$R_{10} = \dfrac{R_1 R_2 v_{o,\text{max}}}{R_3 (v_1 v_2 / v_3)_{\text{max}}}$ (11)
R_{11}	$R_{11} = R_{10}$ (12)
Capacitor values: C_1	$C_1 \approx \dfrac{1}{\pi f_{u1} R_7}$ (13)
C_2	$C_2 \approx \dfrac{1}{\pi F_{u2} R_8}$ (14)
C_3	$C_3 \approx \dfrac{1}{\pi f_{u3} R_9}$ (15)

REFERENCES

1. "Applications Manual for Computing Amplifiers," p. 88, George A. Philbrick Researches, Inc., Nimrod Press, Boston, 1966.
2. Tobey, G. E., J. G. Graeme, and L. P. Huelsman: "Operational Amplifiers: Design and Applications," p. 357, McGraw-Hill, New York, 1971.
3. Millman, J., and C. C. Halkias: "Integrated Electronics: Analog and Digital Circuits and Systems," p. 572, McGraw-Hill, New York, 1972.
4. Smith, J. I.: "Modern Operational Circuit Design," p. 36, Wiley, New York, 1971.
5. Kreeger, R.: AC-to-DC Converters for Low Level Input Signals, *EDN*, Apr. 5, 1973, p. 60.
6. Logarithmic Converters, *National Semiconductor Corp., Appl. Note* AN-30, November 1969.
7. Christie, W. C.: Multiply and Divide with a Dual Photo Resistor, *Electron. Des.,* vol. 21, p. 108, Oct. 10, 1968.
8. Counts, L., and D. Sheingold: Analog Dividers: What Choice Do You Have?, *EDN,* May 5, 1974, p. 55.

Chapter 23

Analog Signal Generation

INTRODUCTION

This chapter differs from Chap. 9 in that linear circuits are used here and digital circuits were used there. Although rectangular-waveform generation is discussed in both chapters, the methods for creating this waveform are substantially different. The present chapter discusses a large variety of waveforms including rectangular, sine, triangle, and saw-tooth waveforms.

The first few sections of this chapter present the theory of operation for monolithic function generators. These chips can simultaneously provide rectangular, triangle, and sine waveforms. They are capable of frequencies down to 0.001 Hz, but their upper frequency is typically only 1 or 2 MHz. The rectangular and triangle waveforms are quite good, but the sine waveform is slightly distorted. The duty factor of the rectangular waveform can be varied from 1 to 99 percent in many devices. The two slopes of the triangle can also be adjusted for a 100:1 difference in either direction.

The second group of circuits presents several of the new developments in low-frequency sine-wave generators. Three or more active devices are required for these circuits, but in most cases multiple-device chips can handle a large part of the circuit.

23.1 ACTIVE-INTEGRATOR FUNCTION GENERATOR

ALTERNATE NAMES Integrator–hysteresis-switch function generator, integrator–dual-comparator function generator.

PRINCIPLES OF OPERATION The active integrator in conjunction with a hysteresis switch has been used as a function generator for many years. It was the first function generator to be successfully mass produced in monolithic form. Although its upper-frequency response is limited by the op-amp integrator slew rate, this circuit is still useful in many applications. Figure 23.1 shows the active-integrator–hysteresis-switch function generator in its most elementary form.

In operation, a rectangular waveform v_1 is present at the output of A1. The instantaneous amplitude of v_1 is compared with the duty-factor voltage V_{DF}. If $v_1 > V_{DF}$, the output of A2 will integrate downward at a rate of $(v_1 - V_{DF})/R_1 C_1$. Likewise, if $V_{DF} > v_1$, v_2 will integrate upward at a rate of $(V_{DF} - v_1)/R_1 C_1$. The resulting v_2 waveform is triangular. If the amplitude of this triangle is to be adjustable, the A3 buffer circuit is installed. This buffer will provide an output amplitude range from zero up to 3 times the maximum value of v_2. The same kind of buffer can be utilized for the rectangular output waveform v_3.

The level detector A4 has hysteresis through use of positive feedback. The size of the hysteresis amplitude is determined by R_6 and R_7. If these resistors are equal in size, the hysteresis amplitude will be 50 percent of the peak-to-peak amplitude of v_3.

The output of A4 will be rectangular as the comparator switches from one saturation level to the other. A portion of v_3 is tapped off R_8 to

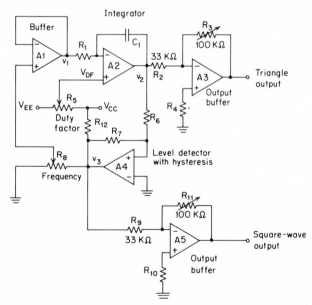

Fig. 23.1 A simplified version of the active-integrator–hysteresis-switch function generator.

drive the integrator. If a small portion is utilized, the integrator will slew at a slow rate and the output frequency will be low. Likewise, a large v_1 produces a high frequency.

DESIGN PARAMETERS

Parameter	Description
A1	Op amp used as buffer so that R_1 is not affected by the location of the R_8 tap point
A2	Op amp used as integrator
A3, A5	Op amp used as variable-gain output buffers
A4	Comparator connected as a noninverting level detector with hysteresis
C_1	Timing capacitor in integrator circuit
DF	Duty factor on waveforms, i.e., ratio of ON time to total period of one cycle
I_{io}	Input offset current of op amp
R_1	Timing resistor in integrator circuit
R_2, R_3	Gain-determining resistors in triangle-output buffer circuit
R_4	Resistor used to minimize output offset of A3
R_5	Variable resistor used to set V_{DF} and duty factor
R_6, R_7	Resistors which set magnitude of hysteresis in A4
R_8	Frequency-determining resistor which delivers adjustable-amplitude rectangular waveform to integrator
R_9, R_{11}	Gain-determining resistors for rectangular-waveform output buffer circuit
R_{10}	Resistor used to minimize output offset in A5
S_{A2}	Slew rate of A2
v_1	Adjustable-amplitude waveform which drives integrator circuit
v_2	Triangle waveform out of A2
v_3	Rectangular waveform out of A4
V_{CC}	Collector supply voltage
V_{DF}	Voltage at noninverting input of integrator which determines duty factor
V_{EE}	Emitter supply voltage (negative)
V_{io}	Input offset voltage of op amp
V_{max}	Maximum output voltage from A4
V_{min}	Minimum output voltage from A4

DESIGN EQUATIONS

Description	Equation	
Slope of triangle waveform	$$\frac{\Delta v_2}{\Delta t} = \frac{V_{DF} - v_1}{R_1 C_1}$$	(1)
Minimum C_1 if integrator drift due to I_{io} is to be minimized	$$C_{1,min} = \frac{I_{io}}{\Delta v_2 / \Delta t \big|_{\text{drift rate, max}}}$$	(2)
Minimum $R_1 C_1$ if slew rate of A2 is not to be exceeded	$$R_1 C_{1,min} = \frac{|v_1 - V_{DF}|}{S_{A2}}$$	(3)
Minimum $R_1 C_1$ if integrator drift due to V_{io} is to be minimized	$$R_1 C_{1,min} = \frac{V_{io}}{\Delta v_2 / \Delta t \big|_{\text{drift rate, max}}}$$	(4)

Description	Equation	
Nominal R_5 to prevent input-bias-current error in A2	$R_5 = 2R_1$	(5)
Upper trip point of level detector; note that $v_{3,\min}$ is negative	$V_U = \dfrac{-R_6 v_{3,\min}}{R_7}$	(6)
Lower trip point of level detector	$V_L = \dfrac{-R_6 v_{3,\max}}{R_7}$	(7)
Integrator op-amp gain required (at frequency f) in order to maintain ramp nonlinearity at less than 1%	$A_v(f) > \dfrac{25}{R_1 C_1 f}$	(8)

23.2 DUAL-CURRENT-SOURCE FUNCTION GENERATOR

ALTERNATE NAMES Switched current-source function generator, source-sink function generator.

PRINCIPLES OF OPERATION In Fig. 23.2 the triangle waveform is generated across capacitor C. The slope of the triangle is positive when

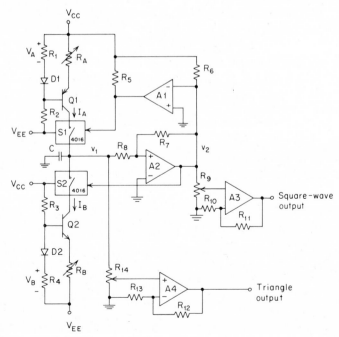

Fig. 23.2 A function generator using a current source and current sink to alternately charge and discharge a timing capacitor.

switch S1 connects current source Q1 to C. During this period S2 is OFF. After the waveform reaches a predetermined positive magnitude, the level detector A2 changes states. This turns S1 OFF and S2 ON. The triangle now decreases in amplitude as current sink Q2 discharges C. When the triangle waveform reaches a predetermined negative amplitude, the level detector again changes states. This causes S1 and S2 to return to the original charging configuration.

The triangle waveform across C will have linear slopes only if $(R_A + R_B)C$ is much larger than the period of the waveform. A larger capacitor is therefore recommended as lower frequencies are approached. The resistors R_A and R_B could be used as slope and fine frequency adjustment. A set of capacitors could then be used as the decade adjustment (coarse frequency-range switch).

If the amplitude of the triangle at v_1 is small, a wide range of slopes using R_A and R_B is possible. This amplitude is made small by using a small hysteresis loop in A2. A small amplitude is achieved by making the ratio R_7/R_8 equal to 10 or more. A ratio of 10 will make the peak-to-peak amplitude one-tenth the peak-to-peak amplitude of v_2.

DESIGN PARAMETERS

Parameter	Description
A1	Comparator or op amp used as inverter
A2	Comparator or op amp which converts triangle waveform into rectangular waveform
A3, A4	Output buffers
C	Timing capacitor which affects both slopes of triangle waveform
D1, D2	Diodes used to cancel thermal drift of emitter-base junction voltages in Q1 and Q2
FG	Function generator
I_A	Current out of Q1 current source
I_B	Current into Q2 current sink
Q1	Active element of current source which charges C
Q2	Active element of current sink which discharges C
R_1, R_2	Biasing resistors for current source
R_3, R_4	Biasing resistors for current sink
R_5, R_6	Pull-up resistors for open-collector outputs of A1 and A2
R_7, R_8	Resistors which provide positive feedback and hysteresis in A2
R_9, R_{14}	Potentiometers to control gain of output buffers
$R_{10}, R_{11}, R_{12}, R_{13}$	Resistors which control gain of output buffers
R_A, R_B	Resistors which control magnitude of source-sink currents in Q1 and Q2
S1, S2	4016 CMOS linear gates
$\Delta v_1/\Delta t$	Slope of capacitor voltage v_1
$v_{2,\text{max}}$	Upper saturation level of v_2 (near V_{CC})
$v_{2,\text{min}}$	Lower saturation level of v_2 (V_{sat} of A2)
V_A, V_B	Voltage developed across R_1 and R_4, respectively
V_{CC}	Collector supply voltage
V_{EE}	Emitter supply voltage
V_L	Lower trip point of level detector
V_U	Upper trip point of level detector

DESIGN EQUATIONS

Description	Equation
Positive slope of triangle at v_1	$$\dfrac{\Delta v_1}{\Delta t} = \dfrac{R_1(V_{CC} - V_{EE} - 0.7)}{CR_A(R_1 + R_2)} \qquad (1)$$
Negative slope of triangle at v_1	$$\dfrac{\Delta v_1}{\Delta t} = \dfrac{-R_4(V_{CC} - V_{EE} - 0.7)}{CR_B(R_3 + R_4)} \qquad (2)$$
Constant current out of Q1 collector	$$I_A = \dfrac{V_A}{R_A} \qquad (3)$$
	where $V_A = \dfrac{R_1(V_{CC} - V_{EE} - 0.7)}{R_1 + R_2}$
Constant current into Q2 collector	$$I_B = \dfrac{V_B}{R_B} \qquad (4)$$
	where $V_B = \dfrac{R_4(V_{CC} - V_{EE} - 0.7)}{R_3 + R_4}$
Upper trip point of level detector	$$V_U = \dfrac{-R_8 v_{2,\min}}{R_7} \qquad (5)$$
Lower trip point of level detector	$$V_L = \dfrac{-R_8 v_{2,\max}}{R_7} \qquad (6)$$

23.3 TRIANGLE–TO–SINE-WAVE CONVERTER

ALTERNATE NAMES   Piecewise-linear sine-wave generator.

PRINCIPLES OF OPERATION   Most monolithic function generators produce a sine-wave output using a piecewise-linear approach. This can be done with a series of diodes or transistors which are biased so that conduction occurs at a different input level for each device. Figure 23.3 shows such a typical circuit using biased transistors. This particular approach is used in the 8038 function-generator chip.

The operation of this circuit is best explained with the functional schematic shown in Fig. 23.4. The eight level detectors shown in Fig. 23.4 are transistors Q1, Q3, Q5, Q7, Q10, Q12, Q14, and Q16 in Fig. 23.3. The trip point of each level detector is established by the resistor divider on the right side of Fig. 23.3. The eight switches of Fig. 23.4 are transistors Q2, Q4, Q6, Q8, Q9, Q11, Q13, and Q15.

The triangle–to–sine-wave converter is, in the simplest terms, a resistive voltage divider. The input resistor R_1 is fixed at 1000Ω. The other eight resistors shown in Fig. 23.4 come into play according to the voltage v_2 present at the right side of R_1. Assuming that $V_{CC} = +5$ V and $V_{EE} = -5$ V, if $|v_2| < 0.53$ V, all eight switches will be in the OFF state. (This also assumes that $R_{12} \| R_{13} = 2600\Omega$ and $R_{25} \| R_{26}$

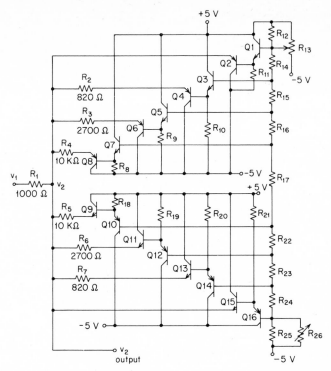

Fig. 23.3 A typical monolithic triangle–to–sine-wave converter (Intersil 8039).

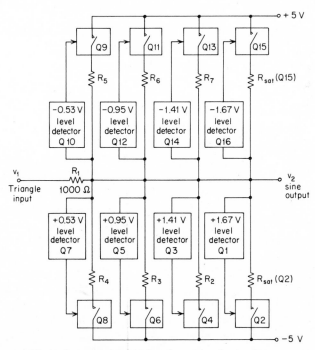

Fig. 23.4 Simplified schematic of the 16-transistor triangle–to–sine-wave converter shown in Fig. 23.3 ($V_{CC} = +5$ V and $V_{EE} = -5$ V are assumed).

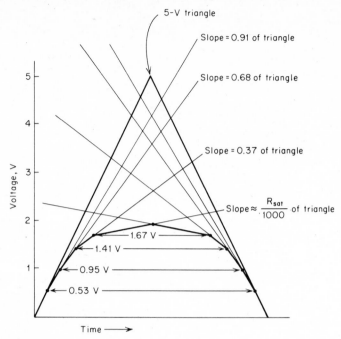

Fig. 23.5 Breakpoints and slopes of the positive portion of the sine wave generated by Fig. 23.3.

$= 2800\Omega$.) As shown in Fig. 23.5, the triangle waveform will be transferred directly to the output for these low-amplitude inputs.

For simplicity we discuss only the positive waveform. The negative side of the waveform is similar in operation. When 0.53 V $< v_2 <$ 0.95 V, level detector Q7 turns on switch Q8. Thus, we have a voltage divider with the output set at ten-elevenths of the input. Figure 23.5 shows how this lower slope deviates from the triangle input.

If 0.95 V $< v_2 < 1.41$ V, we have both R_3 and R_4 forming a voltage divider with R_1. The new slope will now be 0.68 times the triangle slope. Similarly, when 1.41 V $< v_2 < 1.67$ V the 820-Ω resistor R_2 is placed in parallel with R_3 and R_4. The new slope is reduced to 0.37 times the triangle slope. Lastly, when $1.67 < v_2$, the switch Q2 holds v_2 close to 1.67 V through the ON saturation resistance of that switch. The slope of the sine wave is close to zero when v_2 attempts to exceed 1.67 V.

When switches Q2, Q4, Q6, and Q8 are ON, they do not attempt to pull v_2 toward -5 V as one might first think. Instead, the emitter of each switch (in the ON state) is very close in voltage to its corresponding level detector. The large negative V_{EE} is required merely to supply the current needed by the switch.

23.4 STATE-VARIABLE SINE-WAVE OSCILLATOR

ALTERNATE NAMES Phase-shift oscillator, dual integrator-oscillator, stabilized-amplitude sine-wave generator.

PRINCIPLES OF OPERATION An oscillator requires positive feedback with exactly 0 or ±360° phase shift around its feedback loop for proper operation. The state-variable sine-wave oscillator shown in Fig. 23.6 provides a −360° phase shift in op amps A1, A2, and A3. The ac positive feedback loop is from the output of A1 through A2 and A3 and back to the inverting input of A1. Thus, A1 provides a −180° phase shift since it is merely an inverter. The two integrators A2 and A3 each provide a −90° phase shift.

The oscillator also has positive feedback at dc via R_6 to assure oscillation the instant power is applied. Negative feedback is implemented with R_1 and R_4. This feedback stabilizes the amplitude of the oscillator. It provides even better control over amplitude by rectifying one of the

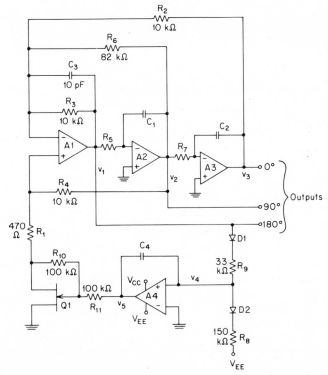

Fig. 23.6 A state-variable sine-wave oscillator utilizing two integrators, one phase inverter, and one level stabilizer.

outputs with D1 and D2 and using this dc voltage to control the amount of negative feedback. For example, if the average output amplitude begins to rise, the dc voltage v_4 rises. This forces v_5 down, causing Q1 to pull less current through R_1. Thus, R_1 is effectively raised in resistance, so that $R_1/(R_1 + R_4)$, the negative-feedback factor, is increased. This increased negative feedback causes the output amplitude v_1 (also v_2 and v_3) to be restored to its original stabilized level.

If we set $R = R_5 = R_7$ and $C = C_1 = C_2$, the frequency of oscillation is simply

$$f_o = \frac{1}{2\pi RC}$$

Ganged potentiometers are recommended for R_5 and R_7. C_1 and C_2 could be changed with a rotary switch to provide coarse frequency adjustment.

The positive and negative feedback loops must be carefully balanced to provide a distortionless sine-wave output. Once the balance is made, a high-quality sine wave is achieved over several decades using the ganged potentiometers. The positive-negative feedback balance can be trimmed by making R_6 or R_1 variable.

Op amp A4 is connected as an integrator to filter out the rectified waveform from D1 and D2. Without good filtering in this location we would have excessive modulation distortion on all outputs. The high dc gain of A4 automatically stabilizes the dc loop and reduces the effect of various parameter variations on the oscillator performance.

DESIGN PARAMETERS

Parameter	Description
A1	Summing amplifier
A2, A3	Integrators
A4	Part of level-control circuit
C_1, C_2	Part of integrator circuits
C_3	Provides phase compensation to prevent feedback instability in A1, A2, A3 loop
C_4	Capacitor to filter output of D1, D2 rectifier so that level stabilizer will not modulate oscillator
D1, D2	Diodes which rectify output waveform to be used in level stabilization
f_o	Output frequency
Q1	FET used as voltage-controlled variable resistor in series with R_1
R_1, R_4	Negative feedback to stabilize output amplitude of oscillator
R_2	Main path for positive feedback which determines center frequency of oscillator
R_3	Gain-determining resistor for A1
R_5, R_7	Part of integrator circuits
R_6	Provides positive feedback at dc to help oscillator turn ON
R_8, R_9	Part of level-controller rectifier circuit
R_{10}, R_{11}	Resistors which provide negative feedback around Q1 so that its voltage-controlled resistance is a linear function of input voltage
R_D	Drain-to-source resistance of Q1

DESIGN EQUATIONS

Description	Equation	
Frequency of oscillation if $R = R_5 = R_7$ and $C = C_1 = C_2$	$$f_o = \frac{1}{2\pi RC}$$	(1)
Frequency of oscillation if $R_2 \neq R_3$, $R_5 \neq R_7$, and $C_1 \neq C_2$	$$f_o = \frac{1}{2\pi} \sqrt{\frac{R_3}{R_2 R_5 R_7 C_1 C_2}}$$	(2)
Necessary balance condition to assure oscillation	$$\frac{R_3}{R_3 + R_6} = \frac{R_1 + R_D}{R_1 + R_D + R_4}$$	(3)

23.5 SAMPLE-AND-HOLD OSCILLATOR

ALTERNATE NAMES Locked oscillator, burst-input sampling oscillator, synchronous sine-wave generator.

PRINCIPLES OF OPERATION The circuit shown in Fig. 23.7 is used
to lock an oscillator to an input reference frequency which need be
present only for short periods of time. Many applications call for a
local clock frequency which must be locked to a remote clock. This
locked condition must be maintained for long intervals even in the absence of the remote clock.

The locked-oscillator function is performed through the use of a sam-

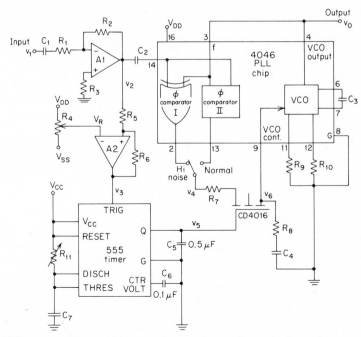

Fig. 23.7 An oscillator which remembers and reproduces the input frequency until a
new input frequency is received.

ple-and-hold device in conjunction with a phase-locked loop. (Since Chap. 27 describes phase-locked loops in detail, we mention their theory of operation only briefly here.) As shown in the figure, the output frequency f_o is provided by the voltage-controlled oscillator (VCO) within the phase-locked loop (PLL) chip. The VCO frequency is determined by the control voltage at pin 9 of the PLL. Since this is a CMOS device, a stored voltage on C_4 can be used to control the VCO for long periods of time.

The input burst frequency is first buffered and amplified in A1. It is then applied to the PLL chip, where it is compared with the VCO output frequency f_o. Phase comparator I can be used if the input signal has a large noise component. Otherwise, comparator II is the optimum circuit for comparing input phases.

The amplified input frequency from A1 is also compared against a reference voltage V_R in A2. Device A2 is connected as a comparator with positive feedback so that its output is rectangular. The adjustable trip level R_4 allows one to reject low-level noise similar to a squelch control on an FM receiver. Each time v_3 makes a negative transition, the 555 timer produces a positive pulse at v_5. The width of this pulse is approximately $0.7 R_{11} C_7$. If this pulse width is longer than the period of the input frequency, v_5 will remain HIGH all during the input burst. With v_5 HIGH the 4016 transmission gate is ON and the PLL loop is closed. Thus, whenever a burst is present at v_1 the PLL is operating and the exact same frequency appears at v_o.

When the input burst disappears, the 555 turns OFF and this likewise turns the 4016 OFF. The PLL loop is broken, but the control voltage is stored on C_4. The output resistance of the 4016 in the OFF state is approximately $5 \times 10^{10} \ \Omega$. Thus, the control voltage v_6 will decay with a time constant of $R_{4016} C_4 = 5 \times 10^{10} \times 10^{-7} = 5000$ s. The error of this voltage will be about 10 percent for 500 s, about 1 percent for 50 s, and about 0.1 percent for 5 s. Under these conditions the VCO will continue to operate until another sample frequency is received at v_1.

DESIGN PARAMETERS

Parameter	Description
A1	Inverting amplifier for boosting input signal
A2	Op amp (or comparator) used to convert input signal into rectangular waveform and also to discriminate against noise
A_v	Voltage gain of A1 stage
C_1	Optional coupling capacitor used to isolate dc levels of signal source from this circuit
C_2	Coupling capacitor required to isolate dc levels of A1 and PLL chip
C_3	VCO timing capacitor

C_4	Part of PLL loop filter to prevent feedback instability; also stores VCO control voltage
C_5	Smoothing filter capacitor at 555 output
C_6	Required by 555 when used as a one-shot
C_7	Helps determine pulse width of 555
f_{max}	Maximum VCO frequency
f_{min}	Minimum VCO frequency
f_O	Nominal VCO frequency
I_{io}	Input offset current of A1
PLL	Phase-locked loop
PW	Pulse width of 555 output
R_1, R_2	Determine voltage gain of A1 stage
R_3	Resistor used to minimize the effect of input bias current in A1
R_4	Potentiometer used to establish discrimination level V_R of A2
R_5, R_6	Resistors used to provide hysteresis in A2, thus improving output rectangular waveform if noise is present in v_2.
R_7, R_8	Parts of PLL loop filter
R_9	Resistor which determines VCO frequency range
R_{10}	Resistor which determines VCO offset frequency
R_{11}	Part of 555 timing circuit which determines duration of PW
R_{4016}	OFF resistance of 4016 switch from drain to source
$v_{1,min}$	Minimum expected peak input voltage
$v_{2,max,PLL}$	Voltage required at PLL input in order to guarantee that phase detectors will operate
V_{DD}	Drain supply voltage
$V_{oo,max}$	Maximum allowable output offset voltage of A1 due to input bias currents
V_R	Reference voltage used to establish discrimination level of A2
VCO	Voltage-controlled oscillator

DESIGN EQUATIONS

Description	Equation	
Pulse width of 555 timer	$PW = 0.7 R_{11} C_7$	(1)
Required value for C_1 to pass lowest input frequency	$C_1 \geq \dfrac{1}{2\pi f_{min} R_1}$	(2)
Voltage gain required for A1 stage at all frequencies of v_1	$A_v = -\dfrac{R_2}{R_1} = -\dfrac{v_{2,max,PLL}}{v_{1,min}}$	(3)
Maximum value for R_2 if $R_3 = R_2$	$R_2 \leq \dfrac{V_{oo,max}}{I_{io}}$	(4)
Nominal value for R_3	$R_3 = R_2$	(5)
Approximate center frequency of VCO assuming $v_6 = V_{DD}/2$ and $R_{10} = \infty$	$f_0 \approx \dfrac{V_{DD}}{10 R_9 C_3}$	(6)
Approximate frequency offset of VCO assuming $v_6 = 0$	$f_{min} \approx \dfrac{V_{DD}}{2 R_{10} C_3}$	(7)
Ratio of maximum to minimum VCO frequency	$\dfrac{f_{max}}{f_{min}} = 1 + 0.84\left(\dfrac{R_{10}}{R_9}\right)^{0.84}$	(8)
Constraint on sizes of R_9 and R_{10}	$10\ k\Omega \leq R_9$ and R_{10}	(9)
Constraint on size of C_3	$100\ pF \leq C_3$ if $5\ V \leq V_{DD}$ $50\ pF \leq C_3$ if $10\ V \leq V_{DD}$	(10)

DESIGN PROCEDURE

One should study the data sheets for the 4046 PLL IC and the 555 timer IC before proceeding with the design. The list of equations is sufficient to produce a design for this sample-and-hold oscillator, but most data sheets contain many subtle points which should be understood. The author has frequently found items in the fine print or footnotes which were critical to his design.

By using a low-frequency PLL such as the 4046 and an op amp for A2 instead of a comparator we have limited this circuit to the audio frequencies. Faster devices will be needed if operation above the audio frequencies is required.

Step 1. The 4046 data sheet says that the maximum ac-coupled input sensitivity on pin 14 is 800 mV (at $V_{DD} = 10$ V). Use this number for $v_{2,max,PLL}$ in conjunction with the minimum input signal v_1 to solve Eq. (3).

Step 2. Find a value for R_2 using Eq. (4). Let the output offset $V_{oo,max}$ be less than $V_{DD}/10$.

Step 3. Now find R_1 using Eq. (3) and set $R_3 = R_2$.

Step 4. A minimum value for C_1 is determined using Eq. (2).

Step 5. Choose a convenient value for C_7 such as 0.1 μF or 0.47 μF. Solve Eq. (1) for R_{11}. The pulse width PW should be at least 10 times longer than the period of the lowest frequency expected at v_1.

Step 6. Choose a practical value for C_3 such as 0.1 μF. Find values for R_9 and R_{10} using Eqs. (6) and (7). Double-check Eq. (8) to see if the frequency range of the VCO is sufficient. Note that f_{max} is as far above f_0 as f_{min} is below f_0.

EXAMPLE SAMPLE-HOLD OSCILLATOR DESIGN

Design Requirements and Device Data

$$f_0 = 1000 \text{ Hz} \qquad f_{min} = 50 \text{ Hz} \qquad f_{max} = 1950 \text{ Hz}$$

$$v_{1,min} = 0.1 \text{ V (peak)} \qquad V_{DD} = 10 \text{ V} \qquad I_{io,LM324} = 50 \text{ nA}$$

Step 1. The gain of the A1 stage must be

$$A_v = -\frac{R_2}{R_2} = -\frac{v_{2,max,PLL}}{v_{1,min}} = -\frac{800 \text{ mV}}{100 \text{ mV}} = -8$$

Step 2. A value for R_2 is

$$R_2 \le \frac{V_{oo,max}}{I_{io}} = \frac{V_{DD}}{10 I_{io}} = \frac{10}{10(5 \times 10^{-8})} = 20 \text{ M}\Omega$$

We will let $R_2 = 1.0$ MΩ.

Step 3. Resistors R_1 and R_3 are found from

$$R_1 = -\frac{R_2}{A_v} = -\frac{10^6}{-8} = 125 \text{ k}\Omega$$

We will let $R_1 = 100 \text{ k}\Omega$.

$$R_3 = R_2 = 1 \text{ M}\Omega$$

Step 4. Equation (2) provides us with a value for C_1

$$C_1 \geq \frac{1}{2\pi f_{min} R_1} = \frac{1}{2\pi(50 \times 10^5)} = 0.032 \text{ }\mu\text{F}$$

We will let $C_1 = 0.1 \text{ }\mu\text{F}$.

Step 5. Let $C_7 = 1.0 \text{ }\mu\text{F}$. We find R_{11} from Eq. (1) as follows:

$$R_{11} = \frac{\text{PW}}{0.7 C_7} = \frac{10/f_{min}}{0.7 C_7} = \frac{10/50}{0.7 \times 10^{-6}} = 280 \text{ k}\Omega$$

Let $R_{11} = 270 \text{ k}\Omega$.

Step 6. We will let $C_3 = 0.1 \text{ }\mu\text{F}$. Equations (6) and (7) are altered to provide us with R_9 and R_{10}:

$$R_9 \approx \frac{V_{DD}}{10 f_o C_3} = \frac{10}{10(1000 \times 10^{-7})} = 10 \text{ k}\Omega$$

$$R_{10} \approx \frac{V_{DD}}{2 f_{min} C_3} = \frac{10}{2(50 \times 10^{-7})} = 1 \text{ M}\Omega$$

We now compute f_{max}/f_{min} using Eq. (8) to make sure our requirements are satisfied:

$$\frac{f_{max}}{f_{min}} = 1 + 0.84 \left(\frac{R_{10}}{R_9}\right)^{0.84} = 1 + 0.84 \left(\frac{1 \text{ M}\Omega}{10 \text{ k}\Omega}\right)^{0.84} \doteq 41$$

Our original requirements were

$$\frac{f_{max}}{f_{min}} = \frac{1950 \text{ Hz}}{50 \text{ Hz}} = 39$$

We cannot expect exact agreement on these results because Eqs. (7) and (8) are approximate. Better correlation can be achieved if one uses the curves shown on the 4046 data sheet.

REFERENCES

1. Feucht, D.: Exploring Function Generator Design Problems and Solutions, *EDN,* June 5, 1975, p. 37.
2. Jung, W. G.: Low-Distortion Oscillator Uses State-variable Filter, *Electronics,* Feb. 5, 1976, p. 90.
3. Anderson, B. D. O.: Oscillator Design Problem, *IEEE J. Solid-State Circuits,* April 1971, p. 89.
4. Reintjes, P.: Self-gating Sample-and-Hold Controls Oscillator Frequency, *Electronics,* June 9, 1977, p. 132.

Chapter 24

Sampling and Multiplexing Circuits

INTRODUCTION

Sample-hold circuits and analog multiplexing circuits are closely allied sections of data-acquisition systems. These circuits are usually found in the analog "front end" of many instruments. In many cases they are the only analog part of the overall system and are therefore the most sensitive, critical, and troublesome part of the system.

Analog multiplexers have many more subtle problems than digital multiplexers. The designer must consider many conflicting topics such as crosstalk, noise, loading of or by other channels, ON-OFF transmission impedances vs. source or load impedances, switching-circuit feedthrough, response time, bandwidth, gain uniformity, and gain drift.

Sample-hold concepts have been around for many years, but the basic problems must be understood by the designer even when using integrated S/H devices. In order for these integrated devices to be generally applicable to a large range of S/H applications they are composed of little more than two op amps and a high-quality analog gate.

Two classes of sampling circuits will be presented in this chapter: (1) real-time sampling circuits, simply called *multiplexers,* and (2) delayed sampling circuits, called sample-and-hold (S/H) multiplexers. The term multiplexer implies that two or more inputs converge to a common output. The number of inputs is determined by system requirements and can conceivably be 100 or more if certain design rules are followed. However, one must pay a price for many inputs. Sampling time per input channel must be lower. Crosstalk and loading from other channels increase with large numbers of channels. These problems are present in both real-time multiplexers and S/H multiplexers.

This chapter contains design data for both types of sampling circuits. An FET real-time multiplexer using CMOS switches will be presented first. The errors expected in an FET multiplexer will be detailed.

A sample-and-hold circuit will also be described in detail. The discussion will explore the conflicting requirements of speed and accuracy. This particular type of sample-and-hold circuit was chosen for presentation since it is available from several manufacturers on a single chip (except for the holding capacitor).

24.1 FET MULTIPLEXER

ALTERNATE NAMES Analog multiplexer, multichannel sampling circuit, analog commutator, multichannel analog switch, CMOS multiplexer, MUX.

PRINCIPLES OF OPERATION Either junction field-effect transistors (JFETs) or metal-oxide semiconductor FETs (MOSFETs) can be used as the switching devices in a multiplexer. If MOSFETs are used, the complementary type, CMOS, is recommended, since it provides the best overall performance. These devices are available in large arrays which are optimized for this application.

Figure 24.1a shows a representative four-channel CMOS multiplexer. The detail of each switch is shown in Fig. 24.1b. Paralleling an n-channel with a p-channel FET minimizes the interaction between the gate voltage and the source voltage. This interaction is explained as follows. The drain-to-source ON resistance R_{DS} in an ideal FET is determined only by the voltage on the gate. In a real FET, however, R_{DS} depends on the voltage between gate and source or between the gate and drain, whichever is smaller. This is not usually a problem in the FET OFF state, since the gate is pulled to a large enough positive or negative voltage to ensure that the source or drain voltage cannot bring the FET into the conductive state. When the FET is ON, however, the drain-to-source resistance is modulated by the drain (or source) voltage. This undesirable feature is called *resistance modulation*. The parallel CMOS switch reduces resistance modulation to a second-order error. This is possible since the curves of resistance vs. V_{GS} for the two devices have opposite slopes. The ON resistance remains almost constant as V_S or V_D varies from zero to maximum.

The errors due to a nonzero $R_{DS,\text{on}}$ are further minimized by using a noninverting op-amp circuit at the output node of the four switches. The input resistance of a noninverting op-amp circuit is approximately

$$R_{\text{in}} \approx A_{vo} R_{id}$$

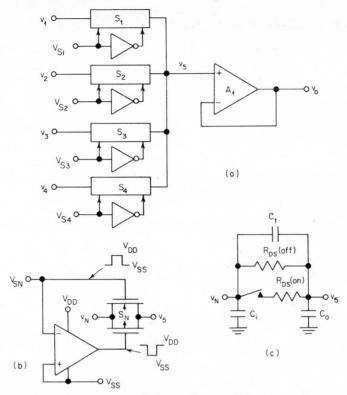

24.1 *(a)* A four-channel CMOS multiplexer; *(b)* a detailed diagram of one switch and its driver; and *(c)* the equivalent circuit of a switch.

The voltage gain of the circuit from a typical input, say v_1, is therefore

$$A_{vc} = \frac{v_o}{v_1} = \frac{R_{in}}{R_{DS1,on} + R_{in}}$$

If a 1 percent accurate multiplexer is required, $R_{in,min}$ must be at least 100 times larger than $R_{DS,on,max}$. Likewise, a 0.1 percent circuit requires an $R_{in,min}/R_{DS,on,max}$ ratio greater than 1000.

Another dc error source which should be considered in a high-accuracy MUX is the leakage current through the OFF switches. These currents can be converted into equivalent leakage resistances to simplify error calculations. Assume that S_1 is ON and all other switches are OFF. If v_2, v_3, and v_4 all equal zero (and all have small source resistances), we assume that the three leakage resistances of S_2, S_3, and S_4 are connected in parallel to ground. In this case it appears that a resistance of

$$R_L = \frac{1}{1/R_{DS2,off} + 1/R_{DS3,off} + 1/R_{DS4,off}}$$

is shunted from v_5 to ground. The voltage gain of the entire circuit, assuming an ideal op amp and nonideal switches, is then

$$A_{vc} = \frac{v_0}{v_1} = \frac{R_L}{R_{DS1,on} + R_L}$$

The ratio $R_{DS,off}/R_{DS,on}$ must be much larger than 1000 if a 0.1 percent MUX is required.

Attenuation of high frequencies through the multiplexer can be caused either by op-amp limitations or by FET output capacitance. The -3-dB frequency of the op amp is simply its unity-gain crossover frequency if the noninverting configuration shown in Fig. 24.1a is used. The -3-dB frequency of each FET switch is $1/(2\pi R_{DS,on}C_o)$. Since the output terminals of all four switches are in parallel, C_o must be multiplied by 4 to determine the actual -3-dB frequency due to this cause.

DESIGN PARAMETERS

Parameter	Description
A_1	Wide-bandwidth operational amplifier which is stable with 100% feedback
A_{vc}	Closed-loop voltage gain of circuit from any particular input to output
A_{vo}	dc open-loop voltage gain of op amp
C_1	Input capacitance of FET switch
C_o	Output capacitance of FET switch
C_t	Transfer capacitance of FET switch
f_{cp}	Pole frequency of circuit (where voltage gain has been reduced 3 dB)
MUX	Multiplexer
$R_{DSN,on}$	ON resistance of Nth FET switch
$R_{DSN,off}$	OFF resistance of Nth FET switch
R_{id}	Differential input resistance of op amp
R_{in}	Closed-loop input resistance of op amp
R_L	Parallel resistance seen at op-amp input due to OFF switches
S_1 to S_4, S_N	FET switches, Nth FET switch
v_0 to v_5	Voltages shown in Fig. 24.1
V_{DD}	Maximum positive voltage applied to CMOS gates, i.e., drain voltage
V_{S1} to V_{S4}, V_{SN}	Drive signals for FET switches, Nth drive signal
V_{SS}	Largest negative voltage applied to CMOS gates, i.e., source voltage

DESIGN EQUATIONS

Description	Equation	
Voltage gain from any input to v_0 assuming ideal op amp and switches	$A_{vc} = \dfrac{v_0}{v_1} = \dfrac{v_0}{v_2} = \dfrac{v_0}{v_3} = \dfrac{v_0}{v_4} = 1$	(1)
Voltage gain from a typical input (say v_1) assuming nonideal op amp and switches	$A_{vc} = \dfrac{v_0}{v_1} = \dfrac{R_L \parallel R_{in}}{R_{DS1,on} + R_L \parallel R_{in}}$	(2)

Approximate input resistance of op amp	$R_{in} \approx A_{vo} R_{id}$	(3)
Shunting resistance of OFF switches assuming S_1 ON and S_2 to S_4 OFF	$R_L = \dfrac{1}{1/R_{DS2.off} + 1/R_{DS3.off} + 1/R_{DS4.off}}$	(4)
Bandwidth of circuit assuming an ideal op amp (S_1 ON and S_2 to S_4 OFF)	$f_{cp} = \dfrac{1}{2\pi R_{DS1.on}(C_{o1} + C_{o2} + C_{o3} + C_{o4})}$	(5)
Maximum signal levels allowed for v_1 (v_2 to v_4 are similar)	$V_{SS} \leqq v_1 \leqq V_{DD}$	(6)

24.2 SAMPLE-AND-HOLD CIRCUIT

ALTERNATE NAMES   Sample-hold circuit, S/H, sampling circuit.

PRINCIPLES OF OPERATION   As shown in Fig. 24.2, an S/H circuit requires a high-output-current op amp A_1, a high-quality switch S, a low-leakage capacitor C, and an output op amp A_2 with a low input bias current. The input op amp must be capable of driving a capacitive load without any hint of instability. The switch must have a high $R_{DS,off}/R_{DS,on}$ ratio so that C can quickly charge to its peak value and maintain that value with minimal droop between sampling times. The switch must also have a small coupling between its digital input and analog output. This coupling would allow switching transients (which occur

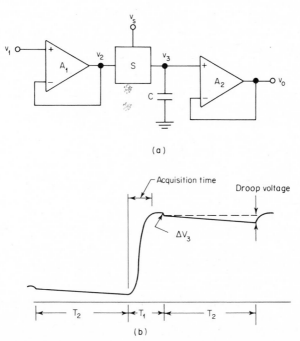

(a)

(b)

Fig. 24.2 (a) A basic sample-and-hold circuit and (b) exaggerated output waveform.

at S turn-off) to change the final voltage stored on C. Lastly, the output op-amp bias and offset currents (and their change with temperature) must be small.

Circuit operation is fairly straightforward. The input op amp A_1 maintains v_2 at the same potential as v_1. During the sampling interval T_1, the voltage v_2 is deposited on C by the closure of S. After T_1 terminates (when S opens), the voltage v_2 is maintained on C for a duration T_2 until the next sampling interval. Op amp A_2 holds the transferred value of v_2 on its output v_0 for the entire T_2 duration.

We will now consider the effect of nonideal parameters on the S/H circuit performance. The holding capacitor must be carefully sized. The following constraints limit the maximum size of C: (1) op amp A_1 must have a current drive capability of at least

$$I_{\max} = \frac{C\,(v_{0,\max} - v_{0,\min})}{T_{1,\min}}$$

and (2) the maximum ON resistance of S and the sampling time T_1 control the upper limit for C. If a 1 percent S/H is required, we need $T_1 > 5 R_{DS,\text{on,max}} C$. If a 0.1 percent system is desired, make $T_1 > 7 R_{DS,\text{on,max}} C$.

Both the above factors tell us that a large C requires a large sampling time T_1. A high current-drive capability from A_1 and a low ON resistance for S are also mandatory if C is large. At the other end of the scale, however, if C is too small, other problems show up: (1) during the hold period T_2, leakage currents through C, S, and A_2 input will easily discharge a small capacitor, and (2) the gate-to-source (or gate-to-drain) capacitance transfers charge to or from C when the gate waveform turns S off. This error adds to, or subtracts from, the sample voltage stored on C. If C is small, the error is more pronounced.

Monolithic circuits are available which contain A_1, A_2, and S on a single chip. Each of these devices is optimized according to the trade-offs itemized above.

DESIGN PARAMETERS

Parameter	Description
A_1	High-current-driver op amp
A_2	Op amp with high input resistance
A_{vc}	Closed-loop voltage gain of entire circuit
C	Holding capacitor
C_{GD}	Gate-to-drain capacitance of switch
$I_{\max}$	Maximum output current available from A1
N	Variable used in accuracy calculations
R_c	Leakage resistance of C
R_{DS}	Drain-to-source resistance of S

R_{ic}	Common-mode input resistance of A2
S	Electronic switch
S/H	Sample and hold
T_1	Sampling time
T_2	Time between samples
v_0 to v_3	Voltages as indicated in Fig. 24.2
v_3 (droop)	Voltage droop on C during hold time T_2
Δv_3	Error voltage added to (or subtracted from) v_3 owing to C_{GD}
Δv_G	Change in gate voltage of S
$V(\pm)$	Power-supply voltages

DESIGN EQUATIONS

Description	Equation	
Voltage gain of circuit during sample time	$A_{vc} = \dfrac{v_0}{v_1} = 1$	(7)
Required output-current capability of A_1	$I_{max} = \dfrac{C(v_{0,max} - v_{0,min})}{T_{1,min}}$	(8)
Required sampling interval T_1	$T_{1,min} = \begin{cases} 5R_{DS,on,max}C & \text{for } 1\% \text{ S/H} \\ 7R_{DS,on,max}C & \text{for } 0.1\% \text{ S/H} \end{cases}$	(9)
Magnitude of error voltage deposited on C when v_G changes by Δv_G	$\Delta v_3 = \Delta v_0 = \dfrac{\Delta v_G C_{GD}}{C + C_{GC}}$	(10)
Maximum droop of voltage v_3 during T_2	$v_{3,droop} = v_{3,max}\left[1 - \exp\left(-\dfrac{T_2}{RC}\right)\right]$ where $R = R_{ic} \parallel R_c \parallel R_{DS,off}$	(11)

DESIGN PROCEDURE

We begin this procedure by assuming that T_2, A_{vc}, $v_{0,max}$, $v_{0,min}$, and Δv_G are fixed by the system into which this S/H circuit is to be installed. The S/H circuit accuracy and droop are also specified. We also assume that A1, A2, S, and the type of holding capacitor have been selected. Our job is to compute the capacitor size, the voltage error of v_3 (and v_0) caused by Δv_G, and the droop of v_3 (and v_0) during T_2.

DESIGN STEPS

Step 1. Equation (11) is rearranged to determine C:

$$C \geq \frac{T_2}{R \ln \left[v_{3,max}/(v_{3,max} - v_{3,droop})\right]}$$

where $R = R_{ic} \parallel R_c \parallel R_{DS,off}$. A value of R_c can be obtained from the specification sheet for the type of capacitor used even though the exact capacitance of the capacitor is not known until the calculation above is performed.

Step 2. Choose a value N from the following table:

Required sampling accuracy, %	N
10	3
1	5
0.1	7
0.01	9

Calculate a first-cut sampling time T_1 using a modified form of Eq. (9):

$$T_{1,\text{min}} = N R_{DS,\text{on},\text{max}} C$$

Determine a second-cut sampling time using a rearranged Eq. (8)

$$T_{1,\text{min}} = \frac{C(v_{0,\text{max}} - v_{0,\text{min}})}{I_{\text{max}}}$$

Use the higher of the two values calculated above for the actual T_1.

Step 3. Use Eq. (10) to find the approximate holding error due to switch capacitance.

S/H CIRCUIT DESIGN EXAMPLE Suppose we want to sample and hold a speech signal before it is sent into an A/D converter. We will assume that the highest audio frequency of interest is 5 kHz. According to the rules of sampling, we must sample at a frequency of at least 10 kHz. Thus $T_2 \approx 1/10^4 = 100$ μs. The maximum input- and output-voltage levels are specified. The gate drive of S is also given.

Design Requirements

$$T_2 = 100 \ \mu\text{s}$$

$$v_{0,\text{max}} = +10 \ \text{V}$$

$$v_{0,\text{min}} = +1 \ \text{V}$$

$$\Delta v_G = 15 \ \text{V}$$

$$v_{3,\text{droop}} = 0.01 \ \text{V}$$

$$V^{(+)} = +15 \ \text{V}$$

$$V^{(-)} = 0.0$$

$$N = 5 \ (1\% \ \text{accuracy})$$

Device Data

$$I_{\text{max}} = 10 \ \text{mA} \ (A_1 = \tfrac{1}{4} \ \text{LM324})$$

$$R_{DS,on,max} = 1000 \ \Omega$$
$$C_{GD} = 4 \ pF$$
$$R_{DS,off,min} = 1.5 \times 10^{11} \ \Omega$$

$$S = \tfrac{1}{4} \ CD4016$$

$$R_{ic} = 1.5 \times 10^{12} \ \Omega \qquad (A2 = CA3130)$$

$$R_c \approx 10^{13} \ \Omega \qquad \text{(polycarbonate capacitor)}$$

Step 1. The holding capacitor is

$$C \geqq \frac{T_2}{R \ \ln \left[v_{3,max}/(v_{3,max} - v_{3,droop}) \right]}$$

$$\geqq \frac{10^{-4}}{(1.5 \times 10^{12}) \| 10^{13} \| (1.5 \times 10^{11}) \ \ln \left[10/(10 - 0.01) \right]}$$

$$\geqq 0.74 \ pF$$

This capacitance is unreasonably small because the leakage resistances are so high. Assume that a capacitance of 100 pF is used for C.

Step 2. We let $N = 5$, since a 1 percent circuit is required. The first-cut $T_{1,min}$ is

$$T_{1,min} = N R_{DS,on,max} C = 5(1000) \times 10^{-10} = 0.5 \ \mu s$$

The second-cut $T_{1,min}$ is

$$T_{1,min} = \frac{C(v_{o,max} - v_{o,min})}{I_{max}} = \frac{10^{-10}(10-1)}{10^{-2}} = 0.09 \ \mu s$$

It appears that the rise time of the voltage on C is constrained more by $R_{DS,on}$ than it is by I_{max}. We therefore let $T_1 = 0.5 \ \mu s$.

Step 3. The approximate holding error due to C_{GD} is

$$\Delta v_o \approx \frac{\Delta v_G \ C_{GD}}{C + C_{GD}} \approx \frac{15(4 \times 10^{-12})}{10^{-10} + 4 \times 10^{-12}} = 0.58 \ V$$

This error is 5.8 percent of $v_{o,max}$ and 58 percent of $v_{o,min}$. If we raise C to 1000 pF, these errors are reduced to 0.58 and 5.8 percent, respectively. Step 2 must also be done over. The new T_1 is 5 μs.

24.3 ALTERNATE S/H CONFIGURATIONS

The basic sample-and-hold circuit shown in Fig. 24.2a is adequate for many applications, but it has a few shortcomings. The input offset voltages associated with each buffer add to give an overall offset which is the algebraic sum of the two offset voltages. The contribution from A_2 can be quite large since it often contains FET input transistors. These errors can be reduced by using overall feedback, as shown in

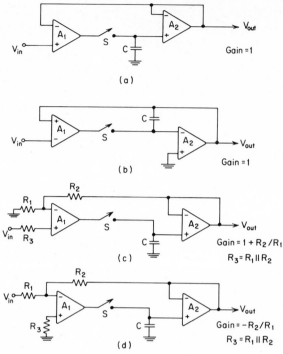

Fig. 24.3 Four alternate sample-and-hold configurations with improved characteristics over the configuration of Fig. 24.2: *(a)* noninverting with unity gain; *(b)* integrator with unity gain; *(c)* noninverting with adjustable gain, and *(d)* inverting with adjustable gain.

Fig. 24.3a. The feedback signal for A_1 is taken from the output of A_2. The dc error caused by the input offset voltage of A_2 is reduced by the gain of A_1.

The circuit of Fig. 24.3a also has difficulties. The FET switch undergoes the same voltage excursions as the input voltage v_{in} when it is closed or opened. Thus, the FET-switch gate-voltage swing must be quite large to keep the switch completely on or off, as required, for all values of v_{in}. This large gate-voltage swing reduces both speed and accuracy. The integrator S/H shown in Fig. 24.3b alleviates this problem. In this case the FET-switch input-output terminals operate at ground potential so that smaller gate drive voltages can be used. This results in lower leakage currents, faster switching time, and reduced parasitic coupling of charge from the FET gate onto the holding capacitor.

If one wants the S/H circuit to provide gain larger than 1, the circuit of Fig. 24.3c can be used. With a slight rearrangement of parts the best features of Fig. 24.3b and c can easily be combined. Likewise, negative adjustable gain is possible using the circuit of Fig. 24.3d. This circuit can also be configured using A_2 as an integrator.

REFERENCES

1. Bergersen, T. B.: Field Effect Transistors in Analog Switching Circuits, *Motorola Appl. Note* AN-220, 1966.
2. Givins, S.: Field Effect Transistors as Analog Switches, *Comput. Des.,* June 1974, p. 106.
3. Fullager, D.: Analog Switches Replace Reed Relays, *Electron. Des.,* June 21, 1973, p. 98.
4. Schmid, H.: Electronic Analog Switches, *Electro-Technology,* June 1968, p. 35.
5. Jones, D.: Applications of a Monolithic Sample-and-Hold/Gated Operational Amplifier, *Harris Semiconductor Appl. Note* 517, March 1974.
6. Buchanan, J. E.: C-MOS Switch Speeds Up Sample-and-Hold Circuit, *Electronics,* Sept. 27, 1973, p. 127.
7. Patstone, W., and C. Dunbar: Choosing a Sample-and-Hold Amplifier Is Not as Simple as It Used to Be, *Electronics,* Aug. 2, 1973, p. 101.
8. Stafford, K. R., P. R. Gray, and R. A. Blanchard: A Complete Monolithic Sample/Hold Amplifier, *IEEE J. Solid-State Circuits,* vol. SC-9, no. 6, p. 381, December 1974.
9. Zuch, E. L.: Designing with a Sample-Hold Won't Be a Problem If You Use the Right Circuit, *Electron. Des.,* Nov. 8, 1978, p. 84.

Chapter 25

Analog-to-Digital Converters

25.1 BASIC DEFINITION OF AN A/D CONVERTER

An analog-to-digital converter changes an analog input voltage into an N-bit digital word. In effect, the input voltage is quantized into 2^N different digital words, one for each state of the N-bit output word.

ALTERNATE NAMES Analog-digital converter, ADC, digitizer, A/D converter.

PRINCIPLES OF OPERATION Hundreds of different approaches to A/D conversion are published in the literature, but only four techniques have been successfully implemented in monolithic form as of 1978. Three of them use an integrator stage and can be classified as high-precision low-speed devices. The last technique uses successive approximation and is several orders of magnitude faster than the integration methods. We will discuss all four conversion techniques in the following pages.

Dual-slope ADC The basic circuit for this technique is shown in Fig. 25.1. The control logic begins each cycle by commanding the input switch to the lower position, where a current proportional to the input voltage is processed by the integrator. Since this current pulls positive charge from the op-amp input, the integrator output slews upward, as shown in Fig. 25.2. This positive slope continues for a fixed time T_1, as established by the control logic. The peak voltage reached by v_I is proportional to the input voltage v_{in}. At the end of T_1 (and the start of T_2) the control logic changes the switch to the reference-current input.

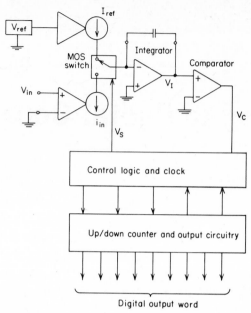

Fig. 25.1 A dual-ramp A/D converter uses an integrator with switched inputs and a comparator.

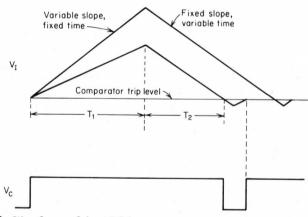

Fig. 25.2 Waveforms of the ADC integrator output v_I and comparator output v_C.

Since the reference current goes into the integrator, the voltage v_I decreases at a rate proportional to V_{ref}. During T_2 a counter begins operating. When the negative-going ramp crosses zero, the comparator output changes state, causing the counter to stop. The magnitude of the number stored in the counter is therefore proportional to the input voltage v_{in}.

We note that during T_1 the integration takes place for a fixed time at a variable slope. During T_2 the integration is accomplished at a fixed slope and a variable time. From Fig. 25.2 we observe that the change in capacitor voltage is equal in magnitude during T_1 and T_2. Thus

$$\frac{1}{C}\int i_{in}\ dt = \frac{1}{C}\int I_{ref}\ dt$$

or

$$i_{in}T_1 = I_{ref}T_2$$

If each voltage-to-current converter is linear and their gains are equal, then

$$v_{in}T_1 = V_{ref}T_2$$

The unknown voltage v_{in} in terms of the three known or measurable parameters is therefore

$$v_{in} = \frac{V_{ref}T_2}{T_1}$$

Both V_{ref} and T_1 are constants, and T_2 is proportional to the digital output word. We conclude that the digital output word is proportional to the magnitude of the input analog voltage.

The principal advantage of a dual-slope ADC is its high resolution for a relatively low cost, accounted for by the absence of an on-chip precision resistor network. Dual-slope A/D converters in monolithic form are easily available with 10 to 12 bits of resolution. Since the input signal is integrated, these converters are nearly immune to noise. They are also highly linear with a guaranteed monotonicity; i.e., missing codes are nearly impossible due to the nature of integration signal processing. The main disadvantage of dual-slope converters is their slow speed. A typical 8-bit dual-slope ADC might require 10 ms per conversion. However, there are many applications such as digital voltmeters where samples are needed only several times per second.

Dual-slope ADC with automatic zero Even though dual-slope A/D converters are highly accurate, they can still be improved using automatic-zero circuitry. A dual slope converter can be considered a two-phase circuit. The first phase is the positive integration of the input signal, and the second phase is the negative integration of the reference source. Automatic-zero ADCs typically use three, four, or five phases per complete cycle. The last two phases operate in the normal dual-slope mode. The automatic-zero operations operate just before the positive signal integration.

The multiphase automatic-zero technique reduces the errors due to offset and drift of the integrator and comparator. The ultimate resolution of the ADC is thereby determined by the noise of the system.

In effect the automatic-zero circuitry measures offset errors just a few milliseconds before the dual-slope measurement and stores the resulting error either on a capacitor or in a digital word. If the error is stored on a capacitor, it is then subtracted from the dual-slope measurement in an analog subtractor circuit, i.e., by using the other input to the integrator. If the error is stored in a digital word, it is subtracted from the digital counter output using digital subtraction techniques. In either case the errors are subtracted from the ADC output on a cycle-by-cycle basis. In this manner the error-correction circuitry tracks errors as they change with time or temperature.

The Analog Devices, Inc., AD7550 A/D converter uses auto-zero circuitry to achieve 13-bit resolution with less than 1 ppm/°C gain and offset sensitivity. This chip is advertised as a quad-slope device, but before the four phases begin, a setup phase ϕ_0 is required. During this time, as shown in Figs. 25.3 and 25.4, switch S1 is closed and switches S2, S3, and S4 are open. This applies the full V_{ref} to the integrator negative input, resulting in a negative-going ramp at the integrator output. When the ramp voltage is equal to the comparator's trip voltage, ϕ_1 begins. This trip voltage is at analog ground.

At the initiation of ϕ_1 a digital counter starts, S3 is closed, and all other switches open. This connects the analog ground to the integrator input to determine whether the ADC's negative supply voltage is at

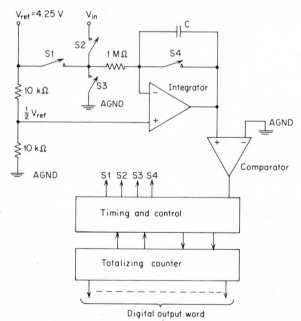

Fig. 25.3 Simplified schematic of Analog Devices AD7550 A/D converter with four-phase auto-zero circuitry.

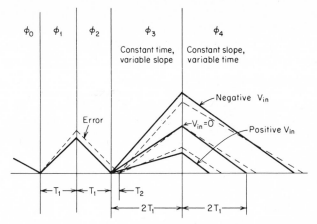

Fig. 25.4 Integrator output voltage during a complete cycle of the auto-zero A/D converter shown in Fig. 25.3.

analog ground potential. If not, an error voltage (dashed line in Fig. 25.4) is produced which is proportional to the difference. This error voltage is stored on the external integrating capacitor.

During ϕ_2 the integrator is connected to V_{ref} by closing S1 and opening all other switches. The output of the integrator falls at a rate proportional to V_{ref} until the comparator trip point is reached. If no errors have been accumulated up to that point, ϕ_2 will last exactly as long as ϕ_1. If errors are found on AGND or V_{ref} with respect to $V_{ref}/2$, the comparator will not trip exactly T_1 after ϕ_2 begins. The trip-time error T_2 can be a positive or negative number about the ideal trip time if an error occurs. T_2 is stored for use during the final two phases of the cycle.

The main integration phase ϕ_3 takes place with S2 closed and all other switches open. This set of conditions lasts for a fixed time $2\,T_1$ plus or minus the error time T_2. As shown with the dashed lines, if the error was positive, the time for the ϕ_3 phase is shortened. This shorter time effectively reduces the gain of the integrator during ϕ_3.

The final phase ϕ_4 is implemented by closing S1 and opening all other switches. This discharges the integration capacitor at a fixed rate determined by V_{ref}. This phase terminates when the capacitor voltage goes through the comparator trip level. Due to the slope correction during ϕ_3 the counter contents at the end of ϕ_4 will contain a corrected A/D conversion number.

Charge-balancing A/D converter This type of conversion process, like the dual-slope process, requires integration of the input signal, but the similarity ends there. Charge balancing, or quantized feedback, as some manufacturers call it, is becoming the preferred technique for LSI A/D converters where high-speed operation is not required.

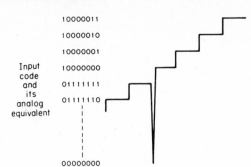

Fig. 26.5 The worst-case glitch can occur when all bits of the DAC input code change simultaneously.

01111111 to **10000000** or vice versa.   If the current-steering switches in the DAC turn OFF faster than they turn ON, for a short period of time the effective input word is **00000000** and the DAC output will be zero.   This large transient, as shown in Fig. 26.5, which may go all the way to zero and back, is referred to as a *glitch.*   This problem can also be caused by improper timing of the code coming into the DAC.

Linearity, differential   This parameter is often merely called linearity, but linearity is classically divided into two types, differential and integral. This parameter is also called *nonlinearity* or *differential nonlinearity.*   The standard definition for differential-linearity error describes the variation in the analog value of transitions between adjacent pairs of digital numbers over the full range.   As shown in Fig. 26.6, a straight line is usually drawn between the full-scale set point and zero.   The normal limit for a good-quality DAC is within ±½ LSB of the normal 1 LSB steps. Since a 3-bit DAC is assumed, ±½ LSB is one-eighth of full scale.

Monotonicity   A DAC which is truly monotonic would never have its

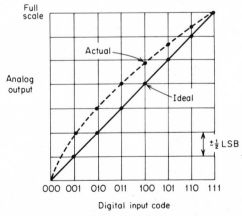

Fig. 26.6 Nonlinearity is defined as the deviation of the DAC output from a straight line drawn between the full-scale output set point and zero.

uses two reference currents summing into the integrator input along with a current proportional to v_{in}. If $2 I_{ref}$ is ON 50 percent of the time, the average reference current will be zero. These would be the operating conditions with zero analog input voltage. With zero v_{in} the up-down counter will fill to the 50 percent point during each conversion period, indicating the midpoint between $-v_{in,max}$ and $+v_{in,max}$. If a positive v_{in} is applied, $2 I_{ref}$ will be ON more than 50 percent of the time to keep the integrator output near the comparator V_{ref}. This will cause a faster accumulation of clock pulses in the up-down counter. Likewise, negative v_{in} voltages create a switch duty cycle less than 50 percent and a corresponding smaller pulse accumulation in the counter.

Successive-approximation ADC   The fastest A/D converter chips use the successive-approximation (S/A) technique. These devices are typically two or three orders of magnitude faster than ADCs using integration, but during any given year the S/A chips lag behind the resolution of integrating chips by 2 to 4 bits. In 1978, for example, the best monolithic integrating ADC provided 14-bit resolution in 20 ms while the best monolithic successive-approximation device was capable of 10-bit resolution in 20 μs.

As indicated in Fig. 25.6, the basic components of an S/A ADC are a voltage comparator, a D/A converter (DAC), a successive-approximation register, and a clock. The successive-approximation register (SAR) enables the bits of the DAC one at a time, beginning with the most significant bit (MSB). As each bit is enabled, the comparator output indicates whether v_{in} is larger or smaller than the DAC output v_D. If $v_{in} < v_D$, the bit is turned OFF. If $v_D < v_{in}$, the bit is left ON. After this test the next lower-order bit is tested. When all bits have been

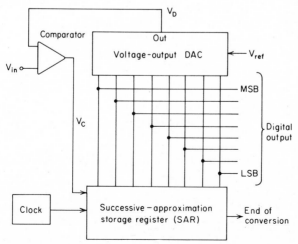

Fig. 25.6 The four basic components of a successive-approximation A/D converter.

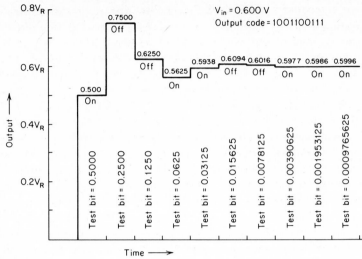

Fig. 25.7   Output voltage from the DAC in a successive-approximation ADC if $V_{in} =$ 0.600 V_R.

tested, the conversion cycle is complete and the lines between the DAC and SAR contain the digital output.   In most devices an end-of-conversion signal is sent to outside circuitry to flag the availability of a complete output word.

Figure 25.7 will help clarify the process just described.   Suppose $v_{in} = 0.600$ V and $V_{ref} = 1.000$ V.   The DAC first tries 0.500 V and finds v_{in} larger, so that bit remains ON.   The next bit causes the DAC output to be $0.500 + 0.25 = 0.75$ V.   This is too large, so the bit is left OFF and the SAR proceeds to the third bit.   After 10 bits have been tested one at a time, the binary number will be **1001100111,** which is equivalent to the following sum:

$$
\begin{array}{r}
0.5000000000 \\
0.0625000000 \\
0.0312500000 \\
0.0039062500 \\
0.0019531250 \\
\underline{0.0009765625} \\
0.5996093755
\end{array}
$$

Of course, since this example concerns a 10-bit ADC the ultimate resolution is 0.1 percent, and so we can only use the first four digits, that is, 0.5996.

25.2   ADC PARAMETER DEFINITIONS

Each manufacturer has a preferred set of parameters to specify its product.   When comparing various ADCs, the designer is often confronted with the problem of evaluating devices from several companies when no standard specifications exist between manufacturers. Before parameters with the same name on different data sheets can be compared, the designer must ascertain that the test conditions for each are identical (or at least similar). Perhaps the most surprising thing resulting from a perusal of a number of ADC data sheets is that some data sheets specify nearly all the parameters listed below but a few data sheets discuss *none*.

Accuracy, absolute This is also called *gain variance* or *full-scale gain set point.* The ability of an ADC to duplicate the transfer function at full scale as established during manufacture is called the absolute accuracy. This parameter can usually be trimmed by the user.

Accuracy, relative Same as integral linearity.

Gain drift This parameter is also called *gain temperature coefficient, gain TC, full-scale drift,* or *full-scale set-point TC.* It is also dependent on drift of the reference voltage. Gain drift specifies the stability of the analog-to-digital transfer function at the maximum input voltage.

Linearity, differential Figure 25.8a clearly illustrates the differential linearity (or nonlinearity) of an ADC. In the case shown only 3 bits of resolution are used, to simplify the drawing. Differential linearity error is the amount of deviation of any quantum step Q from its ideal step size; i.e., it is the deviation in the input analog difference between two adjacent output codes from the ideal value of $FS/2^N$. If an ADC is specified as having a maximum differential linearity error of $\pm\frac{1}{2}$ LSB, the actual size of any quantum step is between $\frac{1}{2}$ and $1\frac{1}{2}$ LSB since the nominal step size is 1 LSB. Figure 25.8a shows differential linearity errors of $+\frac{1}{2}$ and $-\frac{1}{2}$ LSB. The finite step size Q is sometimes called *quantization error.*

Linearity, integral As indicated in Fig. 25.8b, a straight-line transfer function is drawn between the zero input-output point and FS input-output

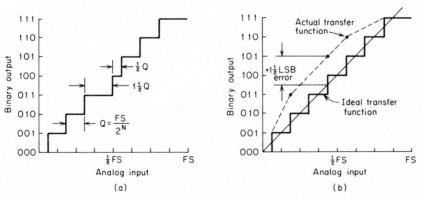

Fig. 25.8 Eight-level examples of (a) differential nonlinearity and (b) integral nonlinearity of an ADC.

point. Integral linearity error is due to curvature of that transfer function. Thus, integral linearity error is defined as the maximum deviation of the actual transfer function from the ideal straight line. This parameter is specified in LSBs or percent of full scale.

Linearity drift Both differential and integral linearity errors are subject to drift induced by temperature and time. This drift is usually specified in ppm per celsius degree.

Noise The input comparator or input op amp of an ADC are the most likely sources of internal noise. Since the output of an ADC is a digital word, any noise measurement must look for false changes of output states when the input is held constant. The noise, however, is referred to the input since noise has traditionally been an analog quantity.

The noise due to quantization of the analog input into 2^N discrete digital output words is an inherent irreducible source of noise. Since an ADC cannot distinguish an analog difference less than Q (see Fig. 25.8), its output at any point may be in error by at least $\pm Q/2$.

Power-supply sensitivity It is desirable for an ADC to maintain a precise transfer function with little dependence on power-supply voltage stability. The power-supply sensitivity specifies the effect of changes in power-supply voltage on the ADC transfer function. Typically, the effect is tabulated at zero input, midrange input, and/or full-scale input. The parameter units are usually percent per percent, percent full scale per percent, or percent per 1 percent. The data sheet also shows the test setup for this measurement.

Some data sheets also separately specify the effect of the power supply on the internal reference voltage. However, any drift in the reference voltage will also show up as drift in the ADC transfer function. The overall power-supply sensitivity parameter should be adequate for most applications.

Resolution The nominal resolution of an ADC is the relative value of the LSB, namely, 2^{-N}. It is expressed as 1 part in 2^N, as a percentage, in parts per million, or simply by N bits. Useful resolution is the smallest uniquely distinguishable change in the transfer function for all operating conditions, i.e., over time and temperature. For example, a 12-bit ADC may have a useful resolution of only 10 bits when all errors are summed together.

Symmetry A/D converters which handle both positive and negative analog inputs have the additional problem of characterizing uniformity of response on both sides of zero. The symmetry parameter is specified at the full-scale positive and full-scale negative inputs. Each of these inputs should produce the exact same digital output word except for the sign bit. Any error is characterized in terms of LSBs or percentage of full scale.

Zero drift Most ADC devices can be trimmed at zero scale and full scale. The gain drift specifies the stability of the full-scale set point. At the other end of the scale, the zero-drift parameter characterizes the stability of the zero set point.

REFERENCES

1. Zuch, E. L.: Interpretation of Data Converter Accuracy Specifications, *Comput. Des.*, September 1978, p. 113.
2. Henry, T.: Successive Approximation A/D Conversion, *Motorola Appl. Note* AN-716, 1974.
3. Sheingold, D. H., and R. A. Ferrero: Understanding A/D and D/A Converters, *IEEE Spectrum*, September 1972, p. 47.
4. Ritmanich, W.: Buffer Improves Converters Small Signal Performance, *Electronics*, September 29, 1977, p. 100.
5. Evans, L: Building Blocks Take the Problem out of A/D Converter Designs, *EDN*, August 5, 1976, p. 68.
6. Kime, R. C., Jr.: The Charge-balancing A/D Converter: An Alternative to Dual-Slope Integration, *Electronics*, May 24, 1973, p. 97.
7. Grandbois, G., and T. Pickerell: Quantized Feedback Takes Its Place in Analog-to-Digital Conversion, *Electronics*, October 13, 1973, p. 103.

Chapter 26

Digital-to-Analog Converters

26.1 BASIC DEFINITION OF A DAC

A digital-to-analog converter (DAC) performs the exact opposite function of an analog-to-digital converter (ADC); i.e., a DAC converts an N-bit binary word into a voltage or current having 2^N possible levels.

ALTERNATE NAMES Digital-analog converter, D/A converter, D/A.

PRINCIPLES OF OPERATION The basic schematic representations of a voltage-output DAC and a current-output DAC are shown in Fig. 26.1. These representations may be misleading, however, because many options are available inside the DAC which are not obvious on the schematic symbol. As indicated in Fig. 26.2, a typical monolithic DAC contains four basic circuits:

1. A resistor network
2. A voltage reference
3. A current-steering transistor-switch network
4. An output circuit-buffer amplifier

Using Ohm's law on Fig. 26.2a, one can determine that if the MSB switch is connected to V_R and all other switches are grounded, the op-amp output will be $-V_R/2$. Likewise, the next MSB switch alone produces an output of $-V_R/4$, the third bit produces $-V_R/8$, and the LSB alone produces an op-amp output of $-V_R/16$. If all switches are connected to V_R, the output is $-15V_R/16$.

Monolithic DACs use $R\text{-}2R$ networks since it is relatively easy to produce resistances within an octave of each other. A popular resistor network used in discrete DACs is shown in Fig. 26.2b. Each resistor

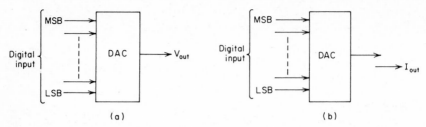

Fig. 26.1   Schematic representation of a DAC with *(a)* a voltage output or *(b)* a current output.

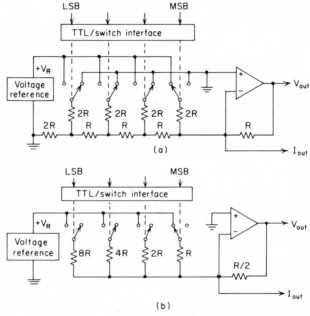

Fig. 26.2   The four basic circuits in the typical monolithic DAC: *(a)* a 4-bit converter using an *R-2R* ladder; *(b)* a 4-bit DAC using binary weighted resistors.

in this ladder is a factor of 2 larger or smaller than adjacent resistors. This technique allows use of a much simpler switching circuit, but it is not popular in monolithic DACs because of the wide spread in resistance values.

The voltage reference must have a very small temperature coefficient if the circuit is to be stable over a wide temperature range.   For example, if a 10-bit DAC is to be $\pm\frac{1}{2}$ LSB stable over the 0 to 75°C temperature range, the complete circuit (V_R, ladder, switches, and op amp) must not drift more than 10 ppm/°C.   One approach used by designers of DAC voltage references utilizes temperature-compensated zener-diode circuits with a constant current source and a buffer-amplifier output.

Even better results have been achieved by replacing the zener diode with a circuit using the band-gap principle, as shown in Fig. 26.3. This circuit uses the positive temperature drift of Q5, Q6, and Q7 to cancel the negative temperature drift of V_{BE} in Q2 and Q3. The resulting stabilized voltage is amplified by the high-gain feedback loop Q1, Q2, Q3, and Q4, which also provides a low output impedance for V_R.

The current-steering switching network which couples the resistor network to the output circuitry represents the greatest challenge to the designers of DAC microcircuits. These switches must be fast, accurate, and uniform and have a large ON-OFF resistance ratio. The current-switching circuit must be uniform to within 0.1 percent in a 10-bit DAC. This is often accomplished with current-mirror circuitry, where the voltage reference controls the current through only one reference current. The other bits use currents which are mirror images of the reference current. A number of transistor switches have been developed for current steering. A simple switch consists of just two transistors, one to switch each resistor network node to ground and another to switch each node to the reference voltage (or current). More exotic switches use up to 10 transistors per cell in a temperature-compensated emitter-coupled configuration.

The output circuit of a DAC is normally the current flowing out of one end of the resistor network. Since most applications require a voltage output, an op-amp current-to-voltage converter (Fig. 26.2) is

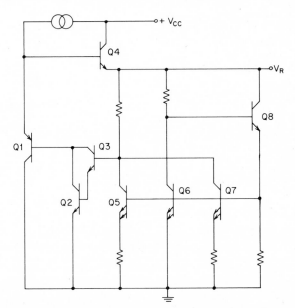

Fig. 26.3 A reference voltage source using the band-gap principle for temperature compensation.

often available on the chip as an option.   The current output is also usually brought out for those who need a more flexible output circuit.

26.2   OPTIONAL DAC CONFIGURATIONS

Several other options are available on various DAC chips in addition to the V/I output option.   If the DAC is to operate as part of a microprocessor system, an input N-bit holding register would be particularly helpful.   Microprocessor systems typically present the data bus to the DAC for a fraction of a microsecond.   The DAC must convert the digital word on the data bus and hold that analog value until the microprocessor presents another word to the DAC.   Since all DAC analog circuitry reacts instantaneously, the digital input is the best location to hold the data temporarily.   In the meantime, the microprocessor is free to perform hundreds or thousands of other tasks using the data bus.   If the DAC input register has a strobe input, the register will be latched with the content of the data bus only when the strobe is activated.   This is illustrated in Fig. 26.4.   We note that this function is the exact opposite of the sample-hold function described in Chap. 24.

DACs with 6, 8, 10, and 12 binary inputs are readily available in monolithic form.   DACs with word sizes less than 6 bits can be implemented using a 6-bit DAC and connecting one or more of the LSBs to ground.   DACs with word sizes above 12 require a technology with accuracies better than 0.025 percent over temperature.   At present

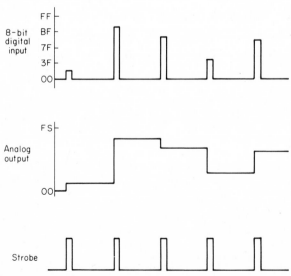

Fig. 26.4   Analog output as a function of digital-word input for a strobed DAC having an input latching register.

TABLE 26.1

Analog output range	Analog output voltage if scaled to ±10 V	Digital input codes				
		Unipolar		Bipolar		
		Binary	BCD	Offset binary	Twos complement	Binary with sign
+FS	+10.0	1111	1001	1111	0111	0111
+FS − 1 bit		1110	1000	1110	0110	0110
+½ FS	+5.0	1000	0100	1100	0100	0100
+1 bit		0001	0001	1001	0001	0001
+0	0.0	0000	0000	1000	0000	0000
−0	0.0					1000
−1 bit				0111	1111	1001
−½ FS	−5.0			0100	1100	1100
−FS + 1 bit				0001	1001	1110
−FS	−10.0			0000	1000	1111

(1979) this capability is just beginning to emerge in monolithic technology although in hybrid form DACs of this resolution have been available for several years.

The five types of input digital codes available in DACs are summarized in Table 26.1. For simplicity of description a 4-bit device is assumed. The left bit of the three bipolar codes is the sign bit. Note that the sign bit for the offset binary code is reversed from the sign bit for the other bipolar codes.

26.3 DAC PARAMETER DEFINITIONS

Nearly two dozen parameters are used by manufacturers of DAC devices to describe their products. No manufacturer uses all the parameters listed below because there is overlap between some of them. Designers must decide which parameters are important for their particular application.

Accuracy, absolute When the full-scale output adjustment is made on a DAC during manufacture, it is set with respect to a standard reference voltage. The ability of the DAC to duplicate that full-scale set point as established during manufacture is called the absolute accuracy. This parameter is often called *gain variance* or *full-scale gain set point*. It is usually a trimmable quantity.

Accuracy, relative Relative accuracy is the difference between any output voltage as a fraction of full scale and its expected value as a fraction of full scale. This parameter is identical to *integral linearity*.

Glitches As a DAC receives small changes in the input code, it passes through *minor* and *major transitions*. The largest major transition is at half scale, where all bits must simultaneously change state, that is,

The dual-slope converter operates by applying an unknown signal to an uncharged capacitor for a fixed length of time and later measuring the time required to discharge the capacitor at a constant rate. In a charge-balancing converter, however, there is no fixed charging period. The charging continues for as long as required to boost the capacitor voltage past a fixed voltage threshold. Each time this crossover occurs, a reference current is subtracted from the input current. When the capacitor discharges past the threshold, the process repeats. This incremental charge balancing repeats many times until the conversion period is complete. During all the conversion period the ADC keeps the integrator output voltage near the level of the reference voltage. In effect, the device measures the amount of charge required to balance the charge developed on the integrating capacitor by the analog input. The net time required by the reference circuitry to hold the capacitor voltage near the threshold is therefore proportional to the input voltage. A counter accumulates clock pulses whenever the reference is switched ON. At the end of the conversion period this counter contains a number proportional to the analog input voltage.

Several variations of the basic charge-balancing technique have been implemented in monolithic form. Figure 25.5 shows a block diagram of the approach used in the Siliconix LD130 converter. This device

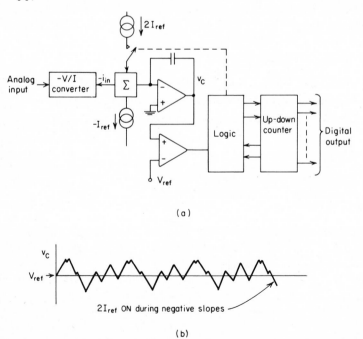

(a)

(b)

Fig. 25.5 A charge-balancing ADC using both positive and negative reference-current sources: (a) block diagram and (b) typical waveform on the integrating capacitor.

output decrease in response to an increased input word. Likewise, a decreased input word should always yield a lower output response. Mathematically speaking, the output is a single-valued function of the input. For a DAC to be monotonic over a wide temperature range, the various error sources in the current-steering switches and resistor network must track each other very closely with temperature.

Noise   The noise in the analog output line of a DAC comes from two types of sources, internally generated noise in the DAC and feed-through noise from the digital inputs. Noise becomes a concern only in high-resolution DACs, that is, 10 bits or more. The noise should not be larger than $\pm\frac{1}{2}$ LSB over the bandwidth in which the DAC output is operating. Noise from glitches and switching transients can be filtered out if they cannot be reduced directly.

Output impedance   This parameter has meaning only for DACs with current-to-voltage-converting op amps in their output circuitry. Since these op amps are a small device on the chip, they often have limited driving capability. When these op amps operate in the closed-loop configuration recommended by the manufacturer, they may still have an output impedance of several ohms.

Power-supply sensitivity   Also called *power-supply rejection*, this term is de-fined as the ratio of the change in full-scale DAC output for a change in any power-supply voltage. A typical DAC might have a power-supply sensitivity of ±0.01 percent per 1 percent V_{CC}. Some data sheets use microvolts per V_{CC}, LSBs per V_{CC}, etc.

Resolution   The resolution of a DAC depends on the number of possi-ble states in its input code. Thus, an 8-bit code has 256 states, etc. The resolution is the size of the smallest state change, i.e., an LSB. The resolution of an 8-bit D/A converter is therefore 1 part in 256.

Settling time   The settling time of a D/A converter is defined as the time required for the analog output to reach its quiescent value within some given margin (such as $\pm\frac{1}{2}$ LSB) following an input code change. Most DACs use a full-scale code change to specify this parameter. The change at one-half full scale where all bits change state is also an important measurement. Settling time is the composite result of many factors: output-amplifier settling time, switching speed, various *RC* time constants, and glitch problems.

Slewing rate   This parameter is often called *slew rate*. Either term refers to the maximum rate at which the DAC output voltage can change in response to a full-scale change in input code. Slew-rate limiting is mostly caused by limited current available to charge a capacitance inside the DAC or in the output circuitry. Slew rate is expressed in volts per microsecond.

Switching time   This parameter is not often used since settling time is a better parameter. Switching time is the time the DAC output takes to switch from one state to another. It includes delay time and rise time (10 to 90 percent) but not settling time.

Symmetry, full-scale   For a bipolar DAC the full-scale symmetry is the difference in magnitudes of +FS and −FS (see Table 26.1). Since most bipolar DACs only have one full-scale adjustment, this parameter is difficult to control. A ±½ LSB error is typical.

Symmetry, zero-scale   In bipolar DACs this parameter is the output-voltage change produced when the sign bit is changed and all magnitude bits are held at zero. In a good-quality DAC this error is much less than ±½ LSB.

Temperature stability   This is also called the *temperature coefficient*. It can be applied to several parameters such as absolute accuracy, linearity, and symmetry. Other parameters are also affected by temperature, but their temperature stability is seldom specified. For any particular variable X_1 at temperature T_1 and X_2 at T_2 the temperature coefficient is defined as follows:

$$TC = \frac{X_1 - X_2}{T_1 - T_2}$$

The full-scale temperature coefficient is affected by the composite stabilities of the voltage reference, resistor network, current switches, and output-amplifier circuit. The zero-scale temperature coefficient of a unipolar DAC is almost entirely controlled by the output op amp. In bipolar DACs the zero is midrange, so many factors affect its zero temperature coefficient.

Zero offset   If all digital inputs are zero, a unipolar DAC should have zero output. Data sheets for most DACs show recommended zero-adjust circuitry.

REFERENCES

1. Kelson, G., H. H. Stellrecht, and D. S. Perloff: A Monolithic 10-b Digital-to-Analog Converter Using Ion Implantation, *IEEE J. Solid State Circuits,* vol. SC-8, no. 6, p. 396, December 1973.
2. Karp, H. R.: Digital-to-Analog Converter: Trading of Bits and Bucks, *Electronics,* Mar. 13, 1972, p. 84.
3. Jung, W. G.: An IC D/A and Three Terminal Voltage Regulator Make a Power D/A, *Electron. Des.,* vol. 13, p. 122, June 21, 1978.
4. *Precision Monolithics Mono* DAC-08 *Tech. Data Sheet,* March 1975.
5. Smith, B. K.: Digital-to-Analog Converters and Their Performance Specifications, *EEE* November 1970, p. 55.

Chapter 27

Phase-Locked-Loop Circuits

INTRODUCTION

Phase-locked loops (PLL) are widely used in communication systems and electromechanical control systems. Within these broad areas we find the PLL in such diverse circuits as

Modulators (AM, FM, FSK, pulse-width)

Demodulators (AM, FM, FSK, synchronous)

Frequency multipliers and dividers

Motor-speed control

Modem transceivers (half and full duplex)

Decoders (tone, ultrasonic, carrier current, FM stereo)

Tracking filters

Generators (function, signal, staircase)

This chapter will present design information on two applications of the PLL. The design equations and design procedures shown for these applications will also be partly applicable to the dozens of other possible uses for the PLL. The chapter begins by discussing, in detail, the two main circuits present in today's single-chip PLL circuits, the phase detector and the voltage-controlled oscillator (VCO). We will then briefly discuss each of the other circuits found in most PLL microcircuits.

The basic elements of a PLL system are shown in Fig. 27.1. In this case a frequency multiplier has been implemented. The phase comparator will be described in Sec. 27.1 while Sec. 27.2 will be devoted to a discussion of the voltage-controlled oscillator. The filter is usually a single-pole low-pass filter or a pole-zero filter. It is commonly of the passive variety although some applications require the more expensive op-amp type. The counter is implemented with a logic family which

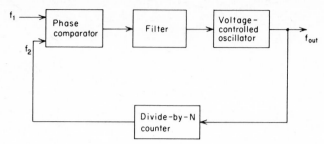

Fig. 27.1   The frequency-multiplier PLL system. One of the most common PLL applications.

is most compatible with the VCO and phase detector.   Some PLL chips require special interface circuits between the VCO and the counter.

27.1   THE PHASE COMPARATOR

ALTERNATE NAMES   Phase detector, phase-difference detector, frequency comparator, frequency detector.

PRINCIPLES OF OPERATION   This device is available either as a separate monolithic chip or as part of a PLL chip.   Many circuits have been designed which will perform phase comparison.   Figure 27.2 shows some of these circuits along with their transfer functions.

The phase comparator shown in Fig. 27.2*a* is sometimes preferred because it requires only one-fourth of a quad XOR.   The transfer function shown in the figure is a plot of average output voltage vs. the phase difference between the inputs.   The output waveform is rectangular with a duty cycle which varies with this phase difference.   The phase-comparator output must be well filtered to provide a smooth dc level for operation of the VCO.   We must usually reach a compromise on good filtering, speed of response, and loop stability.   A filter with a very low frequency pole (large R and C) will allow the VCO to operate with very little jitter (jitter is frequency modulation of the VCO output caused by a noisy control voltage).   Conversely, a filter with this very low frequency pole often makes the PLL system susceptible to feedback instability, manifested as excessive overshoot during capture.   Sustained oscillation of the total loop is even possible under certain conditions.

As might be expected from the periodic transfer function, the XOR phase comparator will produce a valid output when $f_1 = f_2$, $f_1 = 2f_2$, $2f_1 = f_2$, $f_1 = 3f_2$, $3f_1 = f_2$, etc.   This drawback limits its use to applications where the input frequencies will never vary more than ± 20 percent from nominal.   If a VCO generates a feedback signal f_2, as shown in Fig. 27.1, its frequency must be guaranteed to rise immediately to within ± 20 percent of f_1 at system power-up.   If the VCO comes up 25 percent low in frequency, the PLL may slew up to the correct frequency or it may slip down to one-half the correct frequency. If the VCO is not capable of operation at any harmonic or subharmonic of the nominal VCO frequency, this is no problem. Many systems are designed such that the VCO will tune only ±20 percent from nominal as the control voltage goes from absolute minimum to absolute maximum.

The XOR phase comparator requires input signals having exactly 50 percent duty cycles. Failure to implement this requirement will result in a phase comparator with gain which varies as a function of phase angle between f_1 and f_2. This will cause the loop stability also to depend on the phase angle.

The phase comparator shown in Fig. 27.2b utilizes a standard IC comparator. It has nearly the same characteristics as the XOR phase comparator. This circuit has the added advantage of offering some adjustment capability for input trip levels. This is useful in applications where the phase comparator must accept input signals from different logic families and/or signal levels.

Figure 27.2c shows a phase comparator made from a D flip-flop. Each cycle of the reference input frequency resets the flip-flop, and each cycle of the feedback frequency toggles Q to the HIGH state. The transfer function for this phase comparator exhibits good linearity over twice the phase difference range of the previous two detectors. This results in better acquisition, tracking, and locking characteristics. However, the D flip-flop phase detector is also susceptible to phase lock at harmonic frequencies since the transfer function is periodic.

The D flip-flop phase comparator is an edge-sensitive device. This means that 50 percent duty-cycle input waveforms are not required. The asynchronous JK flip-flop and the RS flip-flop shown in Figs. 27.2d and e have performance characteristics identical to those of the D flip-flop in Fig. 27.2c.

Figure 27.2f shows a unique phase comparator which utilizes a sawtooth generator, two sample-and-hold circuits, and three single-shots. The pulse width of all single-shots are approximately equal and several orders of magnitude narrower than the period of the input signals. The triangle generator has a waveform identical to the transfer function shown in the figure. In normal operation the triangle-generator phase leads the feedback-signal triggering edge by θ_D. At this phase angle (delay) the first single-shot causes gate G_1 to sample the triangle voltage level and deposit it on C_1. After an additional delay caused by the second single-shot the third single-shot closes G_2 and transfers part of the voltage from C_1 over to C_2. The second single-shot provides a buffer time so that there will be no chance of G_1 and G_2 coming ON simultaneously. If the capacitors are of equal value, the stored voltage on C_2 at the end of the third single-shot pulse will be one-half the sampled voltage on the triangle.

The dc voltage on C_2 will remain steady only if the feedback signal

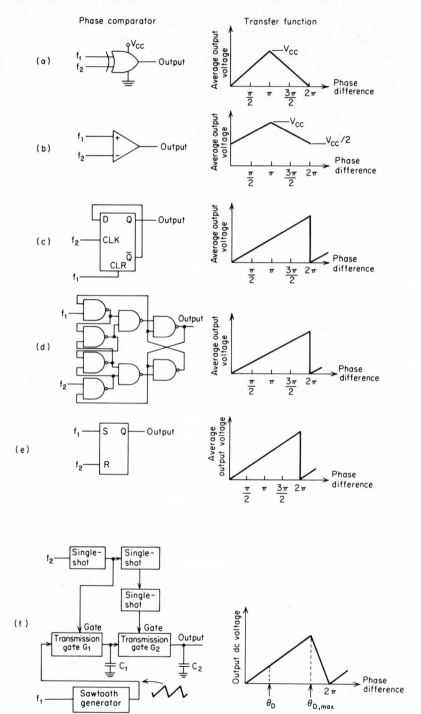

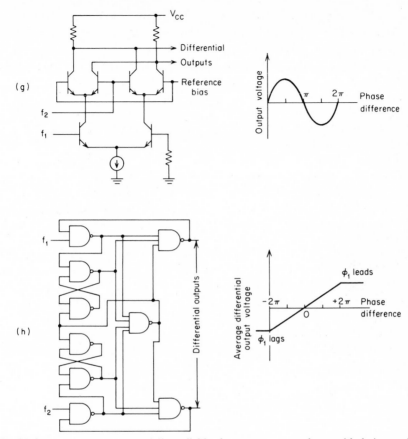

Fig. 27.2   Some of the commercially available phase comparators shown with their transfer functions.

is locked $\theta_D°$ behind the triangle.   A frequency difference between the input signals will cause the single-shots to sample the triangle at random voltages, causing a random output waveform on C_2.   If the two input frequencies are harmonics of each other, a false lock can occur.

The triangle–single-shot–transmission-gate phase comparator does not put out a rectangular waveform when it is in lock.   If the sampling is performed at the same location each cycle, C_2 will maintain a smooth dc voltage level.   This reduces the requirement for filtering which most phase comparators require.

A phase comparator used in several monolithic PLL chips is shown in Fig. 27.2g.   This is a linear circuit useful for low-level signals under 1 mV.   It is sometimes called the double-balanced modulator.   The input signals must have 50 percent duty cycles for proper operation.

Figure 27.2h shows a phase comparator sold as a separate chip

(MC4044) or as part of a PLL chip (MC4046). It has characteristics which overcome most of the problems of other phase comparators. It is both phase- and frequency-sensitive; i.e., its transfer function is not periodic. If the input frequencies are widely separated, the phase comparator will put out a constant high- or low-level signal until the VCO is close to a phase lock. Then the linear part of the transfer function locks the two frequencies at 0° phase difference. Since the output is dc at phase lock, filtering requirements are minimal.

The circuit of Fig. 27.2h is ideal for applications such as PLL motor control. The feedback signal in these systems often comes from a tachometer on the motor shaft. The feedback frequency will therefore start at zero and (relatively) slowly accelerate up to the lock frequency. The phase comparator will put out a steady control voltage until phase lock is near; then it will begin to compare phase to establish a complete phase lock.

The circuit of Fig. 27.2h is sometimes seen in the form shown in Fig. 27.3. (See the CD4046A PLL chip in Ref. 7.) This particular version of the phase comparator has a CMOS three-state output circuit. When both frequency and phase lock are achieved, the output becomes an open circuit. If $f_1 > f_2$, the p-channel output transistor (Q1) is fully on and the n-channel device (Q2) is off. Likewise, when $f_1 = f_2$ and $\phi 1$ leads $\phi 2$, the on time of Q1 is proportional to the phase difference. This on time of Q1 linearly varies from 100 percent when $\phi 1$ leads

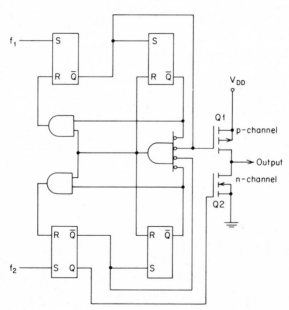

Fig. 27.3 A phase comparator which also compares frequency, thus preventing false lock-up on harmonically related frequencies.

$\phi2$ by 2π rad down to 0 percent when their phases are identical. Q2 remains OFF whenever $\phi1$ leads $\phi2$.

If $f_1 < f_2$, Q1 is fully OFF and Q2 is fully ON. When $f_1 = f_2$ and $\phi1$ lags $\phi2$, the ON time of Q2 is proportional to their phase difference. The ON time of Q2 linearly varies from 100 percent when $\phi1$ lags $\phi2$ by 2π rad down to 0 percent when there is no phase difference.

27.2 THE VOLTAGE-CONTROLLED OSCILLATOR

ALTERNATE NAMES VCO, voltage-to-frequency converter, V/F converter, VFC, voltage-controlled generator, analog-to-frequency converter.

PRINCIPLES OF OPERATION Voltage-controlled oscillator circuits come in a variety of configurations. Most RF VCOs merely use a voltage-variable capacitor (VVC) across a tank circuit, as shown in Fig. 27.4a.

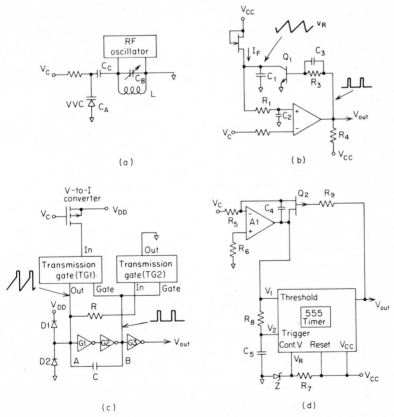

Fig. 27.4 Four common voltage-controlled oscillator circuits: (a) voltage-controlled diode, (b) fixed-current source, (c) variable-current source, and (d) integrator-pulse generator.

The VVC must be reverse-biased for proper operation. At a given control voltage V_C the VVC will exhibit a junction capacitance C_A, which becomes part of the total tank capacitance $C_A + C_B$. The oscillator will resonate at a frequency of

$$f_R = \frac{1}{2\pi \, [L(C_A + C_B)]^{1/2}}$$

The relationship between V_C and C_A is

$$C_A = \frac{C_D}{(1 + V_C/K)^n}$$

where

$$C_D = \text{junction capacitance at zero } V_C$$
$$K \approx 0.6 \text{ for most silicon VVCs}$$
$$n = \begin{cases} \frac{1}{2} & \text{for step-junction diodes} \\ \frac{1}{3} & \text{for graded-junction diodes} \end{cases}$$

The oscillator frequency therefore depends on V_C according to

$$f_R = \frac{1}{2\pi \, [L(C_B + C_D/(1 + V_C/K)^n)]^{1/2}}$$

VCOs for operation below 10 MHz are often designed around the linear-ramp concept. The linear ramp is generated by either charging a capacitor with a current source or by use of an integrator. Figure 27.4b indicates how a constant-current source driving a capacitor is utilized to implement a simple VCO. The current source can be as simple as the junction FET shown or as precise as the dual-transistor current mirrors discussed in the literature [8].

In operation a positive-going linear voltage ramp is generated at v_R. When the ramp voltage slightly exceeds V_C, the comparator output will go to its HIGH state. At this level Q_1 is quickly turned ON through the commutating network $R_3 C_3$. The voltage on C_1 is discharged through Q_1. The discharge time is controlled by the $R_1 C_2$ delay network. The output pulse width is also equal to this discharge time.

If V_C rises to a higher level, the ramp must likewise rise to a higher level before the comparator trips. This results in a lower frequency. The approximate relationship between frequency and V_C is

$$f_R \approx \frac{1}{R_1 C_2 + C_1(V_C - V_{\text{sat}})/I_F}$$

where V_{sat} is the ON saturation voltage of Q_1 and I_F is the current from the FET current source.

Figure 27.4c shows a compact VCO which requires three CMOS chips, a resistor, and a capacitor [4]. At the beginning of each cycle point B is LOW and transmission gate TG1 is ON. This allows the V-to-I converter to generate a ramp voltage on C which has a slope proportional to the magnitude of V_C. When the threshold voltage V_T of G1 has been reached, both G1 and G2 change states. The capacitor now provides positive feedback to the input of G1, causing point A to rise immediately to $V_{DD} + V_D$, where V_D is the diode drop across D1. At this same time TG2 is switched ON, allowing C to discharge through R. When point A falls through the threshold voltage of G1, these two gates again change states. TG2 is switched OFF and TG1 returns to the ON state where the cycle starts over again.

The output pulse width is determined by R and C according to

$$T_P \approx 0.69(R + R_{on})\,C$$

where R_{on} is the ON resistance of TG2.

If the V-to-I converter has a gain of K, then $I_F = KV_C$. The ramp will slew for $CV_T/I_F = CV_T/KV_C$ s before the threshold voltage of G1 is exceeded. Combining this with the retrace time T_P, we get the output frequency

$$f_R \approx \frac{1}{C[0.69(R + R_{on}) + V_T/KV_C]}$$

The circuit shown in Fig. 27.4d is one specific example of a widely used class of VCO circuits [2, 12]. We have an integrator which generates a ramp voltage with a slope proportional to V_C. When the ramp rises to the timer trip level, a pulse is generated which turns the transistor Q_2 ON. This action discharges the integrator capacitor, thereby starting the circuit on a new cycle. The ramp rises at a rate of

$$\frac{\Delta V_1}{\Delta t} = \frac{V_C}{R_5 C_4}$$

As shown in Fig. 27.5, the timer output is normally HIGH. When v_1 rises to V_R, the timer output switches to the LOW state. A LOW voltage at v_{out} causes Q_2 to conduct heavily. The duration of the Q_2 ON time is determined by

$$T_P \approx 0.7 R_8 C_5$$

This delay network ($R_8 C_5$) is required so the Q_2 can fully discharge C_4. The timer's trigger input does not return v_{out} to the HIGH state until

$$v_2 = V_R/2$$

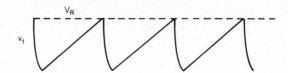

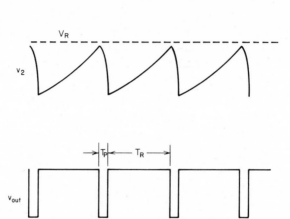

Fig. 27.5   Important waveforms at various locations in the VCO shown in Fig. 27.4d.

Since C_4 must be discharged before the pulse terminates, the following must hold:

$$R_8 C_5 > R_{on} C_4$$

where R_{on} is the ON resistance of Q_2.

The width of T_P determines the upper frequency of the VCO.   In practice the minimum period of a cycle is approximately $2 T_P$.

VCO DESIGN PARAMETERS*

Parameter	Description
C_4	Determines integration time along with R_5
C_5	Controls output pulse width along with R_8
$f_{R,nom}$	Midrange output frequency corresponding to $V_{C,nom}$
I_{DSS}	Drain-to-source saturation current of Q_2 (at $V_{GS} = O$, $V_{DS} = -5$ V)
I_Z	Nominal zener-diode current
R_5	Determines input resistance of circuit and also integration time (along with C_4)
R_6	Cancels effect of op-amp input-bias current
R_7	Controls zener-diode current
R_8	Sets pulse width along with C_5
R_9	Controls drain to gate current in Q_2

R_{on}	ON resistance of Q_2
T_R	Time between pulses (ramp ON time)
T_P	Pulse width
v_1	Ramp waveform out of integrator
v_2	Delayed ramp waveform
v_{out}	Output voltage waveform
V_C	Input control voltage
V_{CC}	Supply voltage
V_P	Pinch-off voltage of Q_2
V_R	Reference voltage established by zener

* Fig. 27.4d.

VCO DESIGN EQUATIONS*

Description	Equation	
Output frequency* if $T_R \gg T_P$	$f_R \approx \dfrac{|V_C|}{V_R R_5 C_4}$ * V_C must be negative.	(1)
Output frequency if $T_R \approx R_P$ (near upper limit of f_R)	$f_R \approx \dfrac{1}{0.7 R_8 C_5 + V_R R_5 C_4/|V_C|}$	(2)
Width of negative output pulse	$T_P \approx 0.7 R_8 C_5$	(3)
Width of positive output pulse	$T_R \approx \dfrac{V_R R_5 C_4}{|V_C|}$	(4)
Resistor values:		
R_5, R_6	$R_5 = R_6 = \dfrac{|V_{C,nom}|}{V_R C_4 f_{R,nom}}$	(5)
R_7	$R_7 \approx \dfrac{V_{CC} - V_R}{I_Z}$	(6)
R_8	$R_8 \approx 10 \text{ k}\Omega$	(7)
R_9	$R_9 \approx 100 \text{ k}\Omega$	(8)
Capacitor values:		
C_4	$C_4 = \dfrac{T_P}{10 R_{on}}$	(9)
C_5	$C_5 = \dfrac{T_P}{0.7 R_8}$	(10)
Q_2 ON resistance (at $V_{GS} = O$)	$R_{on} = \dfrac{V_P}{2 I_{DSS}}$	(11)

* Fig. 27.4d

DESIGN PROCEDURE

These design steps are fairly straightforward due to the simplicity of the circuit. Timer ICs in conjunction with op amps make the design of a large class of circuits greatly simplified. We begin this procedure by assuming that the pulse width T_P and the nominal f_R (and its corresponding $V_{C, nom}$) are specified.

DESIGN STEPS

Step 1. Choose an FET having a V_P less than V_{CC}. Compute R_{on} using Eq. (11).

Step 2. Calculate a nominal value for C_4 using Eq. (9). Use the standard value closest to that calculated.

Step 3. Resistors R_5 and R_6 are now found from Eq. (5). Use a nominal V_C and the corresponding nominal f_R.

Step 4. The selection of R_9 is not critical. For high-speed operation (> 10 kHz) one should probably keep R_9 in the 100-kΩ range. Likewise, a nominal choice for R_8 is 10 kΩ.

Step 5. Calculate an approximate value for R_7 using Eq. (6).

Step 6. Compute a value for C_5 with Eq. (10).

EXAMPLE OF A VCO DESIGN Suppose we want a VCO having a range from nearly dc to 10 kHz. The (negative) pulse width is to be approximately 10 percent of the period at 10 kHz. Let the nominal V_C of -5 V correspond to $f_R = 5$ kHz.

Design Requirements

$$f_{R, \text{nom}} = 5 \text{ kHz} \qquad f_{R, \text{min}} = \text{dc}$$
$$f_{R, \text{max}} = 10 \text{ kHz} \qquad V_{C, \text{nom}} = -5 \text{ V}$$
$$V_{C, \text{min}} = 0 \text{ (shorted to ground)} \qquad V_{C, \text{max}} = -10 \text{ V}$$
$$V_{CC} = \pm 15 \text{ V}$$

Device Data

$$V_{io, \text{max}} = 5 \text{ mV}$$
$$\text{2N2608, measured: } V_P = 3.2 \text{ V} \qquad I_{DSS} = 3.2 \text{ mA}$$
$$\text{1N4566: } I_Z = 0.5 \text{ mA} \qquad V_R = 6.4 \text{ V}$$

Step 1. The 2N2608 FET has a pinch-off voltage of 1 to 4 V. We will use 3.2 V in

$$R_{on} = \frac{V_P}{2 I_{DSS}} = \frac{3.2}{2(3.2 \times 10^{-3})} = 500 \ \Omega$$

Step 2. Equation (9) provides a nominal value for C_4:

$$C_4 = \frac{T_P}{10 R_{on}} = \frac{1}{10 f_{R, \text{max}} \, 10 R_{on}} = \frac{1}{(10 \times 10^4)(10)(500)} = 2000 \text{ pF}$$

[Note that $T_P = 1/(10 f_{R, \text{max}}) = 1/(10 \times 10^4) = 10 \ \mu\text{s}$.]

Step 3. Use Eq. (5) to find R_5 and R_6:

$$R_5 = R_6 = \frac{|V_{C, \text{nom}}|}{V_R C_4 f_{R, \text{nom}}} = \frac{|-5|}{6.4(2 \times 10^{-9})(5000)} = 78 \text{ k}\Omega$$

Step 4.   Let $R_8 = 10 \text{ k}\Omega$ and $R_9 = 100 \text{ k}\Omega$.

Step 5.   The final resistor calculation is

$$R_7 \approx \frac{V_{CC} - V_R}{I_Z} = \frac{15 - 6.4}{5 \times 10^{-4}} = 17{,}200 \,\Omega$$

Step 6.   Capacitor C_5 becomes

$$C_5 = \frac{T_P}{0.7 R_8} = \frac{10^{-5}}{0.7 \times 10^4} = 1430 \text{ pF}$$

27.3   PLL SYSTEMS ON A SINGLE CHIP

ALTERNATE NAMES   Phase-locked loop, PLL FM stereo decoder, monolithic phase-locked loop.

PRINCIPLES OF OPERATION   Single-chip monolithic PLL systems are available which combine some of the better phase comparators of Sec.

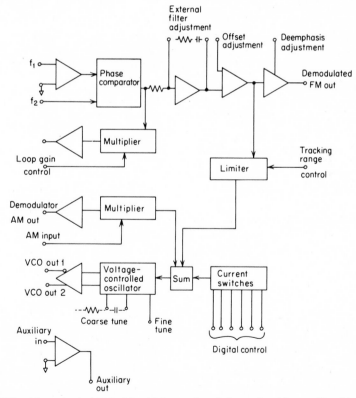

Fig. 27.6   A hypothetical PLL chip which contains all the optional items offered by various manufacturers.

27.1 with some of the better VCOs of Sec. 27.2. These same chips also contain optional items such as phase-lock indicator circuitry, buffer op amp(s), limiter, active filter, auxiliary phase comparator, multiplier, and programmable current switches. The optional items included depend on the particular applications the manufacturer had in mind when the chip was designed. A fictitious PLL chip which contains all these optional items is shown in Fig. 27.6.

In Secs. 27.4 and 27.5 we show design procedures and a numerical example for several widely used applications of the PLL chip.

27.4   A PLL FREQUENCY MULTIPLIER

ALTERNATE NAMES   Harmonic generator, locked oscillator, frequency synthesizer, synchronized generator.

PRINCIPLES OF OPERATION   The frequency multiplier is one of the most basic applications of the PLL. The output frequency, as shown in Fig. 27.7, is exactly N times the input frequency. Not only is the

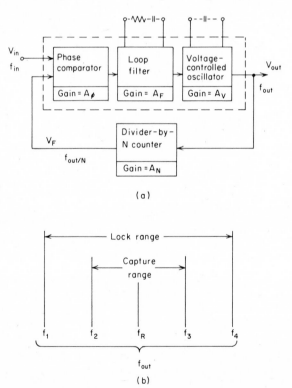

(a)

(b)

Fig. 27.7   (a) A frequency multiplier designed around a standard PLL monolithic circuit; (b) the relationship between VCO free-running frequency, lock range, and capture range.

output frequency precisely controlled, but the phase relationship between the input and output signals is locked at a specific value, depending on the type of phase comparator used.

The operation of this type of feedback circuit is similar to that of voltage or current feedback circuits. In this case, however, the phase of the feedback signal is compared with the phase of the input signal. An incorrect phase relationship between these signals results in an appropriate error voltage from the phase comparator–filter. This error voltage changes the VCO frequency and phase until f_{out}/N is precisely locked to f_{in}.

As shown in Fig. 27.7, three portions of the PLL chip are required for a frequency multiplier.

Phase comparator A frequency-phase comparator, as shown in Fig. 27.2h or Fig. 27.3, is preferred if a wide range of input frequencies is expected. Phase comparators having periodic transfer functions should not be used unless the two input frequencies can be guaranteed to be within ±25 percent of each other. This must also be assured at power

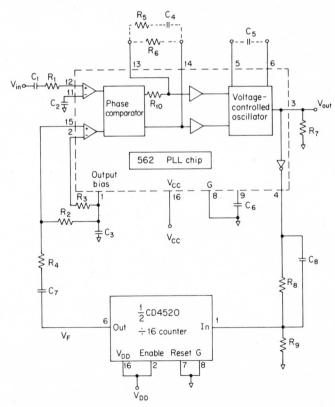

Fig. 27.8 Detailed schematic of a frequency multiplier using the Signetics 562 PLL.

turn-on or the system will come up with an incorrect input-output frequency relationship.

Filter   Either active or passive filters can be used to remove the ripple from the phase-comparator output.   An external *RC* network is used to control the filter response.   The dc voltage produced by the filter is used to control the VCO frequency.   Any ripple on this line will frequency-modulate the VCO.

Voltage-controlled oscillator (VCO)   This is the last major item on the chip normally used in frequency-multiplier applications.   The required range of VCO output frequencies is often the main criterion used for selecting a particular PLL chip.   Most PLL chips, however, contain a VCO capable of operating from less than 1 Hz to higher than 10 MHz.

Not all PLL devices will interface directly with the divide-by-N counter in the feedback loop.   For example, some VCO output signals have a large dc component which must be removed before it can drive the counter.   The designer is advised to evaluate all worst-case PLL input-output parameters carefully to make sure that they will always interface with the worst-case counter input-output parameters.

FREQUENCY-MULTIPLIER DESIGN PARAMETERS

Parameter	Description
A_F	Voltage transfer function of loop filter
A_N	Transfer function of counter in feedback path
A_V	Transfer function of VCO
A_ϕ	Transfer function of phase comparator
C_1, C_2	Provide ac coupling if dc levels are not compatible
C_3	Filter for bias reference output
C_4	Part of loop filter
C_5	Determines VCO free-running frequency
f_1	Lower end of lock-frequency range
f_2	Lower end of capture-frequency range
f_3	Upper end of capture-frequency range
f_4	Upper end of lock-frequency range
f_C	Capture range of frequencies, i.e., from f_2 to f_3
f_{in}	Frequency of input signal
f_L	Lock range of frequencies, i.e., from f_1 to f_4
f_n	Natural frequency of the feedback loop
f_{out}	Frequency of output signal
f_R	VCO free-running frequency
K_A	Loop-gain adjustment factor
K_ϕ	Maximum value of phase-comparator gain
N	Division factor of divider in feedback network
R_1, R_4	Current-limiting resistors required if V_{in} and/or V_F are > 2 V peak to peak
R_2, R_3	Biasing resistors for one of the phase-comparator input amplifiers
R_5	Resistor used to insert zero in loop-filter transfer function
R_6	Optional resistor which is used to adjust total loop gain
R_7	Load resistor required for 562 PLL (note that $R_7 \approx R_8 + R_9$ must be satisfied)

R_8, R_9	Resistors used to adjust PLL average output voltage down to average input required for counter
R_{10}	Output resistance of phase comparator
R_{av}	Average voltage required at counter input
V_{CC}	Collector supply voltage
V_{DD}	Drain supply voltage
V_F	Peak-to-peak feedback voltage
V_{in}	The rms input voltage
$V_{in,0,max}$	Maximum allowable value of LOW logic state driving counter
$V_{in,1,min}$	Minimum allowable value of HIGH logic state driving counter
$V_{out,dc}$	DC level of VCO output waveform
ζ	Damping factor of feedback loop
ϕ	Phase angle between input signal and feedback signal
ω_{out}	Radian frequency of VCO output

FREQUENCY-MULTIPLIER DESIGN EQUATIONS (562)

Description	Equation	
VCO free-running frequency	$f_R = \dfrac{1}{3300\,C_5}$ Hz	(1)
VCO gain	$A_V = \dfrac{2\pi f_{out}}{s} = \dfrac{\omega_{out}}{s}$ rad/s. V	(2)
Phase-comparator gain	$A_\phi = \dfrac{40 V_{in}\cos\phi}{(1 + 625 V_{in}^2)^{1/2}}$ V/rad	(3)
Maximum value of $A_\phi = K_\phi$	$K_\phi = A_\phi\ (\phi = 0°,\ V_{in} > 0.1\ \text{V}) = 1.6$	(3a)
Loop-filter gain for simple lag filter where $R_5 = 0$	$A_F = \dfrac{1}{1 + 2\,sR_{10}\,C_4}$	(4)
Loop-filter gain for pole-zero filter where $R_5 \neq 0$	$A_F = \dfrac{1 + sR_5\,C_4}{1 + s(2\,R_{10} + R_5)\,C_4}$	(5)
Gain constant of digital divider in feedback circuit	$A_N = \dfrac{1}{N}$	(6)
Lock range of frequencies	$f_L = f_4 - f_1 = f_{out}\,\dfrac{K_\phi}{N}$	(7)
Capture range of frequencies if $R_5 = 0$	$f_C = f_3 - f_2 = \left[\dfrac{8\pi K_\phi f_{out}}{NR_{10} C_4}\right]^{1/2}$	(8)
Capture range of frequencies if $R_5 \neq 0$ and assuming $2(f_4 - f_R) = f_4 - f_1$	$f_C = f_3 - f_2 = 2\omega_{out}\left\lvert\dfrac{1 + j\pi f_L R_5 C_4}{1 + j\pi f_L(2\,R_{10} + R_5)\,C_4}\right\rvert$	(9)
Natural frequency of feedback loop if $R_5 = 0$	$f_n = \dfrac{1}{2\pi}\left(\dfrac{K_\phi\,\omega_{out}}{NR_{10}C_4}\right)^{1/2}$	(10)
Natural frequency of feedback loop if $R_5 \neq 0$	$f_n = \dfrac{1}{2\pi}\left[\dfrac{K_\phi\,\omega_{out}}{NC_4(R_{10} + R_5)}\right]^{1/2}$	(11)

Description	Equation	
Damping factor of feedback loop if $R_5 = 0$	$\zeta = \dfrac{\pi f_n N}{K_\phi \, \omega_{out}}$	(12)
Damping factor of feedback loop if $R_5 \neq 0$	$\zeta = \pi f_n \dfrac{N + \omega_{out} K_\phi R_5 C_4}{\omega_{out} K_\phi}$	(13)
Loop gain adjustment factor for finite R_6	$K_A = \dfrac{R_6}{12,000 + R_6}$	(14)
Average voltage required at counter input	$V_{av} = \dfrac{V_{in,1,min} + V_{in,0,max}}{2}$	(15)
Resistor values:		
R_1	$R_1 = \begin{cases} 0 & \text{if } V_{in} \leq 2 \text{ V pp} \\ 1000\,\Omega & \text{if } V_{in} > 2 \text{ V pp} \end{cases}$	(16)
R_2, R_3	$R_2 = R_3 = 1000\,\Omega$	(17)
R_4	$R_4 = \begin{cases} 0 & \text{if } V_F \leq 2 \text{ V pp} \\ 1000\,\Omega & \text{if } V_F > 2 \text{ V pp} \end{cases}$	(18)
R_5	$0 \leq R_5 \leq 200\,\Omega$	(19)
R_6	$R_6 = \dfrac{12,000 K_A}{1 - K_A}$	(20)
R_7	$R_7 = 3000\text{--}12,000\,\Omega$	(21)
R_8	$R_8 = R_7 \dfrac{1 - V_{av}}{V_{out,dc}}$	(22)
R_9	$R_9 = \dfrac{V_{av} R_7}{V_{out,dc}}$	(23)
Capacitor values: C_1, C_2, C_3	$C_1 = C_2 = C_3 = 0.1 \ \mu\text{F}$	(24)
C_4	$C_4 = \dfrac{2 K_\phi f_{out}}{\pi N R_{10} f_C^2}$	(25)
C_5	$C_5 = \dfrac{1}{3300 f_R}$	(26)
C_6	$C_6 = 1000 \text{ pF}$	(27)
C_7, C_8	$C_7 = C_8 = 0.1 \ \mu\text{F}$	(28)

DESIGN PROCEDURE

There is no possible general design procedure which would be applicable to all PLL chips used in frequency-multiplier applications. Each chip has its own peculiarities, design equations, and constraints. The design equations and design steps to follow assume use of the 562 PLL microcircuits.

DESIGN STEPS

Step 1. Calculate the average voltage required at the counter input using Eq. (15).

Step 2. Determine values for R_1 through R_4 with Eqs. (16) to (18).

Step 3. Initially set $R_5 = 0$ and $R_7 = 12 \text{ k}\Omega$.

Step 4. Compute values for R_8 and R_9 using Eqs. (22) and (23).

Step 5. Calculate K_ϕ with Eq. (3a)

Step 6. Determine all capacitor values using Eqs. (24) to (28). Assume $f_R = f_{out}$ for Eq. (26).

Step 7. Determine the lock range of frequencies with Eq. (7).

Step 8. Find the natural resonant frequency f_n of the feedback loop using Eq. (10).

Step 9. Using f_n, we now compute the damping factor ζ with Eq. (12). Use Fig. 27.9 to determine the percentage overshoot of the closed loop. If this overshoot is much above 50 percent, the VCO may never lock onto a multiple of f_{in}. Instead, the VCO output will slowly sweep back and forth at a frequency f_n.

Step 10. If the overshoot is too large, we have two possible remedies: (1) R_5 can be increased from 0 up to 100 or 200 Ω. Equations (11) and (13) are then utilized to find a new f_n and ζ. (2) The loop gain can be reduced by installing R_6. Use Eq. (14) to determine the loop-gain reduction. Equations (10) and (12) can be recomputed using $K_A K_\phi$ in place of the original K_ϕ.

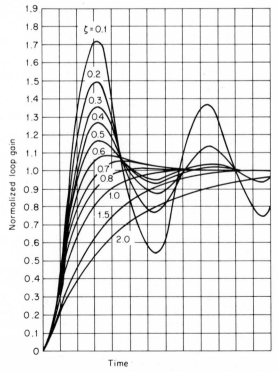

Fig. 27.9 Overshoot of a second-order system as a function of damping factor ζ.

Step 11. A plot of loop gain can be obtained from

$$A_L = \text{loop gain} = \frac{A_V A_\phi A_F}{N}$$

The A_ϕ factor must be lowered by K_A if R$_6$ is installed. The plot of loop gain can be used to accurately determine the gain and phase margins of the PLL (see the following example).

EXAMPLE OF A PLL FREQUENCY MULTIPLIER The design steps and design equations on the previous pages will be illustrated with a numerical example. We will assume that the designer has a precision 100-kHz source for V_{in} and will use the PLL to implement a precision 1600-kHz source.

Design Requirements

$$f_{in} = 100 \text{ kHz} \qquad f_{out} = 1600 \text{ kHz} \qquad f_C = 100 \text{ kHz}$$

$$N = 16 \qquad V_{in} = 1.0 \text{ V rms} = 2.8 \text{ V p–p} \qquad 562 \text{ PLL chip}$$

$$\tfrac{1}{2} \text{ 4520 CMOS dual} \div 16 \text{ counter} \qquad V_{CC} = 18 \text{ V} \qquad V_{DD} = 5 \text{ V}$$

Device Parameters (Typical)

$$562 \text{ PLL: } V_{out,dc} = 12 \text{ V} \qquad R_{10} = 6000 \text{ } \Omega$$

$$4520: V_{in,1,min} = 0.8 V_{DD} \qquad V_{in,0,max} = 0.2 V_{DD}$$

Step 1. The average voltage required at the 4520 input is

$$V_{av} = \frac{V_{in,1,min} + V_{in,0,max}}{2} = \frac{0.8 V_{DD} + 0.2 V_{DD}}{2} = 0.5 V_{DD} = 2.5 \text{ V}$$

Step 2. The CMOS counter output waveform V_f is 5 V p–p. Since both V_f and V_{in} are larger than 2 V p–p, we let

$$R_1 = R_2 = R_3 = R_4 = 1000 \text{ } \Omega$$

Step 3. As recommended, we let $R_5 = 0$ and $R_7 = 12$ kΩ.
Step 4. Values for R$_8$ and R$_9$ are

$$R_8 = R_7 \frac{1 - V_{av}}{V_{out,dc}} = 12,000 \frac{1 - 2.5}{12} = 9500 \text{ } \Omega$$

$$R_9 = \frac{V_{av} R_7}{V_{out,dc}} = \frac{2.5(12,000)}{12} = 2500 \text{ } \Omega$$

Step 5. Equation (3a) provides us with

$$K_\phi = A_\phi \quad (\phi = 0°, V_{in} > 0.1 \text{ V})$$

$$= \frac{40\,V_{\text{in}}\cos\phi}{(1+625\,V_{\text{in}}^2)^{1/2}}\Bigg|_{\substack{\phi=0° \\ V_{\text{in}}>0.1\,\text{V}}} = \frac{40(1)(\cos 0°)}{[1+625(1)^2]^{1/2}} = 1.599$$

Step 6.   Capacitor values are

$$C_1 = C_2 = C_3 = C_7 = C_8 = 0.1\ \mu\text{F}$$

$$C_4 = \frac{2\,K_\phi f_{\text{out}}}{\pi N R_{10} f_{\tilde{c}}^2} = \frac{2.(1.599)(1.6\times 10^6)}{\pi(16)[6000\times(10^5)^2]}$$

$$= 1700\ \text{pF}$$

$$C_5 = \frac{1}{3300\,f_R} = \frac{1}{3300(1600\ \text{kHz})} = 189\ \text{pF} \qquad \text{use 180 pF}$$

$$C_6 = 1000\ \text{pF}$$

Step 7.   The lock range is

$$f_L = f_4 - f_1 = \frac{f_{\text{out}}\,K_\phi}{N} = \frac{(1600\ \text{kHz})(1.599)}{16} = 160\ \text{kHz}$$

Step 8.   The natural resonant frequency of the loop is

$$f_n = \frac{1}{2\pi}\left(\frac{K_\phi\omega_{\text{out}}}{N R_{10}\,C_4}\right)^{1/2}$$

$$= \frac{1}{2\pi}\left[\frac{1.599(2\pi)(1.6\times 10^6)}{(16)(6000)(1700\ \text{pF})}\right]^{1/2}$$

$$= 50\ \text{kHz}$$

Step 9.   The damping factor now becomes

$$\zeta = \frac{\pi f_n N}{K_\phi\omega_{\text{out}}} = \frac{\pi(50\ \text{kHz})(16)}{1.599(2\pi)(1.6\times 10^6)} = 0.156$$

According to Fig. 27.9, a ζ of 0.156 means that the PLL output frequency will momentarily overshoot 60 percent when f_{in} makes an abrupt frequency change.

Step 10.   If the 60 percent overshoot determined in step 9 is too large, we might first try raising R_5 from 0 to 200 Ω. Equation (11) gives us a new f_n

$$f_n\ (R_5 = 200) = \frac{1}{2\pi}\left[\frac{K_\phi\omega_{\text{out}}}{N(R_{10}+R_5)\,C_4}\right]^{1/2}$$

$$= \frac{1}{2\pi}\left[\frac{1.599(2\pi)(1.6\times 10^6)}{16(6000+200)(1.7\times 10^{-9})}\right]^{1/2}$$

$$= 49.1\ \text{kHz}$$

and Eq. (13) a new ζ

$$\zeta = \pi f_n \frac{N + \omega_{out} K_\phi R_5 C_4}{\omega_{out} K_\phi}$$

$$= \pi (49,100) \frac{16 + 2\pi(1.6 \times 10^6)(1.599)(200)(1.7 \times 10^{-9})}{2\pi(1.6 \times 10^6)(1.599)} = 0.206$$

The resistor R_5 does not make a large increase in feedback stability unless it is several kilohms or more. However, when this is done, the

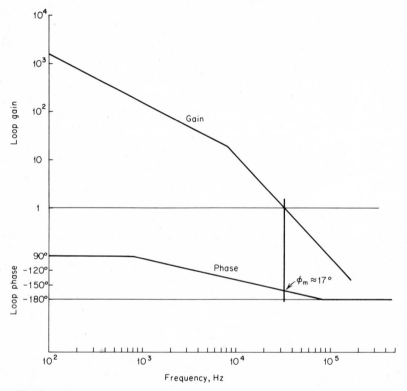

Fig. 27.10 Loop gain and phase as a function of frequency for the PLL frequency multiplier.

filtering action of C_4 is negated to the point where the phase detector drives the VCO with a signal having a large ac component. This frequency-modulates the VCO, which is not desirable in most cases.

We next try increasing stability by lowering the loop gain with R_6. A plot of loop gain as a function of frequency is useful for this approach. Figure 27.10 shows the loop gain, which is merely the product of all block gains in the loop

$$A_L = \frac{A_V K_\phi A_F}{N} = \frac{\omega_{out}}{s} \frac{K_\phi}{N} \frac{1}{1 + 2 s R_{10} C_4}$$

$$= \frac{2\pi(1.6 \times 10^6)}{s} \frac{1.599}{16} \frac{1}{1 + 2\,s(6000)(1.7 \times 10^{-9})}$$

$$= \frac{1.2 \times 10^9}{jf(jf + 7800)}$$

The plot indicates a phase margin of 17°, which corresponds to $\zeta \approx 0.17$. This is found by using the equation [Ref. 11]

$$\phi_m = \tan^{-1} \frac{2\zeta}{[-2\zeta + (2 - 4\zeta^2 + 4\zeta^4)^{1/2}]^{1/2}}$$

We must lower the loop gain until the pole at 8 kHz occurs at unity gain. This will give us a ζ of 0.42 and a phase margin of 45°. The overshoot for $\zeta = 0.42$ is approximately 36 percent.

The loop gain must be lowered by a factor of 20 to put the 8-kHz pole at unity gain. Equation (14) must be made equal to $\frac{1}{20}$ by an appropriate selection of R_6. If Eq. (14) is rearranged, we get

$$R_6 = \frac{12{,}000\,K_A}{1 - K_A} = \frac{12{,}000(\frac{1}{20})}{1 - \frac{1}{20}} = 632 \ \Omega$$

27.5   A PHASE-LOCKED-LOOP FM DETECTOR

ALTERNATE NAMES   Frequency-modulation detector, FM demodulator, FM discriminator, FSK detector, coherent detector.

PRINCIPLES OF OPERATION   The FM detector is probably the most straightforward application of the PLL chip. Figure 27.11 shows an FM detector implemented with the 565 PLL microcircuit. We note that the VCO output signal is coupled directly to the phase-detector feedback input. The other phase-detector input is capacitively coupled to the incoming frequency-modulated signal. The phase-detector output is amplified and low-pass-filtered. This filtered signal is then used for both the FM output and the control voltage for the VCO.

The VCO is tuned to a free-running frequency of

$$f_R = \frac{1}{4\,R_9 C_4}$$

If f_R differs from f_{in}, the phase comparator generates a correction to the VCO control voltage which forces f_R to equal f_{in}. We call this corrected output frequency f_{out}. If the VCO has a linear voltage-to-frequency transfer function, the change in the VCO control voltage will be proportional to the frequency difference between f_R and f_{in}. The ac component of v_c therefore represents the instantaneous FM deviation of f_{in} from the nominal f_R.

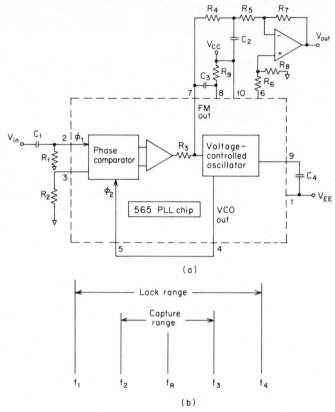

Fig. 27.11 *(a)* An FM demodulator which uses a 565 PLL IC. *(b)* Lock and capture frequency ranges.

The control voltage in the 565 PLL has a large dc offset, which may be undesirable. This offset can be eliminated and some additional gain provided at the same time if the op-amp circuit is included. This circuit is configured as a differential amplifier with the PLL reference output terminal used as the other input. This reference terminal is provided by the PLL for this very reason, and it has a dc level close to that of the FM output terminal.

FM-DEMODULATOR DESIGN PARAMETERS

Parameter	Description
A_F	Voltage transfer function of loop filter
A_V	Transfer function of VCO
A_ϕ	Transfer function of phase comparator

C_1	Provides ac coupling for input signal if dc levels are not compatible
C_2	Part of loop filter
C_3	Recommended by manufacturer to prevent parasitic oscillation
C_4	Controls VCO free-running frequency
F_C	Capture range of frequencies, i.e., from f_2 to f_3
f_1	Lower end of lock frequency range
f_2	Lower end of capture frequency range
f_3	Upper end of capture frequency range
f_4	Upper end of lock frequency range
f_{in}	Frequency of input signal
Δf_{in}	Maximum frequency deviation of input FM signal
$f_{in,min}$	Minimum input frequency $(= f_{in} - \Delta f_{in})$
f_L	Lock range of frequencies, i.e., from f_1 to f_4
f_n	Natural frequency of the feedback loop
f_{out}	Frequency of the output signal
f_R	VCO free-running frequency
K_ϕ	Maximum value of phase-comparator gain
P	Percent modulation of FM input
R_1, R_2	Hold $\phi 1$ input to phase comparator at dc ground potential
R_3	Output resistance of phase comparator
R_4	Resistor used to insert zero in loop-filter transfer function
R_5, R_6	Resistors which control input resistance to the differential amplifier
R_7, R_8	Resistors which control gain of the differential amplifier
R_9	Controls free-running frequency of the VCO
T_p	Pull-in time, i.e., time required to achieve lock from some given frequency offset
v_c	Control voltage to VCO
V_{CC}	Collector supply voltage
V_{EE}	Emitter supply voltage
V_{in}	The value of input voltage
V_{out}	Output from differential amplifier
ζ	Damping factor
θ_e	Peak phase error (maximum value of $\phi 1 - \phi 2$)
$\phi 1$	Phase of input signal
$\phi 2$	Phase of feedback signal
ω_m	FM modulating radian frequency
ω_n	Natural radian frequency of feedback loop

FM-DEMODULATOR DESIGN EQUATIONS (565)

Description	Equation	
VCO free-running frequency	$$f_R = \frac{1}{4\,R_9 C_4} \quad \text{Hz}$$	(1)
VCO gain	$$A_V = \begin{cases} \dfrac{4.1\,f_{out}}{s} & \text{Hz/s}\cdot\text{V} \\[2mm] \dfrac{4.1\,\omega_{out}}{s} & \text{rad/s}\cdot\text{V} \end{cases}$$	(2)
Phase-comparator gain	$$A_\phi = \frac{40\,V_{in}\cos\phi}{(1 + 3460\,V_{in}^2)^{1/2}} \quad \text{V/rad}$$	(3a)
Maximum value of A_ϕ	$$K_\phi = A_\phi\ (\phi = 0°,\ V_{in} > 0.05\ V)$$ $$= 0.68$$	(3b)

Description	Equation
Phase-detector modulation output (control voltage)	$V_C = 0.03\,P$ V pp $\qquad$ (4)
Loop-filter gain for simple lag filter $(R_4 = 0)$	$A_F = \dfrac{1}{1 + sR_3C_2}$ $\qquad$ (5)
Loop-filter gain for pole-zero filter $(R_4 \neq 0)$	$A_F = \dfrac{1 + sR_4C_4}{1 + s(R_3 + R_4)C_2}$ $\qquad$ (6)
Lock range of frequencies (hold-in range)	$f_L = f_4 - f_1 = \dfrac{\zeta\omega_n}{\pi}$ $\qquad$ (7)
Capture range of frequencies	$f_C = f_3 - f_2 \approx \left(\dfrac{8f_{\text{out}}}{\pi\,V_{CC}\,R_3C_2}\right)^{1/2}$ $\qquad$ (8)
Natural frequency of feedback loop $(R_4 = 0)$	$f_n = \dfrac{1}{2\pi}\left(\dfrac{4.1\omega_{\text{out}}K_\phi}{R_3C_2}\right)^{1/2}$ $\qquad$ (9)
Natural frequency of feedback loop $(R_4 \neq 0)$	$f_n = \dfrac{1}{2\pi}\left[\dfrac{4.1f_{\text{out}}K_\phi}{(R_3 + R_4)C_2}\right]^{1/2}$ $\qquad$ (10)
Damping factor of feedback loop $(R_4 = 0)$	$\zeta = \dfrac{1}{2}\left(\dfrac{0.24}{R_3C_2f_{\text{out}}K_\phi}\right)^{1/2}$ $\qquad$ (11)
Damping factor of feedback loop $(R_4 \neq 0)$	$\zeta = \dfrac{1}{2}\left[\dfrac{4.1f_{\text{out}}K_\phi}{(R_3 + R_4)C_2}\right]^{1/2}\left(R_4C_2 + \dfrac{0.24}{f_{\text{out}}K_\phi}\right)$
	$\approx \pi f_n R_4 C_2$ $\qquad$ (12)
Pull-in time	$T_P \approx \dfrac{(\Delta\omega_{\text{in}})^2}{2\zeta\omega_n^3}$ $\qquad$ (13)
Approximate phase error	$\theta_e \approx \dfrac{(\omega_m/\omega_n + 1)^2\,\Delta\omega_{\text{in}}}{(\omega_m/\omega_n - 1)^2 + 8\zeta\omega_n}$ rad $\qquad$ (14)
Resistor values:	
R_1, R_2	$R_1 = R_2 \gg \dfrac{1}{2\pi f_{\text{in,min}}C_1}$ $\qquad$ (15)
R_3	$R_3 = 3600\,\Omega$ (on PLL chip) $\qquad$ (16)
R_4	$R_4 = \dfrac{\zeta\,R_3\,\omega_n}{1.4f_{\text{in}} - \zeta\omega_n}$ $\qquad$ (17)
R_5, R_6	$R_5 = R_6 \gg R_3$ $\qquad$ (18)
R_7, R_8	$R_7 = R_8 = \dfrac{V_{\text{out,pp}}R_5}{v_{c,pp}}$ $\qquad$ (19)
R_9	$R_9 = 4000\,\Omega$ $\qquad$ (20)
Capacitor values:	
C_1	$C_1 > \dfrac{1}{2000\pi f_{\text{in,min}}}$ $\qquad$ (21)
C_2	$C_2 = \dfrac{2\zeta}{R_4\omega_n}$ $\qquad$ (22)
C_3	$C_3 = 0.001\ \mu\text{F}$ $\qquad$ (23)
C_4	$C_4 = \dfrac{1}{4\,f_R\,R_9}$ $\qquad$ (24)

DESIGN PROCEDURE

This procedure is written around the 565 PLL chip. Since not all PLL devices have the same design equations, only a few of the design steps are applicable to other devices. In the following we assume that f_{in}, $V_{in,rms}$, P, f_m, $V_{out,pp}$, V_{CC}, ζ_{min}, and R_3 are initially given. All other parameters are calculated from the given parameters.

Step 1. Compute Δf_{in} and $f_{in,min}$ from

$$\Delta f_{in} = f_{in} P$$

$$f_{in,min} = f_{in} - \Delta f_{in}$$

Step 2. For a start assume that the natural loop frequency ω_n is equal to the maximum modulation frequency ω_m. Solve Eq. (14) to determine the peak phase error. Since the peak phase error must remain less than 90° at all times, several iterations at solving Eq. (14) may be required. The ω_n term can be lowered until θ_e is 1 rad or less.

Step 3. Calculate the peak-to-peak control voltage using Eq. (4).

Step 4. Compute a value for C_1 using Eq. (21).

Step 5. Determine all resistor values using Eq. (15) to (20).

Step 6. The remaining capacitor values are found using Eq. (21) to (24).

Step 7. The phase-comparator gain constant is determined from Eq. (3b).

Step 8. Equations (7) and (8) provide the FM demodulator lock and capture range of frequencies. Verify that $\pm \Delta f_{in}$ does not exceed the f_C range.

Step 9. Use Eqs. (10) and (12) to verify that f_n and ζ are correct.

EXAMPLE OF A PLL FM DEMODULATOR A carrier-current FM intercom demodulator will be designed using the 9 design steps and 24 design equations. The 565 PLL device and the 108 op amp are to be utilized.

Design Requirements and Device Data

$$f_R = f_{in} = f_{out} = 100 \text{ kHz} \qquad V_{in} = 2 \text{ V rms} \qquad P = 5\%$$

$$f_m = 5 \text{ kHz} \qquad V_{out} = 5 \text{ V pp} \qquad V_{CC} = +6 \text{ V}$$

$$\zeta = 0.7 \text{ min} \qquad R_3 = 3600 \text{ }\Omega$$

Step 1. The initial parameter calculations are

$$\Delta f_{in} = f_{in} P = (100 \text{ kHz})(0.05) = 5 \text{ kHz}$$

$$f_{in,min} = f_{in} - \Delta f_{in} = 100 \text{ kHz} - 5 \text{ kHz} = 95 \text{ kHz}$$

Step 2. We start by letting the natural loop frequency equal the maximum modulation frequency

$$\omega_n = 2\pi f_n = \omega_m = 2\pi f_m = 2\pi(5000) = 31,400 \qquad \text{rad/s}$$

The peak phase error in this case is

$$\theta_e \approx \frac{(\omega_m/\omega_n + 1)^2 \, \Delta\omega_{in}}{(\omega_m/\omega_n - 1)^2 + 8\zeta\omega_n} \approx \frac{(1 + 1)^2(2\pi)(5000)}{(1 - 1)^2 + 8(0.7)(31,400)}$$

$$\approx 0.715 \text{ rad} = 41°$$

This phase error is satisfactory, and so no further iteration is required.

Step 3. The peak-to-peak control voltage is

$$v_c = 0.03 \, P = 0.03(5) = 0.15 \text{ V pp}$$

Step 4. Capacitor C_1 is found from

$$C_1 > \frac{1}{2000\pi \, f_{in,min}} = \frac{1}{2000\pi(95,000)}$$

$$> 1675 \text{ pF} \quad \text{use 2000 pF}$$

Step 5. Resistor values are computed:

$$R_1 = R_2 \gg \frac{1}{2\pi \, f_{in,min} C_1} = \frac{1}{2\pi(95,000)(2 \times 10^{-9})}$$

$$\gg 838 \ \Omega \quad \text{use 2 k}\Omega$$

$$R_3 = 3600 \ \Omega \text{ (on PLL chip)}$$

$$R_4 = \frac{\zeta R_3 \omega_n}{1.4 f_{in} - \zeta\omega_n} = \frac{0.7(3600)(31,400)}{1.4 \times 10^5 - 0.7(31,400)}$$

$$= 670 \ \Omega$$

$$R_5 = R_6 \gg R_3 = 3600 \ \Omega \quad \text{use } R_5 = R_6 = 20 \text{ k}\Omega$$

$$R_7 = R_8 = \frac{V_{out,pp} R_5}{v_{c,pp}} = \frac{5(20,000)}{0.15} = 666 \text{ k}\Omega \quad \text{use 680 k}\Omega$$

$$R_9 = 4000 \ \Omega$$

This last resistor may be variable to trim the VCO free-running frequency.

Step 6. The remaining capacitor values are

$$C_2 = \frac{2\zeta}{R_4 \omega_n} = \frac{2(0.7)}{670 \, (31,400)} = 0.066 \ \mu\text{F}$$

$$C_3 = 0.001 \ \mu\text{F (recommended by manufacturer)}$$

$$C_4 = \frac{1}{4 f_R R_9} = \frac{1}{(4 \times 10^5)(4000)} = 625 \text{ pF} \quad \text{use 680 pF}$$

Step 7. The phase-comparator gain constant is

$$K_\phi = A_\phi(\phi = 0°, V_{in} = 2 \text{ V}) = \frac{40 V_{in} \cos \phi}{(1 + 3460 V_{in}^2)^{1/2}}$$

$$= \frac{40(2) \cos 0°}{[1 + 3460(2)^2]^{1/2}} = 0.68 \text{ V/rad}$$

Step 8. The lock range of frequencies is

$$f_L = f_4 - f_1 = \frac{2\zeta\omega_n}{\pi} = \frac{2(0.7)(31,400)}{\pi} = 13,992 \text{ Hz}$$

The capture range of frequencies is

$$f_C = f_3 - f_2 \approx \left(\frac{8 f_{out}}{\pi V_{CC} R_3 C_2}\right)^{1/2} = \left[\frac{8 \times 10^5}{\pi(6)(3600)(0.066 \times 10^{-6})}\right]^{1/2}$$

$$\approx 13,365 \text{ Hz}$$

Since $\pm\Delta f_{in}$ is 10 kHz wide, a capture of the signal is assured.

Step 9. Equations (10) and (12) are used to double-check previous calculations:

$$\omega_n = \left[\frac{4.1 f_{out} K_\phi}{(R_3 + R_4) C_2}\right]^{1/2} = \left[\frac{(4.1 \times 10^5)(0.68)}{(3600 + 670)(0.066 \times 10^{-6})}\right]^{1/2}$$

$$= 31,453 \text{ rad/s}$$

This is in reasonable agreement with step 2.

$$\zeta = \frac{1}{2}\left[\frac{(4.1 f_{out} K_\phi)}{(R_3 + R_4) C_2}\right]^{1/2}\left(R_4 C_2 + \frac{0.24}{f_{out} K_\phi}\right)$$

$$= \frac{1}{2}\left[\frac{(4.1 \times 10^5)(0.68)}{(3600 + 670)(0.066 \times 10^{-6})}\right]^{1/2}\left[670(0.066 \times 10^{-6}) + \frac{0.24}{10^5 \times 0.68}\right]$$

$$= 0.75$$

This result agrees reasonably well with the original 0.7 goal for ζ.

REFERENCES

1. Reed, L. J., and R. J. Treadway: Test Your PLL IQ, *EDN*, Dec. 20, 1974, p. 27.
2. Stout, D. F., and M. Kaufman: "Handbook of Operational Amplifier Circuit Design, McGraw-Hill, New York, 1976, pp. 8-4, 21-5.
3. Renschler, E.: An Integrated Circuit Phase-locked Loop Digital Frequency Synthesizer, *Motorola Appl. Note*, 1972.
4. Schowe, L. F., Jr.: Build a Wideband Phase-locked Loop, *Electron. Des.* 18, Sept. 1, 1973, p. 112.
5. Moore, A. W.: Phase-locked Loops for Motor-Speed Control, *IEEE Spectrum*, April 1973, p. 61.

6. Hager, J. C. Jr.: Edge Triggered Flip-Flops Make 360° Phase Meter, *Electronics,* Aug. 21, 1975, p. 100.
7. RCA Corporation: RCA COS/MOS Integrated Circuits Manual SSD-203C, Somerville, N.J., 1975, p. 472.
8. Cutler, P.: "Linear Electronic Circuits," McGraw-Hill, New York, 1972, p. 82.
9. "Signetics Linear Integrated Circuits," vol. 1, Signetics Corp., Sunnyvale, Calif., 1972.
10. Bryant, J. M.: SL650 and SL651 Applications, Plessey Semiconductors, Santa Ana, Calif.
11. Thaler, G. J.: "Design of Feedback Systems," Dowden, Hutchinson and Ross, Inc., Stroudsburg, Pa., 1973, p. 27.
12. Klement, C.: Voltage-to-Frequency Converter Constructed with Few Components Is Accurate to 0.2%, *Electron. Des.* 13, June 21, 1973, p. 12.
13. Mills, T. B.: The Phase-locked Loop IC as a Communication System Building Block, *National Semiconductor Appl. Note* AN-46, June 1971.

Index